Plastic and Microplastic in the Environment

Plastic and Microplastic in the Environment

Management and Health Risks

Edited by

Arif Ahamad
Daulat Ram College
University of Delhi
New Delhi, India

Pardeep Singh
PGDAV College
University of Delhi
New Delhi, India

Dhanesh Tiwary
Indian Institute of Technology
Banaras Hindu University
Varanasi, India

Registered Offices
John Wiley & Sons, Inc., 111 River Street, Hoboken, NJ 07030, USA
John Wiley & Sons Ltd, The Atrium, Southern Gate, Chichester, West Sussex, PO19 8SQ, UK

Editorial Office
9600 Garsington Road, Oxford, OX4 2DQ, UK

For details of our global editorial offices, customer services, and more information about Wiley products visit us at www.wiley.com.

Wiley also publishes its books in a variety of electronic formats and by print-on-demand. Some content that appears in standard print versions of this book may not be available in other formats.

Library of Congress Cataloging-in-Publication Data

Names: Ahamad, Arif, editor. | Singh, Pardeep, editor. | Tiwary, Dhanesh, editor.
Title: Plastic and microplastic in the environment : management and health risks / edited by Arif Ahamad, Daulat Ram College, University of Delhi, New Delhi, India, Pardeep Singh, PGDAV College, University of Delhi, New Delhi, India, Dhanesh Tiwary, Indian Institute of Technology, Banaras Hindu University, Varanasi, India.
Description: First edition. | Hoboken, NJ : John Wiley & Sons, Inc., 2022. | Includes bibliographical references and index.
Identifiers: LCCN 2021040642 (print) | LCCN 2021040643 (ebook) | ISBN 9781119800781 (hardback) | ISBN 9781119800873 (adobe pdf) | ISBN 9781119800880 (epub)
Subjects: LCSH: Plastic scrap–Environmental aspects. | Microplastics–Environmental aspects. | Plastics–Toxicology.
Classification: LCC TD427.P62 P54 2022 (print) | LCC TD427.P62 (ebook) | DDC 363.738–dc23/eng/20211022
LC record available at https://lccn.loc.gov/2021040642
LC ebook record available at https://lccn.loc.gov/2021040643

Cover Design: Wiley
Cover Image: © alphaspirit.it/Shutterstock

Set in 9.5/12.5pt STIXTwoText by Straive, Pondicherry, India
Printed and bound by CPI Group (UK) Ltd, Croydon, CR0 4YY

C9781119800781_110322

Contents

List of Contributors

Juan Carlos Álvarez-Zeferino
Universidad Autónoma Metropolitana
Mexico City
Mexico

Iqbal Ansari
Department of Earth Sciences and
Environment
Faculty of Science and Technology
Universiti Kebangsaan Malaysia (UKM)
Bangi, Selangor
Malaysia

Charu Arora
Department of Chemistry
Guru Ghasidas University
Bilaspur, Chhattisgarh
India

Sandhya Babel
School of Biochemical Engineering
and Technology
Sirindhorn International Institute of
Technology
Thammasat University
Pathum Thani
Thailand

Kaushik Kumar Bharadwaj
Department of Bioengineering and
Technology
GUIST
University of Guwahati
Guwahati, Assam
India

Yong Chen
School of Environmental Science and
Engineering
Huazhong University of Science and
Technology
Wuhan
China

Bhabesh Kumar Choudhury
Department of Chemistry
University of Guwahati
Guwahati, Assam
India

Arely Areanely Cruz-Salas
Universidad Autónoma de Baja California
Mexicali
Baja California
Mexico

Maha M. El-Kady
Department of Self-pollinated
Vegetables Crops
Horticulture Research Institute (HRI),
Agricultural Research Centre (ARC)
Giza
Egypt

Antima Gupta
Department of Environmental Sciences
University of Jammu
Jammu
India

Shivali Gupta
Department of Environmental Sciences
University of Jammu
Jammu
India

Rupjyoti Haloi
Department of Electrical Engineering
Assam Engineering College
Jalukbari, Guwahati
Assam
India

Marlia Mohd Hanafiah
Department of Earth Sciences and
Environment
Faculty of Science and Technology
Universiti Kebangsaan Malaysia (UKM)
Bangi, Selangor
Malaysia;
Centre for Tropical Climate Change System
Institute of Climate Change
Universiti Kebangsaan Malaysia (UKM)
Bangi, Selangor
Malaysia

Sumi Handique
Department of Environmental Science
Tezpur University
Tezpur, Assam
India

Sumbul Jahan
Department of Biotechnology
Vinoba Bhave University
Hazaribag, Jharkhand
India

Rabindra Kumar
Central Instrumentation Laboratory
Central University of Punjab
Bathinda
India

Rakesh Kumar
Department of Environmental Sciences
University of Jammu
Jammu
India

Sushil Kumar
Department of Chemical Engineering
Motilal Nehru National Institute of
Technology (MNNIT)
Allahabad, Prayagraj
UP
India

Sophayo Mahongnao
Department of Biochemistry
Daulat Ram College
University of Delhi
New Delhi
India

Carolina Martínez-Salvador
Technische Universität Desdren
Dresden
Germany

Ankitendran Mishra
Department of Metallurgical Engineering
IIT
Banaras Hindu University
Varanasi
India

Sarita Nanda
Department of Biochemistry
Daulat Ram College
University of Delhi
New Delhi
India

Sara Ojeda-Benítez
Universidad Autónoma de Baja California
Mexicali
Baja California
Mexico

Ankita Ojha
Department of Chemistry
Maharaja College
Arrah, Bihar
India

Beatriz Pérez-Aragón
Universidad Autónoma Metropolitana
Mexico City
Mexico

Abhay Punia
Department of Zoology Guru Nanak
Dev University
Amritsar, Punjab
India

Sanchayita Rajkhowa
Department of Chemistry
Jorhat Institute of Science and
Technology
Jorhat, Assam
India

Akanksha Rajput
Department of Environmental Sciences
University of Jammu
Jammu
India

Sumeer Razdan
Central Instrumentation Laboratory
Central University of Punjab
Bathinda
India

Jyotirmoy Sarma
Department of Chemistry
Assam Kaziranga University
Jorhat, Assam
India

Steplinpaulselvin Selvinsimpson
School of Environmental Science and
Engineering
Huazhong University of Science and
Technology
Wuhan
China

Soha Shabaka
Hydrobiology Lab
Department of Marine Environment
National Institute of Oceanography and
Fisheries (NIOF)
Cairo
Egypt

Pooja Sharma
Department of Biochemistry
Daulat Ram College
University of Delhi
New Delhi
India

Dharm Pal
Department of Chemical Engineering
National Institute of Technology
Raipur
India

Nalini Singh Chauhan
Department of Zoology
Kanya Maha Vidyalya
Jalandhar, Punjab
India

Richa Singh
Institute of Environment and Sustainable
Development
Banaras Hindu University
Varanasi, Uttar Pradesh
India

Anh Tuan Ta
School of Biochemical Engineering and
Technology
Sirindhorn International Institute of
Technology
Thammasat University
Pathum Thani
Thailand

Jocelyn Tapia-Fuentes
Universidad Autónoma Metropolitana
Mexico City
Mexico

Dhanesh Tiwary
Department of Chemistry
IIT
Banaras Hindu University
Varanasi
India

Hasan Uslu
Food Engineering Department
Engineering Faculty
Niğde Ömer Halisdemir University
Niğde
Turkey

Alethia Vázquez-Morillas
Universidad Autónoma Metropolitana
Mexico City
Mexico

Kailas L. Wasewar
Advance Separation and Analytical
Laboratory (ASAL)
Department of Chemical Engineering
Visvesvaraya National Institute of
Technology (VNIT)
Nagpur, Maharashtra
India

Preface

Humans have modified the environment to better answer their needs, and in the process, they have changed the homeostatic mechanisms of the earth's system. The ecological fabric of the planet has undergone drastic changes due to human-induced modifications. These changes have resulted in certain thresholds of ecological systems being broken. One such human activity is the production of waste, mainly plastic waste. Plastics are a versatile material that provides an inexpensive alternative to many industries. Plastics are a group of synthetic polymers which are now considered as an indicator of the Anthropocene. Of the 300 million tons of plastic produced each year on average, 50% of waste is designed for waste applications, i.e., it is destined to be thrown away. The European Union alone accounts for about 650 thousand tons of flexible plastic wastes. Due to its applicability, potential technology for recycling and degradation of plastic debris has become an essential issue of the 21st century. Renewable raw materials in innovative polymer products with comparable production costs and efficient technology are being explored. Global production of plastic resin has increased from about 1.5 million tons in 1950 to 322 million tons in 2015. Overall, plastic consumption has given rise to the massive load of plastic waste which has disturbed the entire ecosystem. The disposal of plastic waste is a crucial issue regardless of the present awareness and technological capacity. Plastic has been identified to be a global pollutant that now contaminates our aquatic and soil systems. Its durability and use in various industries and as disposables have resulted in a large amount of plastic in landfills and open dumps. Waste disposal is a serious concern, especially in developing countries, due to the non-implementation of rules and lack of public mobilization. Collection, segregation, and treatment or recycling are lacking, and about 90% of the waste is still disposed of through landfills. Because the presence of plastic waste is an unrelenting concern, mainly due to its non-biodegradability and adverse environmental impacts, there is an urgent need to tackle the plastic waste issue. A combination of various techniques is currently used to increase the energy value or reduce the waste volume of plastic wastes. Recycling of plastic waste consists of converting it into fuel or feedstock, which would reduce the volume and the net cost of disposal. Recycling converts plastics into liquids or gases with chemical, thermal, or thermo-chemical procedures. Chemical recycling, including de-polymerization, pyrolysis, and catalytic cracking, are the usual procedures employed by industries for plastic recycling. The omnipresence of microplastics with a size range of <5 mm is increasing worldwide concerns about their implications for human health. Its presence in water, soil, and air poses serious health risks to humans and other organisms via the food chain. However, whether these contaminants pose a substantial risk to human health is far from understood. This book emphasizes the occurrence of plastic and microplastic in the environment, challenges faced, and various management strategies and policies for their management at the global level. It also includes health risk issues related to plastic and microplastic in different componentsof the environment.

This book displays several chapters on microplastic contamination in freshwater, marine water, soil, and air. It also includes a chapter focused on microplastic in various aquatic food chains and its impact on human health. It includes and highlights several techniques related to microplastic detection in different environmental media. Chapters on the distribution of plastic at the global level and challenges related to its recycling, degradation of plastic, and biodegradable plastic are also included. It also includes a chapter on the role of education and society to deal with plastic problems. This book will be helpful for graduate students, researchers, engineers, technologists, NGOs, and government agencies working in plastic and microplastic related issues.

This book is a humble effort to address the plastic and microplastic issues across the globe. We hope that it is an important addition to the available literature. The contributors of the book are from diversebackgrounds that provided holistic information on the topic. We convey our heartfelt gratitude to all the contributors and publishers who helped to produce an incredible and meaningful edited volume on a very relevant theme.

1

Sources, Occurrence, and Analysis of Microplastics in Freshwater Environments: A Review

Anh Tuan Ta and Sandhya Babel

School of Biochemical Engineering and Technology, Sirindhorn International Institute of Technology, Thammasat University, Pathum Thani, Thailand

1.1 Introduction

Plastic products have benefited human life for one hundred years. Plastics are used in almost every sector including construction, packaging textiles, consumer products, transportation, electronics, industry, and medical applications. These products make our life more comfortable, convenient, and safe. The world plastics production in 2018 was about 360 million tonnes (PlasticsEurope 2018). However, due to poor management, about 10% of plastic wastes are discharged into aquatic environments (Cole et al. 2011). Recently, considerable attention is placed on the distribution of micro-sized plastic particles, so-called microplastics (MPs). The ubiquity of MPs has been widely documented in the marine environment, and their possible impact was investigated (Dris et al. 2018). MPs are frequently ingested by organisms, either from ingestion of other organisms containing MPs, or because they cannot distinguish MPs from prey (de Sá et al. 2015). This may cause physical harm for organisms including a disruption of the hormone balance or digestive system, reduced feeding, and impacting reproduction (Carr et al. 2012; Lusher et al. 2013). Another ecological risk relates to the interaction between MPs and toxic chemicals. The small plastic debris has a high surface area to volume ratio that can enhance the interaction of toxic chemicals onto their surface (de Sá et al. 2018). These chemicals may be carried by MPs over long distances and accumulate in organisms after being ingested (Bakir et al. 2016; Lee et al. 2014). Another potential risk is that MPs can be vectors of microorganisms that attach to their surfaces (Viršek et al. 2017). Human pathogens have been found to colonize on MP particles in marine environments (Foulon et al. 2016; Kirstein et al. 2016). However, the existence of MPs in aquatic environments is complex as they have different polymer types, morphologies, sizes, states of degradation, and contain different additives. Therefore, the evaluation of MPs toxicity is still hampered. Until now, most studies on MPs have been conducted on marine environments; while in contrast, the distribution of MPs in freshwater environments has been investigated less. To highlight this, a study by Blettler et al.

(2018) found that 87% of MP studies were related to marine environments, while only 13% of the studies were on freshwater environments. A few studies reported that rivers can transport a huge amount of plastic and MPs into oceans; according to Schmidt et al. (2017), rivers contribute 88–95% of plastic wastes placed into the oceans. Methods for the identification of MPs have been developed by many researchers; however, these methods are not homogeneous and standardized, and this hampers the comparison between different studies (Li et al. 2018; Van Cauwenberghe et al. 2015).

This chapter reviews the current state of MPs in freshwater systems, and different techniques for sample collection, preparation, and analysis of MPs are summarized. Moreover, the potential sources, pathways, and occurrences of MPs into freshwater systems are discussed in the chapter.

1.2 Sources of Microplastic

MPs originate from various sources, but they are mainly sorted into primary and secondary sources. Primary MPs are produced in micro sizes, such as microbeads for personal care products and plastic resins. Secondary MPs are generated from the degradation of large plastics (Andrady 2011). Textile laundering facilities and sandblasting are other sources of MPs (Browne et al. 2011; Napper and Thompson 2016). The pollutants are washed down the drain along with wastewater and enters sewage systems, where they are too small to be removed by WWTPs and so end up in the river systems and finally in the oceans.

1.2.1 Primary Sources

1.2.1.1 Microplastics from Personal Care Products

Plastic materials have been utilized in the cosmetic industry for decades since the 1960s. The particles of plastic used in personal care products can be large enough to see with the naked eye (50–1000 μm), fine particles (lower μm-range), or very fine particles (<2.5 μm) (Leslie 2014). MPs used in the industry may be spherical or irregular morphology as shown in Figure 1.1 (Godoy et al. 2019). MPs used as ingredients in cosmetic products include the two main categories that are typically made from petroleum carbon sources: thermoplastics (e.g. polypropylene (PP), polyethylene (PE), polytetrafluoroethylene (Teflon),

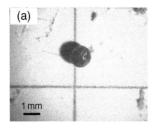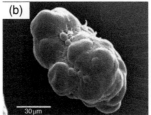

Figure 1.1 (a) Micrograph of a microbead found in Lantau Island, Hong Kong *Source:* Cheung 2016, p. 02 / With permission from Elsevier. (b), (c) Scanning electron microscopy (SEM) of microbeads utilized in cosmetic products *Source:* Napper 2015, p. 05 / With permission from Elsevier; Godoy 2019, p.06 / With permission from Elsevier.

polystyrene (PS), polymethyl methacrylate (PMMA); and thermoset plastics (e.g. polyester, polyurethanes)) (Napper et al. 2015). According to Gouin et al. (2015), around 93% of the microbeads applied in cosmetics are PE, since their smoothness reduces redness and impact on the human skin compared to other polymers.

1.2.1.2 Microplastics from Plastic Resins

Another major source of primary MPs are plastic resins (pellets or powders). The pellets or powder resin (≤ 0.5 mm) are generally cylindrical or disk shaped (Bergmann et al. 2015). These plastics are transported to factories for re-melting and molding into a wide range of commercial plastic products. The plastics are released into the environment due to improper handling, such as accidents during transport or runoff from production processes.

1.2.2 Secondary Sources

1.2.2.1 Microplastics from Degradation of Plastic Debris

Large plastic debris in freshwater environments gradually degrades into smaller particles when exposed to some factors in the environments (Figure 1.2). In the degradation of plastic, the average molecular weight of polymers is drastically reduced (Andrady 2011). The degradation can categorize as agents causing this process: biodegradation is an action of microorganisms; photodegradation is an action of light and sunlight; thermal degradation is the action of high temperatures; thermooxidative degradation is an oxidative breakdown at medium temperatures; hydrolysis degradation is a reaction with water. In the environment, UV radiation from sunlight is the major factor of plastic degradation, which speeds up the oxidative breakdown of polymers (Andrady et al. 1996). Polymers such as high-density polyethylene (HDPE), low-density polyethylene (LDPE), PP, and nylons are degraded mainly by the UV-B radiation in sunlight as they are exposed to environments. As degradation is initiated, the thermooxidative process can continue without further exposure to UV radiation. The autocatalytic degradation may occur as long as the existence of oxygen in the system (Andrady 2011). During the degradation process, plastic wastes typically discolor, evolve surface features, and become weak and brittle, gradually. Other forces such as waves, wind, and human and animal activities can easily crack the embrittled plastics into small particles. The degradation and fragmentation of plastics is the major process for the formation of secondary MPs in aquatic environments (Kershaw 2015).

(a)  (b) 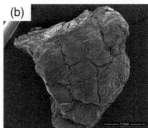 (c)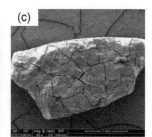

Figure 1.2 (a) Degradation and fragmentation of plastic under environmental factors; (b) and (c) the cracks seen at the surface are caused by photochemical degradation *Source:* Ter Halle 2016, p. 15 / With permission from American Chemical Society.

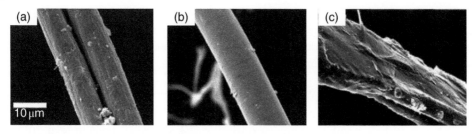

Figure 1.3   SEM of typical fibers: (a) polyester-cotton blend; (b) polyester; (c) acrylic
Source: Napper 2016, p. 03 / with permission from Elsevier.

1.2.2.2   Microplastics from Textile and Domestic Washing

Fibers from synthetic textiles are another source of secondary MPs in the environments (Figure 1.3). Synthetic fibers are made from petroleum through polymerization, polycondensation, or polyaddition processes (Astrom 2016). In 2010, total synthetic fiber production was 49.6 million tons, accounting for 60% of the world's fiber production (Essel et al. 2015). Synthetic textiles can be used in clothes, furniture, geotextiles, cloth, sports, packing, toys, construction, and agriculture. Shedding of textiles relies on the textile types, the yarn and texture type, and the fiber types involved. According to Astrom (2016), most fibers are shed from synthetic fleece and microfleece. A synthetic fleece coat can shed about 1900 fibers with each wash (Browne et al. 2011). These authors also concluded that using detergents causes more shedding compared with using only water. The average size of fibers ranged from 5.0 to 7.8 mm in length, and 11.9 to 17.7 μm in diameter (Napper and Thompson 2016). Compared to other MPs detected in environments, such as pellets or fragments, fibers have a higher surface-to-volume ratio; therefore, they can attract more chemicals than other MPs (Astrom 2016).

1.3   Pathways of Microplastics into Freshwater Environments

MPs can enter freshwater environments by several pathways due to the bulk of plastic wastes in the environment (Figure 1.4). A pathway of MPs entering may be important in one region but less important in another (Lambert and Wagner 2018). For example, MPs applied in personal care products are likely more important in urban than agricultural regions (Lambert et al. 2014). Potential environmental release pathways of MPs can be separated by their primary or secondary sources.

A pathway of primary MPs input to freshwater environments has been found through WWTPs and the utilization of sludge from WWTPs to agricultural lands. Previous studies depicted that 90% of MPs in domestic wastewater are retained within sludge (Magnusson and Norén 2014; Talvitie and Heinonen 2014). In Europe, sewage sludge is normally composted to produce agricultural fertilizer as well as dispose of the sludge to land. The EU countries apply about four to five million tonnes of sludge to agricultural lands, annually (Willén et al. 2017). MPs that cannot be removed in the treatment process will reach the freshwater environments via effluent (Horton et al. 2017). Another pathway of primary MPs could be from the release of industrial products or processes.

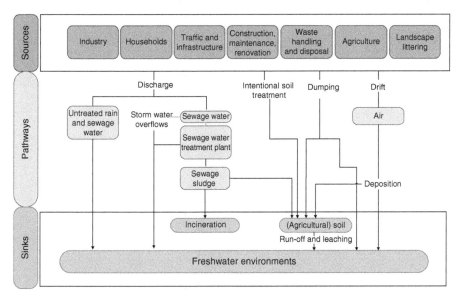

Figure 1.4 Possible exposure pathways of MPs into freshwater environments.

Routes of secondary MPs into freshwater environments are mainly from improper management of plastic wastes. This includes release during collecting, transporting, processing, and landfilling of solid waste. Another route could be the runoff of wastes through drainage canals from agricultural land. Plastic wastes on roads such as vehicle debris, tire wear particles, or fragments of road-marking paints could overflow with storm water into the freshwater environments (Eriksen et al. 2013). Lighter plastic wastes may be transported into freshwater environments by wind (Zylstra 2013), and synthetic fibers could be carried and accumulated by atmospheric fallout (Dris et al. 2015).

1.4 Microplastic Analytical Methods in Freshwater

Until now, there has been no certain methodology for sampling and analysis of MPs in freshwater. Most researchers adopted methods from the marine environment, with modifications. However, homogeneous and standard methods for the determination of MPs are missing. This fact hampers comparability between studies both in freshwater and marine environments. Different techniques for sample collection, preparation, and analysis of MPs are summarized in Figure 1.5.

1.4.1 Sampling of Microplastic

1.4.1.1 Water Samples
As reported in previous studies, the abundance of MPs in water samples is lower than that of sediment samples (Ta and Babel 2020b). Thus, a large quantity of water is usually collected to acquire a representative sample. Volume reduced samples were commonly applied

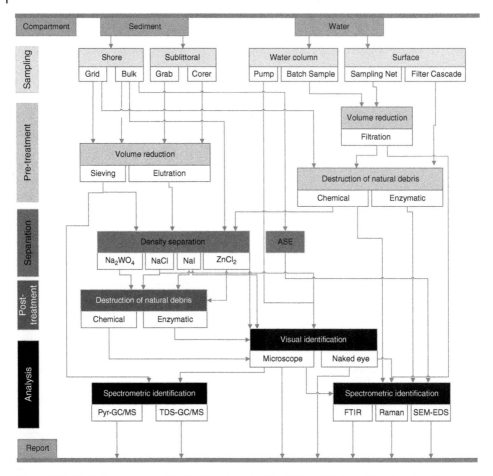

Figure 1.5 Techniques reported in the literature for identifying MPs in sediment and water samples.

for the sampling of surface water. Until now, most studies used manta trawls or plankton nets to collect water samples. Different mesh sizes of these nets are used, ranging from 50 to 3000 µm (Tan et al. 2019; Zhang et al. 2015), while 333 µm is the most commonly used in all studies (Anderson et al. 2017; Free et al. 2014; Su et al. 2016; Ta and Babel 2020b; Wong et al. 2020). Flow meters are attached to plankton nets or manta trawls to determine the filtered water volume. This helps to normalize the number of MPs to the volume of filtered water. Another technique is the measurement of the sampling area by the Global Positioning System (GPS). Thus, results are presented as the number of MPs per surface area. Trawling speed depends on the water velocity or wave action, but usually ranges from one to five knots (Ta and Babel 2020a; Tan et al. 2019; Wong et al. 2020). Another sampling method is direct filtration of water through sieves or by the collection of batch samples (Crew et al. 2020; Yan et al. 2019). In this method, a smaller water volume can be collected, thus the sample representability is reduced (Wang et al. 2017).

1.4.1.2 Sediment Samples

The sampling of sediment samples can be separated into the collection from the shoreline and bottom of the river or lake. For shore sediments, sampling strategies are transected sampling perpendicular, random sampling, and sampling in single squares or parallel to the water. Most studies applied the grid sampling method with depths of 2–5 cm on the surface layer (Jiang et al. 2019; Klein et al. 2018). Frame and corers are usually used to determine the sampling area. Non-plastic tools such as scoop, trowels, or shovels, and non-plastic sampling vessels are required (Alam et al. 2019; Jiang et al. 2019; Peng et al. 2018). Bottom sediment from the riverbed or lakebed can be carried out with grab samplers such as Ekman or Van Veen grabs or corers (Alam et al. 2019; Fan et al. 2019; Ta et al. 2020c; Wang et al. 2017). The sediment samples collected by grab methods are usually disturbed, therefore this is suitable for surface layer (top 5 cm) or bulk sampling. Conversely, sampling by cores allows determining MP depth profiles and undisturbed surface and depth layers. Nevertheless, the number of samples that can be collected is limited. According to Dris et al. (2018), river bottom sediments are mostly collected by grabs, while corers or grabs are used for lake bottom sediments. The number of MPs is usually normalized to the sediment volume or weight, and sampling area.

1.4.2 Sample Preparation

1.4.2.1 Extraction of Microplastics

Due to the complex nature of the sediment, MPs in the samples must be extracted from sample matrices. The density separation is widely applied to extract MPs from sediment samples. The sediment is dried prior to mixing with a concentrated salt solution. After a period of agitation, MPs and light particles float to the surface or stay suspended, whereas heavy particles settle down (Klein et al. 2018). Many studies extract MPs by using sodium chloride (NaCl) solution since this is inexpensive and environmentally friendly (Alam et al. 2019; Campanale et al. 2020; Free et al. 2014; Mani et al. 2015). However, the density of NaCl solution (~1.2 g/cm^3) cannot extract some polymers such as Polyvinyl chloride (PVC), Polyethylene terephthalate (PET), polycarbonate, and polyurethane. Therefore, sodium iodide (NaI), sodium zinc chloride (ZnCl$_2$), and sodium polytungstate (Na$_2$WO$_4$) are viable choices (Ballent et al. 2016; Ta and Babel 2019; Yin et al. 2020). Conversely, MPs in water samples are easily filtered and separated during the sampling step (Dris et al. 2018).

1.4.2.2 Removal of Organic Debris

The detection of MPs is hampered by organic particles that accompany MPs during collecting water samples. Thus, the removal of the organic debris is important to reduce the misidentification or underestimation of MPs. The detection step is conducted before or after the density separation. Organic debris is usually removed by strong acids (Cole et al. 2014; Imhof et al. 2016) or base solutions (Dehaut et al. 2016), or oxidation agents such as hydrogen peroxide (Ta and Babel 2019; Zhao et al. 2017). Some sensitive polymers (i.e. poly(methyl methacrylate) and polycarbonates) can be lost or damaged during the treatment (Dehaut et al. 2016; Li et al. 2018). Alternate chemicals for removing organic debris in samples are enzymes; these chemicals reduce damage to sensitive polymers. (Catarino et al. 2017; Mani et al. 2015). The method can be conducted by using proteinase K or mixtures of technical enzymes including proteinase, lipase, chitinase, amylase, and cellulose (Courtene-Jones et al. 2017; Mani et al. 2015). The enzymatic digestion should be conducted under controlled conditions of pH and temperature. However, several disadvantages of using enzymes are reported. In comparison to chemical treatments, enzymatic treatments are expensive, time-consuming, and may not completely remove the organic material (Courtene-Jones et al. 2017).

1.4.3   Identification of Microplastic

1.4.3.1   Visual Sorting

In most studies, visual sorting is the first step to separate MPs from samples before identification of the polymer type. Large MPs (>1 mm) can be recognized by the naked eye (Anderson et al. 2017), while smaller particles are identified using dissection microscopes (Faure et al. 2015; Mani et al. 2015) or scanning electron microscopy (SEM) (Eriksen et al. 2013; Su et al. 2016). This step requires experienced researchers and good optical quality of the microscope. However, identification of all particles is difficult if they are smaller than a certain size, if they are unable to be distinguished visually or cannot be managed with forceps due to their minuteness. Thus, visual sorting is time-consuming and easy misidentification or underestimation of MPs is possible. Recently, another visual identification method using fluorescence was applied to detect and quantify small MPs. In most studies, Nile Red (NR) was used and dissolved in different solvent solution such as acetone, chloroform, and n-hexane (Crew et al. 2020; Tamminga et al. 2017). Suspected MPs are stained with the NR solution and analyzed with a fluorescence microscope. This technique is inexpensive, can utilize available instruments, and can be semi-automated for large amounts of sample analysis.

1.4.3.2   Identification of Microplastics by Chemical Composition

Pyrolysis-GC/MS   Pyrolysis-gas-chromatography/mass spectrometry (Pyr-GC/MS) can be used to determine the polymer types and additives. In this method, the samples are combusted, and the thermal degradation products of the polymers are used to detect MPs (Fries et al. 2013). The pyrolysis results provide characteristic pyrograms of MPs samples that can be identified by comparing with reference pyrograms of known polymer types. Particles must be inserted in pyrolysis tubes manually, and the technique can analyze only one particle per run; thus the method is unable for analyzing large amounts of samples (Dris et al. 2018; Klein et al. 2018).

Infrared Spectroscopy   The spectroscopic identification techniques consist of Raman spectroscopy and FTIR. These techniques are based on the energy sorption by characteristic functional groups of polymeric materials. Thus, samples must be dried before analysis because water strongly absorbs IR radiation. The purification of samples is important because organic, inorganic–organic, or inorganic may affect the IR spectra of samples. Larger particles (>0.5 mm) may be investigated by FTIR with an attenuated transverse reflection (ATR) unit (Doyle et al. 2011; Klein et al. 2018; Zhang et al. 2015). MPs are manually transferred to the crystal of the ATR unit. The FTIR can be combined with an IR microscope to enhance the identification of small MPs ($>10\,\mu$m) (Sadri and Thompson 2014). In Raman spectroscopy, the sample is exposed to monochromatic laser radiation. The reaction of the molecules and atoms with the laser radiation cause differences in the frequency of the backscattered light compared with the initial laser frequency. This can be computed to create substance-specific Raman spectra. Raman spectroscopy is also able to be connected with microscopy. The Raman micro-spectroscopy can identity very small MPs of sizes down to $1\,\mu$m (Yan et al. 2019). Until now, FTIR is widely used in MPs studies.

1.5 Occurrence of Microplastic in Freshwater Environments

In contrast to research in marine environments, MPs in freshwater environments have received less attention, but in the last few years, research on MPs in freshwater are advancing. This helps to reveal the occurrence of MPs in freshwater environments of several continents.

1.5.1 Microplastic in Lakes

MPs have been reported in lakes of Europe, Africa, Asia, and North America (Table 1.1). In North America, many studies have focused on the Great Lakes system. These studies are clarified by watershed populations, lake areas, and industrial activities. (Anderson et al. 2017; Corcoran et al. 2015; Eriksen et al. 2013). The average number of MPs at the Great Lakes was as high as 193 000 items/km^2 (Anderson et al. 2017). In Europe, research on MPs usually relates to population sizes. It varies from the densely populated lakes in Swiss Lake Geneva (Faure et al. 2015) to the less populated lakes such as Italian Lakes Bolsena, Chiusi, and Garda, and Swiss Lake Brienz (Fischer et al. 2016; Imhof et al. 2013). To the extent of our knowledge, the greatest number of MPs in Europe has been reported in Lake Geneva, Switzerland, with 220 000 items/km^2 (Faure et al. 2015). Most MP research in Asia is conducted in lakes in China. MPs were found in all study areas from the isolated lakes in Tibetan Plateau (Zhang et al. 2016) to lakes in the most developed areas such as the Taihu Lake (Su et al. 2016). The MPs pollution in China is reported to be more serious than in other regions. Su et al. (2016) found MPs contamination in Taihu Lake with the abundance ranging from 0.1×10^6 to 6.8×10^6 items/km^2.

1.5.2 Microplastic in Rivers

The number of MP studies in rivers is mainly in North America and European countries. MPs have been found in rivers in most European countries. MPs were reported in the Tamar River of UK (Sadri and Thompson 2014); in Ofanto River of Italy; in the Rhine River traversing Germany, Netherlands, and Switzerland (Mani et al. 2015); in Swiss rivers (Faure et al. 2015); and in the Seine and Marne rivers of France (Dris et al. 2015). In North America, MPs studies have been conducted in rivers of the United States, such as the Los Angeles basin watershed (Moore et al. 2011), the North Shore Channel in Chicago (McCormick et al. 2014), and the Detroit and Niagara Rivers (Cable et al. 2017), as well as the St. Lawrence

Table 1.1 Summary of selected studies on MP contamination in natural freshwater systems.

Types	Study location	Sampling method	Sample process and analysis	Study finding	References
Lake water	Taihu Lake, China	Plankton net: 333 μm	H_2O_2 (30%); Visual, subset by micro-FIR or SEM/EDS	Max: 6.8×10^6 items/km^2; Min: 0.1×10^6 items/km^2	Su et al. (2016)
	Lake Hovsgol, Mongolia	Manta trawl: 333 μm	Density separation (saltwater, 1.6 g/cm^3), wet peroxide oxidation; Stereomicroscope (visual), subsample with DSC.	Max: 44 400 items/km^2 Mean: 20 264 items/km^2	Free et al. (2014)
	Lake Winnipeg, Canada	Manta trawl: 333 μm	Subsample, wet peroxide oxidation; Visual, subsample with SEM/EDX	Max: 748 000 items/km^2 Mean: 193000 items/km^2	Anderson et al. (2017)
	Great Lakes,USA	Manta trawl: 333 μm	Sieved, hydrochloric acid; Subsamples SEM/EDX	Max: 466 000 items/km^2 Mean: 43 000 items/km^2	Eriksen et al. (2013)
	Lake Bolsena, Italy	Manta trawl: 300 μm	Sieved, density separation (NaCl, 1.2 g/cm^3), HCl digestion, NR staining. Fluorescence microscopy, SEM.	Max: 4.42 particles/m^3 Min: 0.82 particles/m^3	(Fischer et al. 2016)
	Lake Geneva, Switzerland	Manta trawl: 333 μm	Sieved, wet peroxide oxidation; Stereomicroscope (visual), subsample with ATR-FTIR	Mean: 220 000 items/km^2	Faure et al. (2015)

River water	Tamar River, UK	Manta trawl: 300 µm	Sieved; ATR-FTIR and micro- FTIR	Max: 204 pieces of suspected plastic Mean: 0.028 items/m^3	Sadri and Thompson (2014)
	Ofanto River, Italy	Manta trawl: 333 µm	Sieved, H_2O_2 (30%); density separation (salt water, 1.2g/cm^3), Stereomicroscope (visual), Py–GC/MS	Min: 0.9 items/m^3; Max: 13 items/m^3	Campanale et al. (2020)
	Rhine river, Germany, Netherlands, Switzerland	Manta trawl: 300 µm	Enzymatic digestion, density separation (saltwater, 1.2g/cm^3); Stereomicroscope (visual), ATR-FTIR.	Max: 8.9×10^6 items/km^2 Mean: 3.9×10^6 items/km^2	Mani et al. (2015)
	St. Lawrence River, Canada	Grab samples	H_2O_2 (30%); density separation, NR staining; Fluorescent microscope	Mean: 0.12 items/L (upstream) Mean: 0.16 items/L (downstream)	Crew et al. (2020)
	Los Angeles River, USA	Manta trawl: 333 µm	Sieved. Stereomicroscope (visual)	9 items/m^3	Moore et al. (2011)
	Pearl River, China	Grab samples	H_2O_2 (30%); Stereomicroscope (visual), Raman spectroscopy	Mean: 19860 items/m^3	Yan et al. (2019)
	Yangtze River, China	Manta trawl: 112 µm	Sieved, visually screen, density separation; Stereomicroscope (visual), ATR-FTIR	Max: 13.6×10^6 items/km^2 Mean: 8.5×10^6 items/km^2	Zhang et al. (2015)
	Tamsui River, Taiwan	Manta trawl: 300 µm	Sieved, H_2O_2 (30%); ATR-FTIR	Min: 10.1 items/m^3 Max: 70.5 items/m^3	Wong et al. (2020)

River of Canada, (Crew et al. 2020). In Asia, most studies are carried out in the East region of the continent. MPs were found in many river systems of China such as Yangtze and Pearl (Yan et al. 2019; Zhang et al. 2015), and the Tamsui River of Taiwan (Wong et al. 2020). Similar to results from lake studies, the abundance of MPs found in river systems of China was much higher than in other areas. At the Yangtze River, the mean number of MPs was reported at 8.5×10^6 items/km^2. Another study at Pearl River found 19860 items/m^3. In river studies, the abundance of MPs is reported in different units. Some studies presented the number of MPs per filtered volume of water (items/m^3), while others presented it per area of water surface (items/km^2). Thus, the comparison between studies is difficult.

1.6 Conclusions and Recommendations

The number of studies of MP pollution in freshwater environments is growing in the scientific community due to the potential risks to the environment. The results of studies worldwide indicated a high abundance of MPs in freshwater environments. The pollutants were found in all continents of the world, from remote to densely populated areas. Although less data is available, current studies depicted that MPs in freshwater environments are ubiquitous, and concentrations are equivalent to the marine environment. MPs can be discharged into freshwater environments through various pathways such as effluent from WWTPs and during solid waste collection, processing, and landfilling. Therefore, techniques for wastewater treatment and solid waste management need to be improved to mitigate the MP pollution problems. Until now, most studies on MPs in freshwater environments has been conducted in the western hemisphere. Knowledge of the pollutants is insufficient in the other hemispheres such as in South America, Africa, and Asia. Moreover, there are no standard methods for sample collection, preparation, analysis, and reporting of MPs; as a result, data on MPs in freshwater environments cannot be compared easily. This limits further comprehension of MPs and the development of solutions to control the pollutants. Therefore, further studies are required to standardize methods to ensure consistency in monitoring MPs. Furthermore, studies of MPs in freshwater environments need to progress rapidly to fulfill knowledge gaps in the distribution and risks of MP pollution in the environments.

Acknowledgments

This study was funded by the Asia-Pacific Network for Global Change Research (CRRP2018-09MY-Babel). The authors would like to acknowledge a Ph.D. scholarship, provided to the first author by the Thailand Research Fund (PHD/0241/2560).

References

Alam, F.C., Sembiring, E., Muntalif, B.S., and Suendo, V. (2019). *Microplastic distribution in surface water and sediment river around slum and industrial area (Case study: Ciwalengke River, Majalaya district, Indonesia)*. Chemosphere 224: 637–645.

Anderson, P.J., Warrack, S., Langen, V. et al. (2017). *Microplastic contamination in lake Winnipeg, Canada. Environmental Pollution* 225: 223–231.

Andrady, A.L. (2011). *Microplastics in the marine environment. Marine Pollution Bulletin* 62: 1596–1605.

Andrady, A., Pegram, J., and Searle, N. (1996). *Wavelength sensitivity of enhanced photodegradable polyethylenes, ECO, and LDPE/MX. Journal of Applied Polymer Science* 62: 1457–1463.

Astrom, L. (2016). *Shedding of Synthetic Microfibers from Textiles.* University of Gotherburg.

Bakir, A., O'Connor, I.A., Rowland, S.J. et al. (2016). *Relative importance of microplastics as a pathway for the transfer of hydrophobic organic chemicals to marine life. Environmental Pollution* 219: 56–65.

Ballent, A., Corcoran, P.L., Madden, O. et al. (2016). *Sources and sinks of microplastics in Canadian Lake Ontario nearshore, tributary and beach sediments. Marine Pollution Bulletin* 110: 383–395.

Bergmann, M., Gutow, L., and Klages, M. (2015). *Marine Anthropogenic Litter.* Springer.

Blettler, M.C., Abrial, E., Khan, F.R. et al. (2018). *Freshwater plastic pollution: recognizing research biases and identifying knowledge gaps. Water Research* 143: 416–424.

Browne, M.A., Crump, P., Niven, S.J. et al. (2011). *Accumulation of microplastic on shorelines woldwide: sources and sinks. Environmental Science & Technology* 45: 9175–9179.

Cable, R.N., Beletsky, D., Beletsky, R. et al. (2017). *Distribution and modeled transport of plastic pollution in the Great Lakes, the world's largest freshwater resource. Frontiers in Environmental Science* 5: 45.

Campanale, C., Stock, F., Massarelli, C. et al. (2020). *Microplastics and their possible sources: the example of Ofanto river in Southeast Italy. Environmental Pollution* 258: 113284.

Carr, K.E., Smyth, S.H., McCullough, M.T. et al. (2012). *Morphological aspects of interactions between microparticles and mammalian cells: intestinal uptake and onward movement. Progress in Histochemistry and Cytochemistry* 46: 185–252.

Catarino, A.I., Thompson, R., Sanderson, W., and Henry, T.B. (2017). *Development and optimization of a standard method for extraction of microplastics in mussels by enzyme digestion of soft tissues. Environmental Toxicology and Chemistry* 36: 947–951.

Cheung, P.K. and Fok, L. (2016). *Evidence of microbeads from personal care product contaminating the sea. Marine Pollution Bulletin* 109: 582–585.

Cole, M., Lindeque, P., Halsband, C., and Galloway, T.S. (2011). *Microplastics as contaminants in the marine environment: a review. Marine Pollution Bulletin* 62: 2588–2597.

Cole, M., Webb, H., Lindeque, P.K. et al. (2014). *Isolation of microplastics in biota-rich seawater samples and marine organisms. Scientific Reports* 4: 4528.

Corcoran, P.L., Norris, T., Ceccanese, T. et al. (2015). *Hidden plastics of Lake Ontario, Canada and their potential preservation in the sediment record. Environmental Pollution* 204: 17–25.

Courtene-Jones, W., Quinn, B., Murphy, F. et al. (2017). *Optimisation of enzymatic digestion and validation of specimen preservation methods for the analysis of ingested microplastics. Analytical Methods* 9: 1437–1445.

Crew, A., Gregory-Eaves, I., and Ricciardi, A. (2020). *Distribution, abundance, and diversity of microplastics in the upper St. Lawrence River. Environmental Pollution* 260: 113994.

Dehaut, A., Cassone, A.-L., Frère, L. et al. (2016). *Microplastics in seafood: benchmark protocol for their extraction and characterization. Environmental Pollution* 215: 223–233.

Doyle, M.J., Watson, W., Bowlin, N.M., and Sheavly, S.B. (2011). *Plastic particles in coastal pelagic ecosystems of the Northeast Pacific ocean. Marine Environmental Research* 71: 41–52.

Dris, R., Gasperi, J., Rocher, V. et al. (2015). *Microplastic contamination in an urban area: a case study in Greater Paris. Environmental Chemistry* 12: 592–599.

Dris, R., Imhof, H.K., Löder, M.G. et al. (2018). Microplastic contamination in freshwater systems: methodological challenges, occurrence and sources. In: *Microplastic Contamination in Aquatic Environments: An Emerging Matter of Environmental Urgency* (ed. E.Y. Zeng), 51–93. Elsevier.

Eriksen, M., Mason, S., Wilson, S. et al. (2013). *Microplastic pollution in the surface waters of the Laurentian Great Lakes. Marine Pollution Bulletin* 77: 177–182.

Essel, R., Engel, L., Carus, M., and Ahrens, R. (2015). *Sources of microplastics relevant to marine protection in Germany*, Umweltbundesamt Texte, vol. 64, 1219–1226.

Fan, Y., Zheng, K., Zhu, Z. et al. (2019). *Distribution, sedimentary record, and persistence of microplastics in the Pearl River catchment, China. Environmental Pollution* 251: 862–870.

Faure, F., Demars, C., Wieser, O. et al. (2015). *Plastic pollution in Swiss surface waters: nature and concentrations, interaction with pollutants. Environmental Chemistry* 12: 582–591.

Fischer, E.K., Paglialonga, L., Czech, E., and Tamminga, M. (2016). *Microplastic pollution in lakes and Lake shoreline sediments – a case study on Lake Bolsena and Lake Chiusi (Central Italy). Environmental Pollution* 213: 648–657.

Foulon, V., Le Roux, F., Lambert, C. et al. (2016). *Colonization of polystyrene microparticles by Vibrio crassostreae: light and electron microscopic investigation. Environmental Science & Technology* 50: 10988–10996.

Free, C.M., Jensen, O.P., Mason, S.A. et al. (2014). *High-levels of microplastic pollution in a large, remote, mountain lake. Marine Pollution Bulletin* 85: 156–163.

Fries, E., Dekiff, J.H., Willmeyer, J. et al. (2013). *Identification of polymer types and additives in marine microplastic particles using pyrolysis-GC/MS and scanning electron microscopy. Environmental Science: Processes & Impacts* 15: 1949–1956.

Godoy, V., Martín-Lara, M., Calero, M., and Blázquez, G. (2019). *Physical-chemical characterization of microplastics present in some exfoliating products from Spain. Marine Pollution Bulletin* 139: 91–99.

Gouin, T., Avalos, J., Brunning, I. et al. (2015). *Use of micro-plastic beads in cosmetic products in Europe and their estimated emissions to the North Sea environment. SOFW Journal* 3: 40–46.

Horton, A.A., Walton, A., Spurgeon, D.J. et al. (2017). *Microplastics in freshwater and terrestrial environments: evaluating the current understanding to identify the knowledge gaps and future research priorities. Science of the Total Environment* 586: 127–141.

Imhof, H.K., Ivleva, N.P., Schmid, J. et al. (2013). *Contamination of beach sediments of a subalpine lake with microplastic particles. Current Biology* 23: R867–R868.

Imhof, H.K., Laforsch, C., Wiesheu, A.C. et al. (2016). *Pigments and plastic in limnetic ecosystems: a qualitative and quantitative study on microparticles of different size classes. Water Research* 98: 64–74.

Jiang, C., Yin, L., Li, Z. et al. (2019). *Microplastic pollution in the rivers of the Tibet Plateau. Environmental Pollution* 249: 91–98.

Kershaw, P. (2015). Sources, fate and effects of microplastics in the marine environment: a global assessment. Rep. Stud. GESAMP, 90.

Kirstein, I.V., Kirmizi, S., Wichels, A. et al. (2016). *Dangerous hitchhikers? Evidence for potentially pathogenic Vibrio spp. on microplastic particles. Marine Environmental Research* 120: 1–8.

Klein, S., Dimzon, I.K., Eubeler, J., and Knepper, T.P. (2018). *Analysis, occurrence, and degradation of microplastics in the aqueous environment.* In: *Freshwater Microplastics. The Handbook of Environmental Chemistry* (eds. M. Wagner and S. Lambert), 51–58,67. Cham: Springer.

Lambert, S. and Wagner, M. (e.) (2018). *Microplastics are contaminants of emerging concern in freshwater environments: an overview.* In: *Freshwater Microplastics. The Handbook of Environmental Chemistry*, vol. 58, 1–23. Cham: Springer.

Lambert, S., Sinclair, C., and Boxall, A. (2014). *Occurrence, degradation, and effect of polymer-based materials in the environment. Reviews of Environmental Contamination and Toxicology* 227: 1–53.

Lee, H., Shim, W.J., and Kwon, J.-H. (2014). *Sorption capacity of plastic debris for hydrophobic organic chemicals. Science of the Total Environment* 470: 1545–1552.

Leslie, H. (2014). *Review of microplastics in cosmetics. Institute for Environmental Studies [IVM]* 393: 394.

Li, J., Liu, H., and Chen, J.P. (2018). *Microplastics in freshwater systems: a review on occurrence, environmental effects, and methods for microplastics detection. Water Research* 137: 362–374.

Lusher, A., McHugh, M., and Thompson, R. (2013). *Occurrence of microplastics in the gastrointestinal tract of pelagic and demersal fish from the English Channel. Marine Pollution Bulletin* 67: 94–99.

Magnusson, K. and Norén, F. (2014). Screening of microplastic particles in and down-stream a wastewater treatment plant. Technical Report published for IVL Swedish Environmental Research Institute, Stockholm.

Mani, T., Hauk, A., Walter, U., and Burkhardt-Holm, P. (2015). *Microplastics profile along the Rhine River. Scientific Reports* 5: 1–7.

McCormick, A., Hoellein, T.J., Mason, S.A. et al. (2014). *Microplastic is an abundant and distinct microbial habitat in an urban river. Environmental Science & Technology* 48: 11863–11871.

Moore, C., Lattin, G., and Zellers, A. (2011). *Quantity and type of plastic debris flowing from two urban rivers to coastal waters and beaches of Southern California. Revista de Gestão Costeira Integrada-Journal of Integrated Coastal Zone Management* 11: 65–73.

Napper, I.E. and Thompson, R.C. (2016). *Release of synthetic microplastic plastic fibres from domestic washing machines: effects of fabric type and washing conditions. Marine Pollution Bulletin* 112: 39–45.

Napper, I.E., Bakir, A., Rowland, S.J., and Thompson, R.C. (2015). *Characterisation, quantity and sorptive properties of microplastics extracted from cosmetics. Marine Pollution Bulletin* 99: 178–185.

Peng, G., Xu, P., Zhu, B. et al. (2018). *Microplastics in freshwater river sediments in Shanghai, China: a case study of risk assessment in mega-cities. Environmental Pollution* 234: 448–456.

Plasticseurope (2018). Plastics – the Facts 2018. An analysis of European plastics production, demand and waste data. https://www.plasticseurope.org/application/files/6315/4510/9658/Plastics_the_facts_2018_AF_web.pdf (accessed 2019).

de Sá, L.C., Luís, L.G., and Guilhermino, L. (2015). *Effects of microplastics on juveniles of the common goby (Pomatoschistus microps): confusion with prey, reduction of the predatory performance and efficiency, and possible influence of developmental conditions. Environmental Pollution* 196: 359–362.

de Sá, L.C., Oliveira, M., Ribeiro, F. et al. (2018). *Studies of the effects of microplastics on aquatic organisms: what do we know and where should we focus our efforts in the future? Science of the Total Environment* 645: 1029–1039.

Sadri, S.S. and Thompson, R.C. (2014). *On the quantity and composition of floating plastic debris entering and leaving the Tamar Estuary, Southwest England. Marine Pollution Bulletin* 81: 55–60.

Schmidt, C., Krauth, T., and Wagner, S. (2017). *Export of plastic debris by rivers into the sea. Environmental Science & Technology* 51: 12246–12253.

Su, L., Xue, Y., Li, L. et al. (2016). *Microplastics in Taihu Lake, China. Environmental Pollution* 216: 711–719.

Ta, A. and Babel, S. (2019). *Microplastic pollution in surface water of the Chao Phraya River in Ang Thong area. EnvironmentAsia* 12: 48–53.

Ta, A. and Babel, S. (2020a). *Microplastic contamination on the lower Chao Phraya: abundance, characteristic and interaction with heavy metals. Chemosphere* 257: 127234.

Ta, A.T. and Babel, S. (2020b). *Microplastics pollution with heavy metals in the aquaculture zone of the Chao Phraya River Estuary, Thailand. Marine Pollution Bulletin* 161: 111747.

Ta, A.T., Babel, S., and Haarstrick, A. (2020c). *Microplastics contamination in a high population density area of the Chao Phraya River, Bangkok. Journal of Engineering & Technological Sciences* 52: 534–545.

Talvitie, J. and Heinonen, M. (2014). Preliminary study on synthetic microfibers and particles at a municipal waste water treatment plant. *Baltic Marine Environment Protection Commission HELCOM*, 1–14.

Tamminga, M., Hengstmann, E., and Fischer, E. (2017). *Nile Red staining as a subsidiary method for microplastic quantification: a comparison of three solvents and factors influencing application reliability. SDRP Journal of Earth Sciences and Environmental Studies* 2: 165–172.

Tan, X., Yu, X., Cai, L. et al. (2019). *Microplastics and associated PAHS in surface water from the Feilaixia Reservoir in the Beijiang River, China. Chemosphere* 221: 834–840.

Ter Halle, A., Ladirat, L., Gendre, X. et al. (2016). *Understanding the fragmentation pattern of marine plastic debris. Environmental Science & Technology* 50: 5668–5675.

Van Cauwenberghe, L., Devriese, L., Galgani, F. et al. (2015). *Microplastics in sediments: a review of techniques, occurrence and effects. Marine Environmental Research* 111: 5–17.

Viršek, M.K., Lovšin, M.N., Koren, Š. et al. (2017). *Microplastics as a vector for the transport of the bacterial fish pathogen species Aeromonas salmonicida. Marine Pollution Bulletin* 125: 301–309.

Wang, W., Ndungu, A.W., Li, Z., and Wang, J. (2017). *Microplastics pollution in inland freshwaters of China: a case study in urban surface waters of Wuhan, China. Science of the Total Environment* 575: 1369–1374.

Willén, A., Junestedt, C., Rodhe, L. et al. (2017). *Sewage sludge as fertiliser–environmental assessment of storage and land application options. Water Science and Technology* 75: 1034–1050.

Wong, G., Löwemark, L., and Kunz, A. (2020). *Microplastic pollution of the Tamsui River and its tributaries in northern Taiwan: spatial heterogeneity and correlation with precipitation. Environmental Pollution* 260: 113935.

Yan, M., Nie, H., Xu, K. et al. (2019). *Microplastic abundance, distribution and composition in the Pearl River along Guangzhou City and Pearl River estuary, China. Chemosphere* 217: 879–886.

Yin, L., Wen, X., Du, C. et al. (2020). *Comparison of the abundance of microplastics between rural and urban areas: a case study from East Dongting Lake. Chemosphere* 244: 125486.

Zhang, K., Gong, W., Lv, J. et al. (2015). *Accumulation of floating microplastics behind the Three Gorges Dam. Environmental Pollution* 204: 117–123.

Zhang, K., Su, J., Xiong, X. et al. (2016). *Microplastic pollution of lakeshore sediments from remote lakes in Tibet plateau, China. Environmental Pollution* 219: 450–455.

Zhao, S., Danley, M., Ward, J.E. et al. (2017). *An approach for extraction, characterization and quantitation of microplastic in natural marine snow using Raman microscopy. Analytical Methods* 9: 1470–1478.

Zylstra, E. (2013). *Accumulation of wind-dispersed trash in desert environments. Journal of Arid Environments* 89: 13–15.

2

Microplastics in Freshwater Environments – With Special Focus on the Indian Scenario

Sumi Handique

Department of Environmental Science, Tezpur University, Tezpur, Assam, India

2.1 Introduction

Over the last 50 years, plastic has revolutionized the way we live and is now an essential part of our lives. Globally, plastic wastes have increased at a staggering rate in the last few decades, and around 79% are disposed of in landfills or in the surrounding environment with no proper waste management (Geyer et al. 2017). Out of this, a staggering amount of plastic waste, approximately 4.8–12.7 million tons, is estimated to enter the oceans each year, a large quantity of which comes from land-based sources and is transported by fluvial or aeolian processes (Jambeck et al. 2015). Rivers are one of the main contributors of plastic waste to the oceans, and are reported to carry 2 million tons of MPs annually (Lebreton et al. 2017). It has been investigated and found that this river-transported plastic makes up 80% of the plastic debris released from the terrestrial environment to the oceans (Horton et al. 2017; Law & Thompson 2014). With this growing awareness of the importance of the riverine source of plastic wastes in the marine environment in recent years, several studies have been carried out in various world rivers. These include the Los Angeles River (Moore et al. 2005), Danube (Lechner et al. 2014), Yangtze Estuary (Zhao et al. 2014), Rhine (Mani et al. 2015), Selenga River (Battulga et al. 2019), Beijiang River (Tan et al. 2019), Ciwalengke River (Alam et al. 2019), and others. This chapter discusses the various works done with freshwater MPs across the globe, with a special focus on studies in India.

2.2 The Nature and Production of Microplastics

The Joint Group of Experts on the Scientific Aspects of Marine Environmental Protection (GESAMP) (2015) defines plastic as a synthetic, water-insoluble polymer, generally of petrochemical origin, that can be molded on heating and designed into various shapes to be maintained during use (Arthur et al. 2009; Lassen et al. 2015).

Plastic and Microplastic in the Environment: Management and Health Risks, First Edition.
Edited by Arif Ahamad, Pardeep Singh, and Dhanesh Tiwary.
© 2022 John Wiley & Sons Ltd. Published 2022 by John Wiley & Sons Ltd.

The plastic pollutants encompass a wide range of synthetic polymers, some of which are polyethylene terephthalate, high-density polyethene, polyvinyl chloride, polyethene, polypropylene, and polystyrene. These present a wide range of different sizes of plastic polymer materials that are present (meter to micrometers) in our surroundings. According to National Oceanic and Atmospheric Administration (NOAA) and the European Chemicals Agency, MPs are small plastic pieces less than five millimeters.

After entering the ecosystems, these plastic particles undergo degradation and fragmentation processes and it becomes difficult to identify and remove them, particularly the smaller size fractions. These particles are water-insoluble and not easily biodegradable, and are chemically durable over long periods. These MP pollutants easily move long distances through aeolian transport (Gasperi et al. 2018) and water, and accumulate in the environment. Thus, due to the validation of the long-range transport of plastics via air and water, the misconception of plastics being a local junk or waste as thought of a few decades ago is now being rebuffed, and plastics are now acknowledged as a serious threat to the global environment.

The threat of plastic pollution can be managed early through efficient source identification. In recent years many researchers have reported potential MP sources, fluxes, and sinks of these pollutants using theoretical models (Figure 2.1) across a wide variety of hydrological reservoirs (Alimi et al. 2018; Browne 2015; de Souza Machado et al. 2018; Horton et al. 2017; Horton & Dixon 2018; Nizzetto et al. 2016; Wagner et al. 2014). Extensive risk management and assessment of this emerging contaminant requires a proper and exhaustive understanding and quantification of the sources and emissions pathways across the world, spatially as well as temporally, with a special focus on freshwater sources and fluxes, This will help in the understanding of the sources, fluxes, and sinks, and contribute immensely to a proper estimation of the global budget of plastics input to the oceans, which can be useful for source mitigation schemes and for planning long-standing monitoring and assessment strategies.

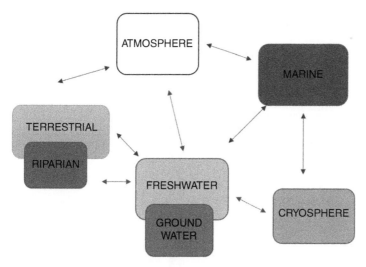

Figure 2.1 Showing the pathways of plastic fluxes across the various hydrological reservoirs, indicated by brown arrows, which represent the fluxes where extensive research is going on. Blue arrows represent plastic fluxes (theoretical) that have yet to be studied in detail *Source:* Modified from Windsor et al. 2019.

2.3 Global Ecological Impacts of Plastic Pollution

The harmful effect of plastics on the various life forms and the ecology may be due to various feedback mechanisms (Figure 2.2). While available literature reports mostly physical effects due to exposure to this pollutant on organisms or ecosystem function, chemical effects due to various processes such as adsorption and bioaccumulation of toxic chemicals, and the impacts due to leaching of harmful additives in plastics have not been investigated.

One of the most observed hazardous physical effects of plastic pollution that has been found in various studies is entanglement and external physical damage to larger organisms from plastic items such as fishing nets and rope (Jacobsen et al. 2010). The smaller organisms also face problems such as zooplanktons exposed to MPs, which suffered from antennal and carapace deformities (Ziajahromi et al. 2017). Moreover, due to the similarity in size of these debris with that of the larvae of several organisms and planktons, many aquatic life forms can suffer by ingesting these plastics (Besseling et al. 2014; Boerger et al. 2010; Browne et al. 2008; Kaposi et al. 2014; Tanaka & Takada 2016). Recently, several studies have investigated that the hazardous chemicals can be transported into organisms

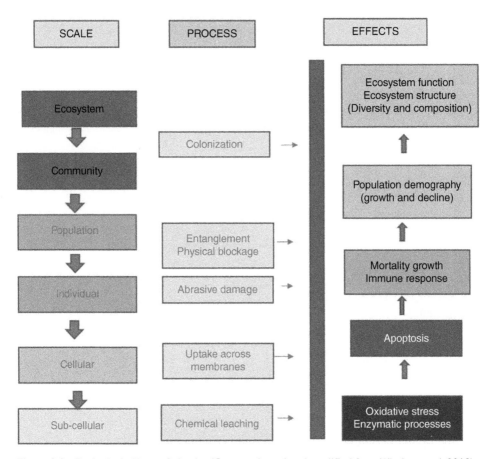

Figure 2.2 Ecological effects of plastics (*Source:* adapted and modified from Windsor et al. 2019).

through various pathways, and these can eventually cause health hazards in humans by reaching through the food chain, which is another concern (Besseling et al. 2017; Browne et al. 2013; Koelmans et al. 2013; Koelmans et al. 2016; Rochman et al. 2015; Tanaka et al. 2013, 2015; Thompson et al. 2009; Van Cauwenberghe & Janssen 2014; Wright & Kelly 2017). A recent study (Wilson et al. 2020) revealed that plastics represent the most diverse habitat for invertebrates in some rivers, which is a sad state of affairs.

2.4 Socio-Economic Impacts of Plastic Pollution

Plastic delivers many benefits to society and is responsible for the promotion of a wide range of technological advances which have revolutionized our day-to-day lives. However, increasing awareness and available records of potential environmental impacts, at present mostly on the marine world (Thompson 2017), are forcing us to stress upon the potential negative impact on the economy and various industries, such as fishing and tourism. Very scattered and scarce data are available, mostly restricted to local impacts, yet these are indicative of widespread global socio-economic effects due to plastic pollution. The fishing industry, in particular, is likely to suffer detrimental impacts of plastic pollution due to reducing amount and damaged catches (Thompson 2017); one such study surveyed fishing vessels in Scotland and revealed that 86% of the fishing vessels reported that plastic pollution in the sea resulted in a reduction in fish catches (Mouat et al. 2010). In addition, another serious problem is entanglement reported within marinas and harbors, with 70% of the marinas and harbors that were surveyed reported that they experience various problems due to plastic litter (Mouat et al. 2010).

Contaminated catches may also result in significant economic loss due to the high concentration of plastics in the fish stocks (Foekema et al. 2013; Lusher et al. 2013). Moreover, it may have a detrimental effect on the salability of commercial fishes due to the negative public perception of these contaminated supplies (GESAMP 2016). Another industry likely to be affected is tourism, as public perceptions of esthetically pleasing and clean sites are likely to influence people's choices of places to visit. Tourists visiting the coasts may not have positive feedback about the locations where they found litter, affecting the market value of the place as a tourist destination (Brouwer et al. 2017), and litter on beaches can cause physical injury (Werner et al. 2016). Thus, to boost the tourism sector and for environmental reasons, the local authorities execute cleanliness drives (Mouat et al. 2010). Plastic wastes are causing huge problems in tourist places, e.g. blockage in sewage drains and the combined costs of removing these plastics as well as the loss in tourists will have a negative effect on the tourism sector (Drinkwater & Moy 2017).

2.5 Freshwater Plastic Pollution

Freshwater plastic pollution is an emerging hazard. To overcome this global challenge, reliable data on river plastic transport and accumulation is key. Approximately 4 million tons of plastic waste enters the oceans via rivers every year. River health is intrinsically linked to marine health; there is no separating the two. Why are people more concerned about plastic in the ocean than they are about plastic in rivers? Most ocean plastic 'litter' originates on land, with major rivers providing important source-to-sink pathways into the ocean. How effectively the problem of plastic pollution in the ocean can be resolved is dependent on establishing effective methods for surveying, quantifying, and modeling pollution pathways on land and examining how these can vary through time. Because it will be a herculean task to clean up the plastic debris that have already filled up the oceans, we must take necessary steps and precautions to reduce the amount of plastic entering the marine environment. To prevent the plastic debris from entering the sea, the path taken by the plastic has to be elucidated so that immediate and efficient steps can be taken.

To date, there is very little available data on the quantification and characterization of riverine MPs. As the impact of plastic pollution increases with decreasing particle size, the investigation of MPs (particles <5 mm) is particularly relevant. MPs particles, ranging from microns to millimeters in size, pose a significant risk to natural ecosystems and habitats. However, despite the potential ecological impacts from MPs pollution, the ability to accurately predict MP transport by environmental flows (e.g. in rivers, estuaries, and coastal currents) is limited. It is important to understand how they are transported to predict their dispersion and behavior, and ultimately understand their impact on ecological and human health. The extensive quantification of major rivers as sources of MPs pollution is yet to be established, and extensive research is needed to focus on transport pathways, fluxes, and fate of this emerging pollutant to understand the threat it poses to human health and ecosystems across the world (Alimi et al. 2018; Browne 2015; de Souza Machado et al. 2018; Horton et al. 2017; Horton and Dixon 2018; Nizzetto et al. 2016; Wagner et al. 2014).

2.5.1 Sources of Freshwater Microplastics

According to literature, MPs sources can be categorized as primary sources or secondary sources (Andrady 2011; Cole et al. 2011). We can refer to primary sources as those MP particles released into the environment which were manufactured in the range of micrometers, whereas secondary sources are those sources that produce the MPs by the process of disintegration and/or breakdown of larger plastic particles due to various physical and chemical mechanisms (e.g. exposure to sun rays, weathering, mechanical wear and tear, etc.). Faure et al. (2012) and Eriksen et al. (2013) propose three major sources: (i) effluent from wastewater treatment plants (WWTP); (ii) sewage treatment overflow during high-volume rain events; and (iii) runoff from agricultural or public lands. In recent literature, sources of MPs in personal care and cosmetic products and clothing have been reported (Gouin et al. 2015). Sundt et al. (2014) tried to attempt a detailed study of source apportionment for Norway; primary as well as secondary sources that release MPs and in the study, and made the conclusion that the MPs from tire dust are the sources for the largest contribution in the Baltic Sea with a small contribution from other consumer products. Data for other terrestrial sources of MPs are almost nonexistent, for example, agricultural and urban soils (Lwanga et al. 2017). Literature suggests that potential sources of MPs from agricultural equipment or use of WWTP products in agriculture along with landfill waste disposal sites soil pose a threat to the environment, which have not been assessed in detail (Wagner et al. 2014). As per literature, polythene (PE) and polyethylene terephthalate (PET) are found to be the major polymer type found in freshwater samples (Figure 2.3, Li et al. 2020).

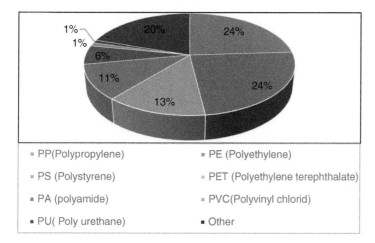

Figure 2.3 Composition of microplastics found in freshwater samples *Source:* Modified from Li et al. 2020.

2.5.2 Studies on Freshwater Plastic Pollution from around the World

As we are becoming increasingly aware of the problems caused by plastic to marine life, the terrestrial environments and freshwaters are now being recognized as the source and transport pathways of plastics to the oceans (Horton et al. 2017), yet how MPs move from terrestrial to marine environments is poorly understood due to scarcity of data (Eerkes-Medrano et al. 2015). Thus, there is an ardent need to focus on the freshwater sources of plastic pollution and discover ways to minimize it before it reaches the oceans. Studies in freshwater environments have received some focus in recent years, for example, recent studies include lakes (Horton et al. 2017; Imhof et al. 2016; Fischer et al. 2016); tributaries of the Great Lakes (Baldwin et al. 2016), the Seine River (Dris et al. 2015, 2018), various rivers in Switzerland (Faure et al. 2015), the Rhine River (Mani et al. 2015), various river sites near Chicago (McCormick et al. 2014), and the Danube River (Klein et al. 2015; Lechner et al. 2014). Some researchers working on river plastic pollution (Blettler & Wantzen 2019; Islam & Tanaka 2004; Jambeck et al. 2015; Lebreton et al. 2017; Lechner et al. 2014; Nollkaemper 1994; Schmidt et al. 2017) have identified riverine source to be an important contributor to the ocean environment. It is now a widely accepted fact that rivers are major carriers of plastic debris to the marine environment (Carr et al. 2016; Jambeck et al. 2015; Lebreton et al. 2017) for primary as well as secondary MPs (Andrady 2011). Castro-Jiménez et al. (2019) describe the transport of macro-plastics from the Rhone to the Mediterranean, and research from Indonesia (van Emmerik et al. 2019a) studied the movement of plastic wastes through rivers to the oceans. The ongoing research on freshwater plastic pollution includes topics such as monitoring and analyzing concentrations of MP river debris (Dris et al. 2015; Mani et al. 2015; Verster et al. 2017; Yonkos et al. 2014), outflow flux estimations to the marine environment (Lebreton et al. 2017; Mani et al. 2015; Nizzetto et al. 2016; Schmidt et al. 2017; Siegfried et al. 2017), MP source apportionment and identification (Carr et al. 2016; Leslie et al. 2013; McCormick et al. 2016; Murphy et al. 2016; Yonkos et al. 2014; Ziajahromi et al. 2016), and the effect of MPs on freshwater life forms (Besseling et al. 2017; Dris et al. 2015; Hoellein et al. 2017; Wagner et al. 2014). Researchers suggest that population, land use, basin characteristics, and hydrology (Lebreton et al. 2017) have a significant relationship with riverine plastic concentrations, and have found that MPs are concentrated in rivers near areas with high population density (Mani et al. 2015; Yonkos et al. 2014). Some researchers have used various mathematical models to theoretically estimate the transport of MPs in watersheds (Nizzetto et al. 2016).

Two of the areas in riverine plastic pollution studies where there is almost no data is how MPs transport changes along the river, and that temporal variation remains unknown. More studies can help in increasing our understanding of the origins, sinks, and accumulation zones in catchments. Most studies that we have come across focused on the movement of MPs plastic in specific river cross-sections (Crosti et al. 2018; van Emmerik et al. 2018) or the output from complete river systems (Tramoy et al. 2019). A study conducted in the Los Angeles rivers indicated significant temporal variations in plastic transport within one year (Moore et al. 2011). A study in rivers of Indonesia even suggested monthly variations on plastic transport (van Emmerik et al. 2019a,b).

The plastic emission from Asian rivers is estimated to be significantly high, which may be due to various factors such as high population density, a large quantity of primary MPs production, and hydrological regimes with heavy rainfalls. This results in huge MP waste transport from Asian continent to the oceans; 86% of the total global input, with an estimated annual input of 1.21 million tons (Lebreton et al. 2017). Lebreton et al. (2017) estimated that the Chinese Yangtze River catchment is the largest contributor, followed by the Ganges River catchment. With the growing awareness in recent years, some studies are focusing on Asian rivers (Blettler et al. 2018; van Emmerik et al. 2019a; Jambeck et al. 2015; OCMCBE 2015). Surface samplings at the Chinese Yangtze River mouth showed considerably higher plastic concentrations than any other sampled river worldwide (Zhao et al. 2014), with a reported 4137 particles per cubic meter. The significant differences between sampled estuarine concentrations and nearshore monitoring in the area confirmed that the Yangtze River is a major regional source of plastic input into the marine environment. Kataoka et al. (2019) reported of the MP concentrations on 29 Japanese river surfaces, which may be a source of MPs for the MP hotspot in the East Asian seas. They found MPs in 31 of the 36 sites and demonstrated that concentration of MPs in the river basins were dependent on population density, urbanization, and biological oxygen demand (BOD), which suggested that river water quality and plastic pollution in rivers are related.

2.5.3 The Problem of Freshwater Microplastics in Developing Countries

The problem of freshwater plastic pollution occurs more in developing countries due to the lack of proper waste management systems in these countries. As a result, the waste is directly disposed of into open areas or water bodies like rivers, from which they reach the oceans (OCMCBE 2015; Kaza et al. 2018). As observed by van Emmerik et al. (2019a), rivers in South East Asia are known to be major contributors of plastic waste into the marine ecosystem, but with no proper observation and monitoring of these riverine plastic wastes, the sources and fate of these MPs remain unknown. The current research on freshwater plastic pollution is mainly focused on the developed countries with almost no proper monitoring in the developing countries (Blettler et al. 2018; Blettler & Wantzen 2019). This disparity is mainly because the countries doing quality research in sciences (especially experimental) are countries with a high-income economy (OECD 2017). From an environmental perspective this cannot be sustainable, as the developing countries are generating the largest amount of solid wastes due to high population growth and high consumption rate, with no proper solid waste management (UNHSP 2016). Furthermore, many developed countries are dumping their solid wastes by exporting to these low-income countries, adding to the environmental issues in these countries. Lastly, the largest river drainage systems are located in these countries (Latrubesse 2008) and they are now recognized as the largest potential contributors of plastic to the marine ecosystem. Thus, due to these reasons, extensive quantification and monitoring of sources and transport pathways, as well as fluxes of riverine MPs in these countries is urgently required.

2.5.4 Status of India's Freshwater Plastic Problem

The plastic consumption rate among the Indian population is among the highest in the world, and results in the generation of approximately 5.6 million tons of plastic waste per year (Toxics Link 2014). Not surprisingly, India is described as the twelfth largest contributor of marine plastic pollution in the world (Jambeck et al. 2015), but very few studies to date have investigated the presence of MPs in the water or sediment in freshwater or estuarine environments in India. Recently there have been a few studies on MPs pollution in India but most of these studies are on coastal or marine sediments (Jayasiri et al. 2013; Reddy et al. 2006; Robin et al. 2020; Suman et al. 2020) and studies on fresh water environments are almost non-existent (Table 2.1). Owens & Kamil (2020) carried out a study of plastic pollution in a river in Kerela, India (Karamana River) and Indonesia (Tukad Badung River).

   In one of the first reports on MPs in the freshwater environments in India, Sruthy & Ramasamy (2017) studied the sediments of Vembanad Lake, a Ramsar site in India with low-density polythene dominating the sediment samples. The authors pointed out that as the locals consume the aquatic fishes and clams from this lake, the fate of MPs entering humans via the food web is a potential threat. Ram & Kumar (2020) studied MPs from Sabarmati River sediments, where they reported that higher amount of MPs were observed in the river in areas near landfill sites from where the surface runoff might have carried the plastic debris to the river. Sarkar et al. (2019) estimated distribution of meso- and microplastics in the sediments of the lower reaches of the river Ganga, where they observed a relation between MPs abundance and other water quality parameters such as BOD. Karthik et al. (2018) studied MPs particles at beaches along the southeast coastal region of India, where they found the highest abundance of MPs on beaches adjacent to the river mouth. They also found the MPs in 10.1% of the 79 fishes they studied. Reddy et al. (2006) reported the observed plastic debris in the marine sediments on the coast of Gujarat, and a group of researchers reported plastic particles in the beaches of Mumbai (Jayasiri et al. 2013). Veerasingam et al. (2016) studied the MPs in surface sediments along the Chennai coast during March 2015 (pre-Chennai flood) and November 2015 (post-Chennai flood) and found that the MPs in the sediments increased threefold in post flooding, which may be due to huge input of MPs through the Cooum and Adyar rivers during the flood. This study highlights the importance of rivers as sources of plastic pollution to the marine environment.

Table 2.1 Concentrations and sizes of microplastics reported in samples from freshwater environments (studies from India).

Sl. No.	Location	Average concentration	Method	Sample type	Size	Polymer type	References
1	River Ganga	11.48–63.79 ng/g	FT-IR	Sediments	63–850 μm, 850 μm–5 mm	PET, PE	Sarkar et al. (2019)
2	Sabarmati River	134.53–581.70 mg/kg	SEM	Sediments	4 mm–75 μm	Plastic debris and fibers	Ram & Kumar (2020)
3	Vembanad Lake	0.27 g/l	Raman	Sediments	0.2–1 mm	*HDPE, LDPE, PP, PS	Sruthy & Ramasamy (2017)
4.	Netravathi River	288 pieces/m^3 (water), 96 pieces/kg (sediment) 84.45 pieces /kg (soil)	–	Water, sediments, and soil	5–0.3 mm	PE, PET	Amrutha & Warrier (2020)

* Low-density polyethylene (LDPE) High-density polyethylene (HDPE).

In India, plastic litter is documented to have caused serious damage to biodiversity in places like Cochin, Lakshadweep, Sutrapada, Vembanad Lake, Chilika Lake, Mandapam, Kilakkarai, Erwadi and Periyapattinam. There are multiple cases of ingestion and entanglement from plastic debris leading to mortality of marine mammals and birds in the country. The tourism and fishing industries are significant contributors of plastic pollution in India, especially in the coastal areas, as these areas are being affected by plastic pollution. Recently, eight beaches in India have been recommended for the coveted "Blue Flag" international eco-label, and we should be able to reduce plastic pollution to maintain the clean status of all our tourist places.[†]

[†] The 'Blue Flag' is a certification that can be obtained by a beach, marina, or sustainable boating tourism operator, and serves as an eco-label. The certification is awarded by the Denmark-based non-profit Foundation for Environmental Education, which sets stringent environmental, educational, safety-related, and access-related criteria that applicants must meet and maintain. It is awarded annually to beaches and marinas in FEE member countries.

Amrutha & Warrier (2020) presented a detailed source-to-sink characterization of freshwater MPs collected from the water, sediment, and soil samples of the Netravathi River catchment, a tropical Indian river. The authors reported that sampling sites close to important religious places were observed to have higher concentration of MPs fibers, which may be due to garments washing.

2.6 Conclusion and Future Prospects

Although the awareness of plastic pollution is increasing, the knowledge of the scale and severity of impacts on humans and ecosystems is limited. With the growing awareness of the potential threat and subsequent consequences of plastic pollution to human populations, more detailed investigations of these emerging pollutants are needed for effective management action and risk assessment to reduce the detrimental social, ecological, and economic impacts. Detailed studies on qualitative and quantitative estimation of macro- and microplastics with strong focus on predicting current and future trends across spatial and temporal scales, using portable sensors, etc., is imperative to address the global plastic wastes. Together with a detailed understanding and quantification of the origins, transport pathways, fluxes, and fate of these emerging pollutants, urgent steps to manage plastic wastes such as making laws banning plastic littering more stringent, providing for plastic waste management infrastructure, and making waste segregation at source mandatory, are required to determine and reduce the risks to marine ecosystem as well as all other ecosystems and humans as a whole.

References

Alam, F.C., Sembiring, E., Muntalif, B.S., and Suendo, V. (2019). *Microplastic distribution in surface water and sediment river around slum and industrial area (case study: Ciwalengke River, Majalaya district, Indonesia)*. Chemosphere 224: 637–645.

Alimi, O.S., Farner Budarz, J., Hernandez, L.M., and Tufenkji, N. (2018). *Microplastics and nanoplastics in aquatic environments: aggregation, deposition, and enhanced contaminant transport*. Environmental Science & Technology 52 (4): 1704–1724.

Amrutha, K. and Warrier, A.K. (2020). *The first report on the source-to-sink characterization of microplastic pollution from a riverine environment in tropical India*. Science of the Total Environment 739: 140377.

Andrady, A.L. (2011). *Microplastics in the marine environment*. Marine Pollution Bulletin 62 (8): 1596–1605.

Arthur, C., Baker, J.E. and Bamford, H.A. (2009). *Proceedings of the International Research Workshop on the Occurrence, Effects, and Fate of Microplastic Marine Debris* (September 9–11, 2008), University of Washington Tacoma, Tacoma, WA, USA. NOAA Technical Memorandum.

Baldwin, A.K., Corsi, S.R., and Mason, S.A. (2016). *Plastic debris in 29 Great Lakes tributaries: relations to watershed attributes and hydrology*. Environmental Science & Technology 50 (19): 10377–10385.

Battulga, B., Kawahigashi, M., and Oyuntsetseg, B. (2019). *Distribution and composition of plastic debris along the river shore in the Selenga River basin in Mongolia. Environmental Science and Pollution Research* 26 (14): 14059–14072.

Besseling, E., Wang, B., Lürling, M., and Koelmans, A.A. (2014). *Nanoplastic affects growth of S. obliquus and reproduction of D. magna. Environmental Science & Technology* 48 (20): 12336–12343.

Besseling, E., Quik, J.T., Sun, M., and Koelmans, A.A. (2017). *Fate of nano-and microplastic in freshwater systems: a modeling study. Environmental Pollution* 220: 540–548.

Blettler, M.C. and Wantzen, K.M. (2019). *Threats underestimated in freshwater plastic pollution: mini-review. Water, Air, & Soil Pollution* 230 (7): 174.

Blettler, M.C., Abrial, E., Khan, F.R. et al. (2018). *Freshwater plastic pollution: recognizing research biases and identifying knowledge gaps. Water Research* 143: 416–424.

Boerger, C.M., Lattin, G.L., Moore, S.L., and Moore, C.J. (2010). *Plastic ingestion by planktivorous fishes in the North Pacific Central Gyre. Marine Pollution Bulletin* 60 (12): 2275–2278.

Brouwer, R., Hadzhiyska, D., Ioakeimidis, C., and Ouderdorp, H. (2017). *The social costs of marine litter along European coasts. Ocean & Coastal Management* 138: 38–49. https://doi.org/10.1016/J.OCECOAMAN.2017.01.011.

Browne, M.A. (2015). Sources and pathways of microplastics to habitats. In: *Marine Anthropogenic Litter* (eds. M. Bergmann, L. Gutow and M. Klages), 229–244. Cham: Springer https://doi.org/10.1007/978-3-319-16510-3_9.

Browne, M.A., Dissanayake, A., Galloway, T.S. et al. (2008). *Ingested microscopic plastic translocates to the circulatory system of the mussel, Mytilusedulis (L.). Environmental Science & Technology* 42 (13): 5026–5031.

Browne, M.A., Niven, S.J., Galloway, T.S. et al. (2013). *Microplastic moves pollutants and additives to worms, reducing functions linked to health and biodiversity. Current Biology* 23 (23): 2388–2392.

Carr, S.A., Liu, J., and Tesoro, A.G. (2016). *Transport and fate of microplastic particles in wastewater treatment plants. Water Research* 91: 174–182.

Castro-Jiménez, J., González-Fernández, D., Fornier, M. et al. (2019). *Macro-litter in surface waters from the Rhone River: plastic pollution and loading to the NW Mediterranean Sea. Mar. Pollut. Bull* 146: 60–66. https://doi.org/10.1016/j.marpolbul.2019.05.067.

Cole, M., Lindeque, P., Halsband, C., and Galloway, T.S. (2011). *Microplastics as contaminants in the marine environment: a review. Marine pollution bulletin* 62 (12): 2588–2597.

Crosti, R., Arcangeli, A., Campana, I. et al. (2018). *'Down to the river': amount, composition, and economic sector of litter entering the marine compartment, through the Tiber river in the Western Mediterranean Sea. RendicontiLincei. ScienzeFisiche e Naturali* 29 (4): 859–866.

Drinkwater, A. and Moy, F. (2017). *Wipes in sewer blockage study – Final report.* Report Ref. No. http://21CDP.WS4.WS. Water UK, London, UK (accessed 26 October 2021).

Dris, R., Imhof, H., Sanchez, W. et al. (2015). *Beyond the ocean: contamination of freshwater ecosystems with (micro-) plastic particles. Environmental Chemistry* 12 (5): 539–550.

Dris, R., Johnny, G., and Bruno, T. (2018). Sources and fate of microplastics in urban areas: a focus on Paris megacity. In: *Freshwater Microplastics the Handbook of Environmental Chemistry* (eds. M. Wagner and S. Lambert), 69–83. Cham: Springer https://doi.org/10.1007/978-3-319-61615-5_4.

Eerkes-Medrano, D., Thompson, R.C., and Aldridge, D.C. (2015). *Microplastics in freshwater systems: a review of the emerging threats, identification of knowledge gaps and prioritisation of research needs. Water research* 75: 63–82.

van Emmerik, T., Kieu-Le, T.C., Loozen, M. et al. (2018). *A methodology to characterize riverine macroplastic emission into the ocean. Frontiers in Marine Science* 5: 372.

van Emmerik, T., Loozen, M., Van Oeveren, K., and Buschman, F. (2019a). *Riverine plastic emission from Jakarta into the ocean. Environmental Research Letters* 14 (8): 084033.

van Emmerik, T., Tramoy, R., van Calcar, C. et al. (2019b). *Seine plastic debris transport tenfolded during increased river discharge. Front. Mar. Sci.* 6: 642. https://doi.org/10.3389/fmars.2019.00642.

Eriksen, M., Mason, S., Wilson, S. et al. (2013). *Microplastic pollution in the surface waters of the Laurentian Great Lakes. Marine Pollution Bulletin* 77 (1–2): 177–182.

Faure, F., Corbaz, M., Baecher, H., and De Alencastro, L.F. (2012). *Pollution due to plastics and microplastics in Lake Geneva and in the Mediterranean Sea. Archives des Sciences* 65: 157–164.

Faure, F., Demars, C., Wieser, O. et al. (2015). *Plastic pollution in Swiss surface waters: nature and concentrations, interaction with pollutants. Environmental Chemistry* 12 (5): 582–591.

Fischer, E.K., Paglialonga, L., Czech, E., and Tamminga, M. (2016). *Microplastic pollution in lakes and Lake shoreline sediments – a case study on Lake Bolsena and Lake Chiusi (Central Italy). Environmental Pollution* 213: 648–657.

Foekema, E.M., De Gruijter, C., Mergia, M.T. et al. (2013). *Plastic in North Sea fish. Environmental Science & Technology* 47 (15): 8818–8824.

Gasperi, J., Wright, S.L., Dris, R. et al. (2018). *Microplastics in air: are we breathing it in? Current Opinion in Environmental Science & Health* 1: 1–5.

GESAMP (2015). Sources, fate and effects of microplastics in the marine environment: a global assessment. In: *IMO/FAO/UNESCO-IOC/UNIDO/WMO/IAEA/UN/UNEP/UNDP Joint Group of Experts on the Scientific Aspects of Marine Environmental Protection*, Rep. Stud.–GESAMP, No. 90 (ed. P.J. Kershaw), 96.

GESAMP (2016). Sources, fate and effects of microplastics in the marine environment: part two of a global assessment. In: *IMO/FAO/UNESCO-IOC/WMO/WHO/IAEA/UN/UNEP Joint Group of Experts on the Scientific Aspects of Marine Environmental Protection* (eds. P.J. Kershaw and C.M. Rochman), 220. London, UK: International Maritime Organization.

Geyer, R., Jambeck, J.R., and Law, K.L. (2017). *Production, use, and fate of all plastics ever made. Science Advances* 3 (7): e1700782.

Gouin, T., Avalos, J., Brunning, I. et al. (2015). *Use of micro-plastic beads in cosmetic products in Europe and their estimated emissions to the North Sea environment. SOFW Journal* 141 (3): 40–46.

Hoellein, T.J., McCormick, A.R., Hittie, J. et al. (2017). *Longitudinal patterns of microplastic concentration and bacterial assemblages in surface and benthic habitats of an urban river. Freshwater Science* 36 (3): 491–507.

Horton, A.A. and Dixon, S.J. (2018). *Microplastics: an introduction to environmental transport processes. Wiley Interdisciplinary Reviews: Water* 5 (2): e1268.

Horton, A.A., Walton, A., Spurgeon, D.J. et al. (2017). *Microplastics in freshwater and terrestrial environments: evaluating the current understanding to identify the knowledge gaps and future research priorities. Science of the Total Environment* 586: 127–141.

Imhof, H.K., Laforsch, C., Wiesheu, A.C. et al. (2016). *Pigments and plastic in limnetic ecosystems: a qualitative and quantitative study on microparticles of different size classes.* Water Research 98: 64–74.

Islam, M.S. and Tanaka, M. (2004). *Impacts of pollution on coastal and marine ecosystems including coastal and marine fisheries and approach for management: a review and synthesis.* Marine Pollution Bulletin 48 (7–8): 624–649.

Jacobsen, J.K., Massey, L., and Gulland, F. (2010). *Fatal ingestion of floating net debris by two sperm whales (Physetermacrocephalus).* Marine Pollution Bulletin 60 (5): 765–767.

Jambeck, J.R., Geyer, R., Wilcox, C. et al. (2015). *Plastic waste inputs from land into the ocean.* Science 347 (6223): 768–771.

Jayasiri, H.B., Purushothaman, C.S., and Vennila, A. (2013). *Quantitative analysis of plastic debris on recreational beaches in Mumbai, India.* Marine Pollution Bulletin 77 (1–2): 107–112.

Kaposi, K.L., Mos, B., Kelaher, B.P., and Dworjanyn, S.A. (2014). *Ingestion of microplastic has limited impact on a marine larva.* Environmental Science & Technology 48 (3): 1638–1645.

Karthik, R., Robin, R.S., Purvaja, R. et al. (2018). *Microplastics along the beaches of southeast coast of India.* Science of The Total Environment 645: 1388–1399.

Kataoka, T., Nihei, Y., Kudou, K., and Hinata, H. (2019). *Assessment of the sources and inflow processes of microplastics in the river environments of Japan.* Environmental Pollution 244: 958–965.

Kaza, S., Yao, L., Bhada-Tata, P., and Van Woerden, F. (2018). *What a Waste 2.0: A Global Snapshot of Solid Waste Management to 2050.* The World Bank.

Klein, S., Worch, E., and Knepper, T.P. (2015). *Occurrence and spatial distribution of microplastics in river shore sediments of the Rhine-Main area in Germany.* Environmental Science & Technology 49 (10): 6070–6076.

Koelmans, A.A., Besseling, E., Wegner, A., and Foekema, E.M. (2013). *Plastic as a carrier of POPs to aquatic organisms: a model analysis.* Environmental Science & Technology 47 (14): 7812–7820.

Koelmans, A.A., Bakir, A., Burton, G.A., and Janssen, C.R. (2016). *Microplastic as a vector for chemicals in the aquatic environment: critical review and model-supported reinterpretation of empirical studies.* Environmental Science & Technology 50 (7): 3315–3326.

Lassen, C., Foss Hansen, S., Magnusson, K. et al. (2015). *Microplastics – Occurrence, effects and sources of releases to the environment in Denmark.* Environmental project No. 1793, Danish Ministry of the Environment–Environmental Protection Agency (Denmark): 204.

Latrubesse, E.M. (2008). *Patterns of anabranching channels: the ultimate end-member adjustment of mega rivers.* Geomorphology 101 (1–2): 130–145.

Law, K.L. and Thompson, R.C. (2014). *Microplastics in the seas.* Science 345 (6193): 144–145.

Lebreton, L.C., Van Der Zwet, J., Damsteeg, J.W. et al. (2017). *River plastic emissions to the world's oceans.* Nature Communications 8: 15611.

Lechner, A., Keckeis, H., Lumesberger-Loisl, F. et al. (2014). *The Danube so colourful: a potpourri of plastic litter outnumbers fish larvae in Europe's second largest river.* Environmental Pollution 188: 177–181. https://doi.org/10.1016/j.envpol.2014.02.006.

Leslie, H.A., Van Velzen, M.J.M. and Vethaak, A.D. (2013). Microplastic survey of the Dutch environment. Novel data set of microplastics in North Sea sediments, treated wastewater effluents and marine biota, The Netherlands.

Li, C., Busquets, R., and Campos, L.C. (2020). *Assessment of microplastics in freshwater systems: a review. Science of The Total Environment* 707: 135578.

Lusher, A.L., McHugh, M., and Thompson, R.C. (2013). *Occurrence of microplastics in the gastrointestinal tract of pelagic and demersal fish from the English Channel. Marine Pollution Bulletin* 67 (1–2): 94–99.

Lwanga, E.H., Vega, J.M., Quej, V.K. et al. (2017). *Field evidence for transfer of plastic debris along a terrestrial food chain. Scientific Reports* 7 (1): 1–7.

Mani, T., Hauk, A., Walter, U., and Burkhardt-Holm, P. (2015). *Microplastics profile along the Rhine River. Scientific Reports* 5 (1): 1–7.

McCormick, A., Hoellein, T.J., Mason, S.A. et al. (2014). *Microplastic is an abundant and distinct microbial habitat in an urban river. Environmental Science & Technology* 48 (20): 11863–11871.

McCormick, A.R., Hoellein, T.J., London, M.G. et al. (2016). *Microplastic in surface waters of urban rivers: concentration, sources, and associated bacterial assemblages. Ecosphere* 7 (11): e01556.

Moore, C.J., Lattin, G.L., and Zellers, A.F. (2005). A brief analysis of organic pollutants sorbed to pre- and post-production plastic particles from the Los Angeles and San Gabriel River Watersheds. In *2005 Conference Focusing on the Land-Based Sources of Marine Debris, Redondo Beach*, California, USA.

Moore, C.J., Lattin, G.L., and Zellers, A.F. (2011). *Quantity and type of plastic debris flowing from two urban rivers to coastal waters and beaches of Southern California. Revista de GestãoCosteiraIntegrada-Journal of Integrated Coastal Zone Management* 11 (1): 65–73.

Mouat, T., Lopez-Lozano, R., and Bateson, H. (2010). *Economic Impacts of Marine Litter.* KommunenesInternasjonaleMiljøorganisasjon.

Murphy, F., Ewins, C., Carbonnier, F., and Quinn, B. (2016). *Wastewater treatment works (WwTW) as a source of microplastics in the aquatic environment. Environmental Science & Technology* 50 (11): 5800–5808.

Nizzetto, L., Bussi, G., Futter, M.N. et al. (2016). *A theoretical assessment of microplastic transport in river catchments and their retention by soils and river sediments. Environmental Science: Processes & Impacts* 18 (8): 1050–1059.

Nollkaemper, A. (1994). *Land-based discharges of marine debris: from local to global regulation. Marine Pollution Bulletin* 28 (11): 649–652.

OCMCBE (2015). *Stemming the Tide: Land-Based Strategies for a Plastic-Free Ocean.* Ocean Conservancy and McKinsey Center for Business and Environment.

OECD (2017, 2017). OECD science, technology and industry scoreboard. In: *The Digital Transformation.* Paris: OECD Publishing http://dx.doi.org/10.1787/9789264268821-en.

Owens, K.A. and Kamil, P.I. (2020). *Adapting coastal collection methods for river assessment to increase data on global plastic pollution: examples from India and Indonesia. Frontiers in Environmental Science* 7: 208.

Ram, B. and Kumar, M. (2020). *Correlation appraisal of antibiotic resistance with fecal, metal and microplastic contamination in a tropical Indian river, lakes and sewage. NPJ Clean Water* 3 (1): 1–12.

Reddy, M.S., Basha, S., Adimurthy, S., and Ramachandraiah, G. (2006). *Description of the small plastics fragments in marine sediments along the Alang-Sosiya ship-breaking yard, India. Estuarine, Coastal and Shelf Science* 68 (3–4): 656–660.

Robin, R.S., Karthik, R., Purvaja, R. et al. (2020). *Holistic assessment of microplastics in various coastal environmental matrices, southwest coast of India. Science of the Total Environment* 703: 134947.

Rochman, C.M., Tahir, A., Williams, S.L. et al. (2015). *Anthropogenic debris in seafood: plastic debris and fibers from textiles in fish and bivalves sold for human consumption. Scientific Reports* 5: 14340.

Sarkar, D.J., Sarkar, S.D., Das, B.K. et al. (2019). *Spatial distribution of meso and microplastics in the sediments of river ganga at eastern India. Science of the Total Environment* 694: 133712.

Schmidt, C., Krauth, T., and Wagner, S. (2017). *Export of plastic debris by rivers into the sea. Environmental Science & Technology* 51 (21): 12246–12253.

Siegfried, M., Koelmans, A.A., Besseling, E., and Kroeze, C. (2017). *Export of microplastics from land to sea. A modelling approach. Water Research* 127: 249–257.

de Souza Machado, A.A., Kloas, W., Zarfl, C. et al. (2018). *Microplastics as an emerging threat to terrestrial ecosystems. Global Change Biology* 24 (4): 1405–1416.

Sruthy, S. and Ramasamy, E.V. (2017). *Microplastic pollution in Vembanad Lake, Kerala, India: the first report of microplastics in lake and estuarine sediments in India. Environmental Pollution* 222: 315–322.

Suman, T.Y., Li, W.G., Alif, S. et al. (2020). *Characterization of petroleum-based plastics and their absorbed trace metals from the sediments of the Marina Beach in Chennai, India. Environmental Sciences Europe* 32 (1): 1–10.

Sundt, P., Schulze, P.-E., and Syversen, F. (2014). *Sources of microplastic pollution to the marine environment.* Mepex for the Norwegian Environment Agency (Miljødirektoratet): 86.

Tan, X., Yu, X., Cai, L. et al. (2019). *Microplastics and associated PAHs in surface water from the Feilaixia reservoir in the Beijiang River, China. Chemosphere* 221: 834–840.

Tanaka, K. and Takada, H. (2016). *Microplastic fragments and microbeads in digestive tracts of planktivorous fish from urban coastal waters. Scientific Reports* 6 (1): 1–8.

Tanaka, K., Takada, H., Yamashita, R. et al. (2013). *Accumulation of plastic-derived chemicals in tissues of seabirds ingesting marine plastics. Marine Pollution Bulletin* 69 (1–2): 219–222.

Tanaka, K., Takada, H., Yamashita, R. et al. (2015). *Facilitated leaching of additive-derived PBDEs from plastic by seabirds' stomach oil and accumulation in tissues. Environmental Science and Technology* 49: 11799–11807. https://doi.org/10.1021/acs.est.5b01376.

Thompson, R. (2017). *Environment: a journey on plastic seas. Nature* 547 (7663): 278.

Thompson, R.C., Swan, S.H., Moore, C.J. and Vom Saal, F.S. (2009). *Our plastic age.* royalsocietypublishing.org (accessed 26 October 2021).

Toxics Link (2014). *Plastics and the Environment, Assessing the Impact of the Complete Ban on Plastic Carry Bag.* New Delhi, India: Central Pollution Control Board (CPCB) http://toxicslink.org/docs/Full-Report-Plastic-and-the-Environment.pdf.

Tramoy, R., Gasperi, J., Dris, R. et al. (2019). *Assessment of the plastic inputs from the seine basin to the sea using statistical and field approaches. Frontiers in Marine Science* 6: 151.

UNHSP (United Nations Human Settlements Programme) (2016). World Cities Report. Nairobi, Kenya: United Nations Human Settlements Programme.

Van Cauwenberghe, L. and Janssen, C.R. (2014). *Microplastics in bivalves cultured for human consumption. Environmental Pollution* 193: 65–70.

Veerasingam, S., Mugilarasan, M., Venkatachalapathy, R., and Vethamony, P. (2016). *Influence of 2015 flood on the distribution and occurrence of microplastic pellets along the Chennai coast, India. Marine Pollution Bulletin* 109 (1): 196–204.

Verster, C., Minnaar, K., and Bouwman, H. (2017). *Marine and freshwater microplastic research in South Africa. Integrated Environmental Assessment and Management* 13 (3): 533–535.

Wagner, M., Scherer, C., Alvarez-Muñoz, D. et al. (2014). *Microplastics in freshwater ecosystems: what we know and what we need to know. Environmental Sciences Europe* 26 (1): 1–9.

Werner, S., Budziak, A., van Franeker, J.A. et al. (2016). *Harm caused by Marine Litter*. MSFD GES TG Marine Litter – Thematic Report EUR 28317. Luxembourg.

Wilson, H.L., Johnson, M.F., Wood, P.J. et al. (2020). *Anthropogenic litter is a novel habitat for aquatic macroinvertebrates in urban rivers. Freshwater Biology* 66: 524–534.

Windsor, F.M., Durance, I., Horton, A.A. et al. (2019). *A catchment-scale perspective of plastic pollution. Global Change Biology* 25 (4): 1207–1221.

Wright, S.L. and Kelly, F.J. (2017). Threat to human health from environmental plastics. *BMJ* 358: j4334. https://doi.org/10.1136/bmj.j4334.

Yonkos, L.T., Friedel, E.A., Perez-Reyes, A.C. et al. (2014). *Microplastics in four estuarine rivers in the Chesapeake Bay, USA. Environmental Science & Technology* 48 (24): 14195–14202.

Zhao, S., Zhu, L., Wang, T., and Li, D. (2014). *Suspended microplastics in the surface water of the Yangtze estuary system, China: first observations on occurrence, distribution. Marine Pollution Bulletin* 86 (1–2): 562–568.

Ziajahromi, S., Neale, P.A., and Leusch, F.D. (2016). *Wastewater treatment plant effluent as a source of microplastics: review of the fate, chemical interactions and potential risks to aquatic organisms. Water Science and Technology* 74 (10): 2253–2269.

Ziajahromi, S., Kumar, A., Neale, P.A., and Leusch, F.D. (2017). *Impact of microplastic beads and fibers on waterflea (Ceriodaphniadubia) survival, growth, and reproduction: implications of single and mixture exposures. Environmental Science & Technology* 51 (22): 13397–13406.

3

Microplastic Contamination in the Marine Food Web: Its Impact on Human Health

Richa Singh

Institute of Environment and Sustainable Development, Banaras Hindu University, Varanasi, Uttar Pradesh, India

3.1 Introduction

Plastic or synthetic polymers are artificially made from petroleum products, and due to their versatile nature of being lightweight, strong, durable, transparent, and waterproof they are now part of everything we human beings use. Thus, they are widely distributed to the entire segment of human life, from waking up in the morning until going to bed at night; modern humans are surrounded by plastic. Plastics have proliferated into food packaging industries; stationary, electric, and electronic goods; vehicles, private and public transport; medical appliances; fishing nets and more. In today's era, one cannot imagine a day without plastic. Toothpastes, shaving creams, and soaps all have synthetic polymers in the form of microbeads (Sun et al. 2020). Plastics are highly resistant to microbial degradation as they are of artificial (human-made) petroleum-based products; therefore, their remediation by natural processes is difficult enough and takes a long time that can vary in the range of hundreds of years (Wierckx et al. 2018). Due to chemical (acid rain) and physical (temperature, pressure, moisture) processes, they degrade and break down into smaller fragments and, in smaller forms which do not degrade completely and pose negative impacts on the environment, these smaller forms have high potential to enter into the water or air matrices in invisible forms. MPs have a size ranging from 100 nm to 5 mm (Zhang et al. 2020). The microbeads and nanoparticles that are intentionally added into facewash, shaving creams, soaps, etc. are termed "primary" MPs; however, those which are added to the environment after the fragmentation of larger size plastic particles by impact of natural phenomenon, as well as anthropogenic activities in the environment, are considered "secondary" MPs (Lei et al. 2017). Due to poor waste management practices and lack of knowledge about proper disposal of such wastes among the consumers of such plastic-based products, plastic debris is dumped directly into the ocean every year in huge quantities. Varying sizes of plastic particles, including larger and smaller beads of MPs, ultimately reach the water column of the ocean, and their presence is significantly reported in sediments (Harris 2020). They are found on the deep floor of the sea, in underground water

Plastic and Microplastic in the Environment: Management and Health Risks, First Edition.
Edited by Arif Ahamad, Pardeep Singh, and Dhanesh Tiwary.
© 2022 John Wiley & Sons Ltd. Published 2022 by John Wiley & Sons Ltd.

tables, and in soils. Global plastic production has significantly boomed from 300 to 360 million metric tons in the last five years (*Deccan Herald*, 2020). As they are highly resistant to microbial degradation, they persist for a longer time in the ecosystem. MPs are a matter of great concern because of their high potential to make any organism unfit, as they have huge impacts on their metabolism.

According to Boucher & Friot (2017), approximately 1.53 million tons/year of primary MPs enter the ocean via different pathways. These pathways include flushed water from our washroom's containing the MPs in microbead form from face scrubs, toothpaste, detergents, facewash, shampoo, cosmetic cream, etc., which goes into rivers through the drainage systems, and later become part of the ocean, as sewage treatment plants are not made for such efficiency in developing countries like India. MPs are extensively distributed throughout our ocean ecosystem; from zooplankton, bivalves, crustacean, fish, and seabirds, and ultimately reach humans due to their extensive consumption and dependency on seafood (Santillo et al. 2017). In aquatic organisms, the uptake pathways of MPs are through their gills and gastrointestinal tracts (Franzellitti et al. 2019). The organisms are often confused as these colorful MP fragments look similar to plankton species on which they feed, and thus a significant portion of MPs reach and are accumulated in consumer organisms (Setälä et al. 2014). Sometimes these MPs are deposited on seaweed or algal blooms, and become part of the food to the organisms, where they enter the gastrointestinal tracts of organisms (Walkinshaw et al. 2020). Once these MPs are mistakenly consumed by smaller organisms in confusion of phytoplanktons, they make their way to successive trophic levels, as predators consume prey already having MPs in their guts. Polymers of rope, usually old fishing gear left deliberately or mistakenly in the ocean, can entangle marine creatures, suffocating them by restricting their mobility and unintentional killing them; this is referred to as "ghost fishing" (Gilman 2015). These polymers may also undergo reduction in size due to natural forces acting on them in the ocean, such as waves, water temperature, contact with other floating debris, or larger marine creatures nibbling the pieces into smaller fragments, which then contributes to MPs. Recent studies show the presence of MPs of an array of shapes and sizes in various organs of different organisms such as gills, liver, gut, muscles, etc. They cause disturbances in processes of metamorphosis, metabolism disorder, behavioral change, oxidative stress, genotoxicity, etc. (Rahman et al. 2021). This is becoming a food safety threat as these organisms are heavily consumed as food by humans as seafood, and contribute to a significant enough proportion of the daily diet of people in coastal areas; for example, India has a huge coastline of 7516.6 km.

MPs have a high density, and become settled on the ocean floor, both after entering into the oceans and through the feces of organisms. There it causes oxygen-deficient conditions like anoxia and hypoxia, in which there is less availability of oxygen and nutrients. In this way, it causes harm to seaweed, corals, and planktons (Seeley et al. 2020). Deposition of layers of MPs on body surfaces of corals (in the tropics) cause their degeneration by lowering the absorption of essential nutrients from the surrounding environment.

However, plastics play an important role in food safety and security by providing safe and durable packaging, and is a great contribution to the pharmaceutical and medical industries for packing medicine and providing disposal medical equipment (Hui et al. 2020). Due to a lack of proper waste management, over 250 000 tons of plastic pieces are dumped into the oceans (Hahladakis 2020). Larger plastics sizes are undergoing slow degradation by integrated physical–chemical and biological processes. They mainly degrade due to photo- and thermo-oxidative processes (Mierzwa-Hersztek et al. 2019).

MPs are generally found in the form of pellets, fragments, or fibers. Some of them are denser than seawater and settle at the seafloor like polyamide, polyester, polyvinyl chloride (PVC), and acyclic, etc. In contrast, those found throughout the water column and floating on the sea surfaces are a lighter density than the sea surface, e.g. polyethylene, polypropylene, polystyrene (Hidalgo-Ruz et al. 2012). Although plastic is treated as non-toxic because of its less reactive nature (Hwang et al. 2020).

3.1.1 Microplastic in the Marine Food Web

Due to wide distribution and unmanaged dumping (some through rivers) in the ocean, MP is a common ailment. However, European countries are more prone to this problem because they have busy sea routes as well as industries near coastal areas. The MPs are added to the ocean from terrestrial sources along with the secondary MPs from larger submerged plastics. Ballast water from ships release huge amounts of MPs. Those MPs are hotspots of toxic chemicals, pathogens, harmful algal bloom, etc. (Naik et al. 2019). Microplastic is of serious concern due to its wide distribution from pelagic to benthic marine biota (Thompson et al. 2009). These are a great matter of concern as they are affecting every segment of the ecosystem. They become attached to the planktons and cause disturbances in performing photosynthesis, and make a film over the water surface and provide the breeding ground for bacterial pathogens. The transfer of MPs from one trophic level to the next is a big concern. Another serious threat is the biomagnification of MPs along with the associated chemicals to the successive trophic level (Walkinshaw et al. 2020). The associated chemicals of MPs have a large area-to-volume ratio, which absorbs hydrophobic pollutants from the surrounding marine environment (Figure 3.1) (Smith et al. 2018). The most serious threat is of bioaccumulation of heavy metal in the presence of micro MPs at every trophic level of food chain, and which may lead to biomagnification of toxic heavy metals among the higher-level organism in food chain, which have a high probability of being eaten by the human population.

3.1.2 Toxic Impacts on Primary Producers

Phytoplankton are considered the main contributor to the primary productivity in the oceans, which fixes almost half of the carbon dioxide of the earth during photosynthesis process by using photosynthetic active radiation (PAR) from Sun and carbon dioxide (Uitz et al. 2010). The MP deposition over phytoplankton decreases the chlorophyll concentration, photosynthesis, cell growth, and morphology of phytoplankton. Microplastics deposited upon the phytoplankton penetrate the cell walls and interfere with the chlorophyll mechanisms in green algae (Nerland Bråte et al. 2014). Phytoplankton absorb persistent organic pollutants released during the degradation of MPs which is further transferred along the marine food web (Chandra et al. 2020). This hazardous chemical has the property of bioaccumulation in successive trophic levels, which causes toxicity to them. When MPs are deposited over harmful alga, they release phycotoxin, which is transferred to phytoplankton, bivalves, and crustaceans (Sharma & Chatterjee 2017). The toxins are then

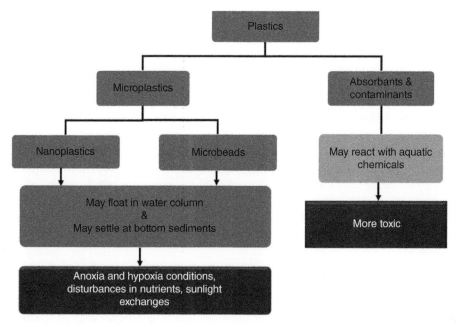

Figure 3.1 Flow diagram of the fate of plastic entering the environment.

bioaccumulated in their bodies and move to the next trophic level. These are then consumed by humans, which may result in many health issues. The coral reef, which has the highest biodiversity in the tropical shallow parts of marine realm, is also badly impacted by MPs. These coral reefs have the mutual collaboration of algae and fungi. However, most of the time, algal partners depend on phytoplankton, zooplankton, etc., for their food requirements, but they are confused by the colorful MPs and consume them. The digestive tracts of coral reefs (coral polyps) cannot deal with MPs and they have very harmful impacts on their health.

These phytoplankton are trapped as part of the marine snow and an important constituent of marine organic matter, this organic matter is taken as food by benthos and nektons. Marine algae aggregates over the floating MPs and settle down to the sediment water interface. This reduces the residence time of floating organic matter in the water column, which in turns lowers the food availability to those organisms residing in the water column. Furthermore, hetero-aggregates of MP and phytoplankton are consumed by zooplankton and have harmful impacts to them. These MP contaminated zooplankton are bioavailable to the predator and transferred to successive trophic levels. In these ways MP potentially disturbs the food transfer and, most importantly, reduces the energy flow from primary producers level.

3.1.3 Toxic Impacts on Consumers

Free-floating zooplankton are key members of the marine food web, as they are the connecting link in transferring energy from producer to consumer level in the food chain, and then to the higher trophic levels. They feed upon phytoplankton and the MPs present over the surface of the phytoplankton are accumulated into their bodies (Cózar et al. 2014). Zooplankton consist of many species and have different life cycle stages with a wide range of feeding mechanisms (Wirtz 2012). Studies from lab experiments revealed that these zooplankton are capable of absorbing tiny plastic latex beads (Cole et al. 2013) as MPs of <5 mm have been found in 15 different taxa of zooplankton, from copepods to jellyfish. Ciliated heterotrophic planktons engulf MPs by phagocytosis (Laist 1987). The zooplankton ingests MPs particles, which may either pass through their digestive tract or get stuck in the gut, causing disturbed digestive health, such as lack of hunger due to feeling of filled stomach by MPs, and behavioral changes, which ultimately leads to their death. Sometimes these MPs are successfully excreted from the zooplankton body in the form of pellets (and become part of marine snow), and are distributed to the water column and ultimately settle to the bottom. These MP-contaminated pellets are then available for benthic organisms. The benthic invertebrates comprise 98% of overall marine biota, which includes oysters, blue mussels, barnacles, lobsters, etc., and these have all been reported to have MPs in them (Nerland Bråte et al. 2014).

According to Possatto et al. (2011) and Lusher et al. (2013), 30% of the fish species that humans extensively consume, including sea bass, are contaminated with MPs. The main exposure route of the MPs in fishes is ingestion during feeding upon MP-contaminated phytoplankton and zooplankton, or direct ingestion due to confusion with prey. Gills are another exposure root to MPs for marine creatures. This organ serves the purpose of osmoregulation, acid–base regulation, gaseous exchange, and nitrogenous exchange. Interaction of MPs to the gills may cause partial or total blockage, which disrupts these functions and causes fatal harm to the organism (Watts et al. 2016). These MPs often accumulate in their bodies, causing starvation, hormonal imbalance, behavioral changes, and malnourishment, ultimately leading to fatality (Welden & Cowie 2016). Sometimes MPs stuck in the gills of organisms, cause suffocation, and even death of organism. Microplastics of larger size (5 mm) are more harmful as they remain for a longer time in the fish body compared with smaller size (2 mm). Smaller sized MPs are easily excreted with feces. Many of the organisms extensively consumed by humans are reported with MP contamination during field studies, some of them are mentioned in Table 3.1.

Sea birds feed upon fishes. Sometimes they take contaminated fishes with MPs accumulated in them. Sea birds like Albatross and Shearwater feed on the ocean surface and in that way, they take in a huge amount of floating MPs inside their gastrointestinal tract. Ryan (2008) found the presence of MPs in south Atlantic birds. These MPs have many negative impacts on their bodies, including starvation due to lack of hunger caused by MPs accumulated in their gastro-intestinal tract, and blockage of respiratory organs like gills leading to suffocation and ultimately death. The sediment water interface is the main hub of these artificial polymers due to sinking and sedimentation. MPs have lower density than the oceanic water; however due to biofouling (deposition and colonization of microorganisms on any surface exposed to them in water) by microorganism they settle down. Biofouling increases the density of MP and make it more dense than the oceanic water, although this may take up to a week, a month, or more than year (Fazey & Ryan 2016). These phenomena are highly surface area-to-volume ratio dependent, as smaller fragments have more surface area-to-volume ratio than the larger debris (Kowalski et al. 2016). Therefore, smaller sized particles lose their bounce earlier and settle in benthic environments. The fecal matter of

Table 3.1 The occurrence of MPs in marine organisms reported in some studies.

Marine organisms contaminated with MPs	Sampling location	Occurrence of MPs	Specific detail	References
Caretta caretta (54 sea turtle samples)	Adriatic Sea	35%	Fatality seen in juvenile turtle due to debris ingestion	Lazar & Gračan (2011)
Lampris sp. (595 samples)	North Pacific	19%	Highest debris ingestion seen in mesopelagic (rarely comes in contact with surface water)	Choy & Drazen (2013)
Mesoplodon mirus	North and west coasts of Ireland	85%	MPs detected throughout the digestive tracts	Lusher et al. (2015)
26 different fish species (178 individuals sampled)	Saudi Arabian Red Sea coast	15%	Highest contamination was found in Parascolopsis eriomma species which feed on benthic organisms	Baalkhuyur et al. (2018)
Oysters	China coast	84%	Average conc. of MPs: 0.62 items/g (wet weight) or 2.93 items/individual	Teng et al. (2019)
150 analyzed fish (50 per species)	Northeast Atlantic Ocean	49%	Lipid oxidative damage found in gills and muscle which cause neurotoxicity	Barboza et al. (2020)
European Sardine	Northwestern Mediterranean Sea	58%	Positive relation between MPs and parasite ingestion	Pennino et al. (2020)
Anchovies (45 samples)	Madura Strait, Indonesia	335 plastic particles: 63% fibers, 34% fragments	2.98% of total MPs found in all anchovy samples	Guntur et al. (2021)

zooplanktons also contain MP pellets; they ingest the MPs and are unable to digest to any further, simpler form of nutrition, so what is ingested is excreted. These also settle to the bottom. MPs in fecal materials speed up its sinking rate in the water column (Cole et al. 2016). These are then bioavailable to the benthos and become the part of benthic food system.

Other sea organisms like turtle, whale, seal, etc., are also at a high risk of MP accumulation and toxicity (Egbeocha et al. 2018). Whales have high lipid content in their body; therefore, they are more prone to accumulate MPs in their blabber, stomach, and intestines. Polar bears in the arctic region are also highly infected with MPs (Singh et al. 2020). Plastic debris reaching oceans from populated landmasses travels long distances along the water current and are distributed to every part of ocean over time, during which fragmentation of plastics also occurs, generating MPs, today their presence is reported in fish bodies in polar areas of the Arctic.

According to Food and Agriculture Organization (FAO, 2002) there are four pillars of food security (introduced in World Food Summit 1996); accessibility, availability,

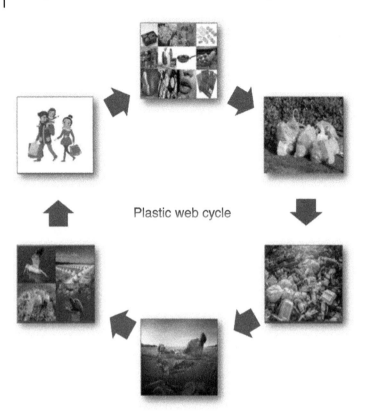

Figure 3.2 Plastic web cycle: Plastic starting its journey from human and returning back to them. *Source:* chaiyapruek/Adobe Stock.

utilization, and stability. Although, due to the presence of MPs in every segment of our ecosystem, two pillars, i.e. food availability and utilization, are compromised in the case of marine food (De-la-Torre 2020). Organisms consume these MPs but their enzymatic actions are unable to break down the polymeric particles of the MPs, and this hinders the process of assimilation of other available nutrients. This causes starvation and lack of nutrition and ultimately leaves them undernourished. These organisms which are contaminated with MPs reach to humans in their food sources as those food sources are already nutrient deficient, so they are not sufficient for a human diet. Instead of nutrients, humans may meet with potential health threats due to MPs (Figure 3.2).

3.1.4 Associated Risk

MPs can readily absorb harmful chemicals from the atmosphere and pathogenic contaminants due to its surface deposition (Verla et al. 2019). Along with their own harmful impacts, this MP has more associated risk when exposed to the environment, as they are breeding grounds for pathogens (Lu et al. 2019).

Plastic polymers are manufactured using chemical additives and other smaller monomer units, which give them the desired shape, structure, strength, and durability. These

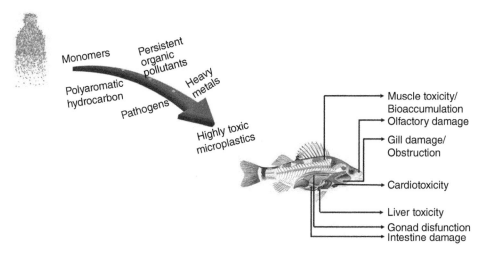

Figure 3.3 Associated chemical toxicants entering marine organisms have potential health impacts.

additives include wide ranges of chemicals like heavy metals; e.g. mercury, plasticizers, flame-retardants, pigments, heat stabilizer filler, UV stabilizers, and many more, and this accounts for 4% of their total weight (Ambrogi et al. 2017). These chemicals tend to deposit over the MPs rather than dissolve in the water bodies (Ziccardi et al. 2016). When these chemicals leach into the environment due to photolysis chemical, and physical breakdown processes, they exhibit harmful impacts in every possible environment (air, water, or soil) as most of them are carcinogenic (Abdel-Shafy and Mansour 2016). These MPs slowly degrade, and their surface area to volume ratio increases, which increases the chemical leaching (Chamas et al. 2020). These leached chemicals are mixed with the surrounding matrix and are transferred to the organism, causing bioaccumulation of these chemicals in them (Figure 3.3). If these MPs are directly eaten by primary consumers like zooplankton and then become prey by high trophic level, there is a high chance of bioaccumulation (Ziccardi et al. 2016).

Fecal matter of zooplankton is an important component of marine organic matter and plays a significant role in the biological pump. The biological carbon pump is helpful in the transportation of carbon, nutrients, and energy to the deeper water sediments (Cisternas-Novoa et al. 2019). Due to feeding on the MPs contaminated meals, their fecal pellets are highly susceptible to the persistent pollutants, hydrocarbons, and petroleum residue. Moreover, benthos feed upon these contaminated meals and are highly prone to bioaccumulation and biomagnification of these harmful chemicals.

3.2 Human Health Implication

The route of exposure of MPs to humans is mainly from inhalation, ingestion, and dermal contact (Prata et al. 2020). The plastics provide the food security by being less reactive and less expensive, therefore they are extensively used in the packaging of food, water, medicine, and the packaging industry in general. Studies reveal the presence of MPs in bottled mineral water (Weisser et al. 2021) and honey, among other foods. The main route of exposure to MPs are inhalation from the airborne MPs coming from the construction and demolition areas, industrial emissions, waste storage and disposal sites, such as landfill and waste reduction plants, where incineration of waste is done. These MPs are ingested along with food and water into the human body via dermal contacts, mainly by applying the cosmetics containing microbeads in them. However, fishes are a potential pathway for the entrance of MPs to human bodies (Barboza et al. 2020). The ocean is the largest producer of biomass and provides varieties of products which we use in our day-to-day life. There is a chance of toxicity due to xenobiotic compounds as well, which provide an important iodide source, i.e. salt. People are using cosmetics, toothpaste, and face scrubs, which have MPs of >1 μm, where the absorption of PE and PP particles in tissues can cause skin damage (Sharma & Chatterjee 2017). While our skin may provide a direct entry of MPs and other contaminants to our bodies, there are other routes which can cause direct exposure to them, such as sweat glands, open injury, and hair follicles. The microbeads present in the toothpaste are unintentionally swallowed, reach the gut, and may absorbed into the blood, which can cause chromosomal alteration, hormonal imbalance (which can lead to infertility), or cancer (Usman et al. 2020).

Plastic polymer products are pervasive in human life; therefore, their particle exposure is inevitable for humans. Humans are highly exposed to the MPs problem as they are widely present in the air, water, and soil (Campanale et al. 2020).

As foreign particles, MPs are resistant to the natural degradation process or defense mechanisms naturally present in one's body. They may cause particle toxicity, oxidative stress, disruption of immune function, or neurotoxicity (Prata et al. 2020).

If MPs enter into the gut, they are not easily excreted by the body and may cause blockage of the gastrointestinal tract, which may disturb the function of the digestive system and lead to death (Wright et al. 2013). The immune system is unable to remove them, and this leads to chronic inflammation, which may cause neoplasia; a tumorous growth potentially capable of turning into a cancerous state (Prata et al. 2020). MPs, in sizes ranging from 0.2–150 μm have a high potency of translocation in humans across the cells to the lymphatic and circulatory systems, possibly through Peyer's patches of the intestine (Hussain et al. 2001). However, this process is not well known and needs further studies.

Although human bodies are capable of excreting more than 90% of the MPs (Schwabl et al. 2019), its fate in the human body is not yet fully understood as it has associated risk of toxicity of chemicals like heavy metals (cadmium, lead, chromium etc.) (Massos & Turner 2017), hydrophobic organic pollutants (organo-chloride, polyaromatic hydrocarbons, polychlorinated biphenyls etc.), and additive compounds (Wright & Kelly 2017). These chemicals are highly carcinogenic and easily transported to the lymphatic system (Wright & Kelly 2017). These chemicals are hydrophobic and easily adsorbed by MPs, and even a very small dose is sufficient to adversely affect the biological metabolism of humans and animals.

Due to its property of being an essential component in nutrition, common salt is extensively consumed by humans globally. Because it is also used in food preservatives, the human community consumes a small quantity of salt in almost every food item, which includes both the freshly cooked and preserved items like chips, aerated drinks, and packaged juice, etc. Salts are also found in cosmetics, pharmaceuticals, toothpaste, and other personal care products. Salts are deracinated from seawater or saline lakes, rocks, and wells. Sea salts are extracted by constructing shallow basins over salty and mineral-rich lakes, generally referred to as salt evaporation ponds, where salts are obtained by the natural evaporation process. Prior to crystallization, the seawater is mixed with fresh water and then passed through a number of successive ponds to maintain a salinity gradient. These areas are highly prone to anthropogenic contaminants, mainly with MPs from marine debris, and this water circulation process adds contaminants into the water in the form of microbeads. Sea salts have higher MP contamination than a lake or rock salts (Peixoto et al. 2019). All these types of exposure to MP contaminants to the human body are still under investigation.

3.3 Conclusion and Future Perspective

Microplastic waste is a matter of concern for researchers, if we don't find a remediation technique then the time is not far when our green planet would be wrapped in colorful plastics. This is quite evident from the oceans, where MPs have close interaction with all segments including biotic and abiotic. They enter into the organisms bodies via food and water as per their feeding habits. However, their harmful impacts on the body metabolism of organisms largely depend upon the particles size, and the biological processes governing their presence in organisms' bodies. The organisms' bodies have different development stages such as egg, larvae, adults etc., which would define their accumulation, fate, and impacts. Some creatures have a self-defense mechanism that restricts them to feed upon these harmful MPs that others confuse with prey and consume in huge amounts. Sometimes, particles of size <5 mm can be successfully excreted out of the body, while sometimes they cause blockage of the digestive system and are fatal to organisms.

Studies revealed the presence of MPs in sea products like common salts, which is the main source of dietary iodine to the human community, so they have potential harmful impacts on human health. Their significant presence has been seen in seafood like crustacean, mollusk, fishes, and birds, etc. Therefore, it is a matter of great concern toward food and nutrient safety and security, as human population is exposed to high concentration of these sea products.

There is a vast knowledge gap of MP toxicity in the human body due to limited research. However, the main factors behind its toxicity are largely influenced by exposure route, chemical composition, adsorbed contaminants, and susceptibility of organisms. The research on human health is ethically restricted because of biosecurity measures in human-sample handling. Therefore, we can conclude the information of exposure and possible toxicity by reviewing the possible impacts on other organisms. However, these plastics being smaller size, generally do not pass through the biological system.

To reduce this problem, the first major step taken should be to reduce the length of our food chain, which can reduce the huge consumption of plastic used in packaging and processing the raw foods. This means fresh vegetable and fruits should be directly taken from garden to the kitchen or to the table, which will reduce the environmental exposure as well as the additional wrapping of plastics for their future use. Instead of using mineral water or packaged drinking water which is filled in plastic bottles, metal or glass bottles should be used for this purpose, as this might reduce the additional MPs in them due to leaching from plastic packagings.

Acknowledgements

The author would like to thank the editors of the book for the kind invitation to write this chapter. I would also like to place on record my thanks to the Department of Environmental Sciences, Banaras Hindu University, Varanasi, for giving me an opportunity to work in the lab of the Department.

References

Abdel-Shafy, H.I. and Mansour, M.S.M. (2016). *A review on polycyclic aromatic hydrocarbons: source, environmental impact, effect on human health and remediation. Egyptian Journal of Petroleum* 25 (1): 107–123. Egyptian Petroleum Research Institute. https://doi.org/10.1016/j.ejpe.2015.03.011.

Ambrogi, V., Carfagna, C., Cerruti, P., and Marturano, V. (2017). Additives in polymers. In: *Modification of Polymer Properties*, 87–108. Elsevier Inc. https://doi.org/10.1016/B978-0-323-44353-1.00004-X.

Baalkhuyur, F.M., Bin Dohaish, E.J.A., Elhalwagy, M.E.A. et al. (2018). *Microplastic in the gastrointestinal tract of fishes along the Saudi Arabian Red Sea coast. Marine Pollution Bulletin* 131: 407–415. https://doi.org/10.1016/j.marpolbul.2018.04.040.

Barboza, L.G.A., Lopes, C., Oliveira, P. et al. (2020). *Microplastics in wild fish from North East Atlantic Ocean and its potential for causing neurotoxic effects, lipid oxidative damage, and human health risks associated with ingestion exposure. Science of the Total Environment* 717: 134625. https://doi.org/10.1016/j.scitotenv.2019.134625.

Boucher, J. and Friot, D. (2017). Primary microplastics in the oceans: A global evaluation of sources. In: *Primary Microplastics in The Oceans: A Global Evaluation of Sources.* International Union for Conservation of Nature (IUCN) https://doi.org/10.2305/iucn.ch.2017.01.en.

Campanale, C., Massarelli, C., Savino, I. et al. (2020). *A detailed review study on potential effects of microplastics and additives of concern on human health. International Journal of Environmental Research and Public Health* 17 (4). MDPI AG. https://doi.org/10.3390/ijerph17041212.

Chamas, A., Moon, H., Zheng, J. et al. (2020). Degradation rates of plastics in the environment. *ACS Sustainable Chemistry and Engineering* 8 (9): 3494–3511. https://doi.org/10.1021/acssuschemeng.9b06635.

Chandra, P., Enespa, and Singh, D.P. (2020). Microplastic degradation by bacteria in aquatic ecosystem. In: *Microorganisms for Sustainable Environment and Health* (eds. P. Chowdhary et al.), 431–467. India: Elsevier https://doi.org/10.1016/b978-0-12-819001-2.00022-x.

Choy, C.A. and Drazen, J.C. (2013). *Plastic for dinner? Observations of frequent debris ingestion by pelagic predatory fishes from the central North Pacific. Marine Ecology Progress Series* 485: 155–163. https://doi.org/10.3354/meps10342.

Cisternas-Novoa, C., Le Moigne, F.A.C., and Engel, A. (2019). *Composition and vertical flux of particulate organic matter to the oxygen minimum zone of the Central Baltic Sea: impact of a sporadic North Sea inflow. Biogeosciences* 16 (4): 927–947. https://doi.org/10.5194/bg-16-927-2019.

Cole, M., Lindeque, P., Fileman, E. et al. (2013). *Microplastic ingestion by zooplankton. Environmental Science and Technology* 47 (12): 6646–6655. https://doi.org/10.1021/es400663f.

Cole, M., Lindeque, P.K., Fileman, E. et al. (2016). *Microplastics Alter the properties and sinking rates of zooplankton faecal pellets. Environmental Science and Technology* 50 (6): 3239–3246. https://doi.org/10.1021/acs.est.5b05905.

Cózar, A., Echevarría, F., González-Gordillo, J.I. et al. (2014). *Plastic debris in the open ocean. Proceedings of the National Academy of Sciences of the United States of America* 111 (28): 10239–10244. https://doi.org/10.1073/pnas.1314705111.

Deccan Herald. (2020). *Global plastic production up by 60 million metric tonnes in five years.* https://www.deccanherald.com/science-and-environment/global-plastic-production-up-by-60-million-metric-tonnes-in-five-years-890311.html (accessed 26 October 2021).

De-la-Torre, G. E. (2020). *Microplastics: an emerging threat to food security and human health. Journal of Food Science and Technology* 57(5), 1601–1608). Springer. https://doi.org/10.1007/s13197-019-04138-1.

Egbeocha, C., Malek, S., Emenike, C., and Milow, P. (2018). *Feasting on microplastics: ingestion by and effects on marine organisms. Aquatic Biology* 27: 93–106. https://doi.org/10.3354/ab00701.

Fazey, F.M.C. and Ryan, P.G. (2016). *Biofouling on buoyant marine plastics: an experimental study into the effect of size on surface longevity. Environmental Pollution* 210: 354–360. https://doi.org/10.1016/j.envpol.2016.01.026.

Food and Agriculture Organization (FAO) (2002). *The state of food insecurity in the world 2001.* Rome: Food and Agriculture Organization.

Food and Agriculture Organization (FAO). (2009). Declaration of the World Summit on Food security. WSFS 2009/2, Rome.

Franzellitti, S., Canesi, L., Auguste, M. et al. (2019). *Microplastic exposure and effects in aquatic organisms: a physiological perspective. Environmental Toxicology and Pharmacology* 68: 37–51. Elsevier B.V. https://doi.org/10.1016/j.etap.2019.03.009.

Gilman, E. (2015). *Status of international monitoring and management of abandoned, lost and discarded fishing gear and ghost fishing. Marine Policy* 60: 225–239. https://doi.org/10.1016/j.marpol.2015.06.016.

Guntur, G., Asadi, M.A. and Purba, K. (2021). *Ingestion of microplastics by anchovies of the Madura Strait,* Indonesia. (Vol. 14).

Hahladakis, J.N. (2020). *Delineating the global plastic marine litter challenge: clarifying the misconceptions. Environmental Monitoring and Assessment* 192 (5): 1–11. https://doi.org/10.1007/s10661-020-8202-9.

Harris, P.T. (2020). *The fate of microplastic in marine sedimentary environments: a review and synthesis. Marine Pollution Bulletin* 158: 111398. Elsevier Ltd. https://doi.org/10.1016/j.marpolbul.2020.111398.

Hidalgo-Ruz, V., Gutow, L., Thompson, R.C., and Thiel, M. (2012). *Microplastics in the marine environment: a review of the methods used for identification and quantification. Environmental Science and Technology* 46 (6): 3060–3075. https://doi.org/10.1021/es2031505.

Honingh, D., van Emmerik, T., Uijttewaal, W. et al. (2020). *Urban River water level increase through plastic waste accumulation at a rack structure. Frontiers in Earth Science* 8 https://doi.org/10.3389/feart.2020.00028.

Hui, T.K.L., Mohammed, B., Donyai, P. et al. (2020). *Enhancing pharmaceutical packaging through a technology ecosystem to facilitate the reuse of medicines and reduce medicinal waste. Pharmacy* 8 (2): 58. https://doi.org/10.3390/pharmacy8020058.

Hussain, N., Jaitley, V., and Florence, A.T. (2001). *Recent advances in the understanding of uptake of microparticulates across the gastrointestinal lymphatics. Advanced Drug Delivery Reviews* 50 (1–2): 107–142. https://doi.org/10.1016/S0169-409X(01)00152-1.

Hwang, J., Choi, D., Han, S. et al. (2020). *Potential toxicity of polystyrene microplastic particles. Scientific Reports* 10 (1): 1–12. https://doi.org/10.1038/s41598-020-64464-9.

Knaeps, E., Sterckx, S., Moshtaghi, M. and Meire, D. (n.d.). *Hyperspectral Remote Sensing of Marine Plastics (HYPER).*

Kowalski, N., Reichardt, A.M., and Waniek, J.J. (2016). *Sinking rates of microplastics and potential implications of their alteration by physical, biological, and chemical factors. Marine Pollution Bulletin* 109 (1): 310–319. https://doi.org/10.1016/j.marpolbul.2016.05.064.

Laist, D.W. (1987). *Overview of the biological effects of lost and discarded plastic debris in the marine environment. Marine Pollution Bulletin* 18 (6, Suppl. B): 319–326. https://doi.org/10.1016/S0025-326X(87)80019-X.

Lazar, B. and Gračan, R. (2011). *Ingestion of marine debris by loggerhead sea turtles, Caretta caretta, in the Adriatic Sea. Marine Pollution Bulletin* 62 (1): 43–47. https://doi.org/10.1016/j.marpolbul.2010.09.013.

Lei, K., Qiao, F., Liu, Q. et al. (2017). *Microplastics releasing from personal care and cosmetic products in China. Marine Pollution Bulletin* 123 (1–2): 122–126. https://doi.org/10.1016/j.marpolbul.2017.09.016.

Lu, L., Luo, T., Zhao, Y. et al. (2019). *Interaction between microplastics and microorganism as well as gut microbiota: a consideration on environmental animal and human health. Science of the Total Environment* 667: 94–100. Elsevier B.V. https://doi.org/10.1016/j.scitotenv.2019.02.380.

Lusher, A.L., McHugh, M., and Thompson, R.C. (2013). *Occurrence of microplastics in the gastrointestinal tract of pelagic and demersal fish from the English Channel. Marine Pollution Bulletin* 67 (1–2): 94–99. https://doi.org/10.1016/j.marpolbul.2012.11.028.

Lusher, A.L., Hernandez-Milian, G., O'Brien, J. et al. (2015). *Microplastic and macroplastic ingestion by a deep diving, oceanic cetacean: the True's beaked whale Mesoplodon mirus. Environmental Pollution* 199: 185–191. https://doi.org/10.1016/j.envpol.2015.01.023.

Massos, A. and Turner, A. (2017). *Cadmium, lead and bromine in beached microplastics. Environmental Pollution* 227: 139–145. https://doi.org/10.1016/j.envpol.2017.04.034.

Mierzwa-Hersztek, M., Gondek, K., and Kopeć, M. (2019). *Degradation of polyethylene and biocomponent-derived polymer materials: an overview. Journal of Polymers and the Environment* 27 (3): 600–611. https://doi.org/10.1007/s10924-019-01368-4.

Naik, R.K., Naik, M.M., D'Costa, P.M., and Shaikh, F. (2019). *Microplastics in ballast water as an emerging source and vector for harmful chemicals, antibiotics, metals, bacterial pathogens and HAB species: a potential risk to the marine environment and human health. Marine*

Pollution Bulletin 149: 110525. Elsevier Ltd. doi:https://doi.org/10.1016/j.marpolbul. 2019.110525.

Nerland Bråte, I. L, Halsband, C., Allan, I. et al. (2014). Norwegian institute for water research negative environmental impact. 47, 55.

Peixoto, D., Pinheiro, C., Amorim, J. et al. (2019). *Microplastic pollution in commercial salt for human consumption: a review*. Estuarine, Coastal and Shelf Science 219: 161–168. Academic Press. https://doi.org/10.1016/j.ecss.2019.02.018.

Pennino, M.G., Bachiller, E., Lloret-Lloret, E. et al. (n.d.). *Basque Research and Technology Alliance (BRTA)*.

Possatto, F.E., Barletta, M., Costa, M.F. et al. (2011). *Plastic debris ingestion by marine catfish: An unexpected fisheries impact*. Marine Pollution Bulletin 62 (5): 1098–1102. https://doi. org/10.1016/j.marpolbul.2011.01.036.

Prata, J.C., da Costa, J.P., Lopes, I. et al. (2020). *Environmental exposure to microplastics: An overview on possible human health effects*. Science of the Total Environment 702: 134455. Elsevier B.V. https://doi.org/10.1016/j.scitotenv.2019.134455.

Rahman, A., Sarkar, A., Yadav, O.P. et al. (2021). Potential human health risks due to environmental exposure to nano- and microplastics and knowledge gaps: a scoping review. *Science of the Total Environment* 757: 143872. Elsevier B.V. https://doi.org/10.1016/ j.scitotenv.2020.143872.

Ryan, P.G. (2008). *Seabirds indicate changes in the composition of plastic litter in the Atlantic and south-western Indian oceans*. Marine Pollution Bulletin 56 (8): 1406–1409. https://doi. org/10.1016/j.marpolbul.2008.05.004.

Santillo, D., Miller, K. & Johnston, P. (2017). *Microplastics as contaminants in commercially important seafood species*. Integrated Environmental Assessment and Management, 13(3), 516–521). Wiley–Blackwell. https://doi.org/10.1002/ieam.1909.

Schwabl, P., Koppel, S., Konigshofer, P. et al. (2019). *Detection of various microplastics in human stool: a prospective case series*. Annals of Internal Medicine 171 (7): 453–457. https:// doi.org/10.7326/M19-0618.

Seeley, M.E., Song, B., Passie, R., and Hale, R.C. (2020). *Microplastics affect sedimentary microbial communities and nitrogen cycling*. Nature Communications 11 (1): 1–10. https:// doi.org/10.1038/s41467-020-16235-3.

Setälä, O., Fleming-Lehtinen, V., and Lehtiniemi, M. (2014). *Ingestion and transfer of microplastics in the planktonic food web*. Environmental Pollution 185: 77–83. https://doi. org/10.1016/j.envpol.2013.10.013.

Sharma, S. and Chatterjee, S. (2017). *Microplastic pollution, a threat to marine ecosystem and human health: a short review*. Environmental Science and Pollution Research 24 (27): 21530–21547. https://doi.org/10.1007/s11356-017-9910-8.

Singh, N., Granberg, M., Caruso, G. et al. (2020). *5 118 SESS Report 2020-The State of Environmental Science in Svalbard*. http://dx.doi.org/10.5281/zenodo.4293836.

Smith, M., Love, D.C., Rochman, C.M., and Neff, R.A. (2018). *Microplastics in seafood and the implications for human health*. Current Environmental Health Reports 5 (3): 375–386. Springer. https://doi.org/10.1007/s40572-018-0206-z.

Sun, Q., Ren, S.Y., and Ni, H.G. (2020). *Incidence of microplastics in personal care products: An appreciable part of plastic pollution*. Science of the Total Environment 742 https://doi. org/10.1016/j.scitotenv.2020.140218.

Teng, J., Wang, Q., Ran, W. et al. (2019). *Microplastic in cultured oysters from different coastal areas of China. Science of the Total Environment* 653: 1282–1292. https://doi.org/10.1016/j.scitotenv.2018.11.057.

Thompson, R.C., Swan, S.H., Moore, C.J., and Vom Saal, F.S. (2009). *Our plastic age. Philosophical Transactions of the Royal Society B: Biological Sciences* 364 (1526): 1973–1976. Royal Society. https://doi.org/10.1098/rstb.2009.0054.

Uitz, J., Claustre, H., Gentili, B., and Stramski, D. (2010). *Phytoplankton class-specific primary production in the world's oceans: seasonal and interannual variability from satellite observations. Global Biogeochemical Cycles* 24 (3) https://doi.org/10.1029/2009GB003680.

UNEP. (2018). *The state of plastics: world environment day outlook 2018.* Combating marine plastic litter and microplastics: an assessment of the effectiveness of relevant International, regional and subregional governance strategies and approaches – summary for policy makers' UN Doc UNEP/AHEG/2018/1/INF/3 (20 April 2018) 5.

Usman, S., Razis, A.F.A., Shaari, K. et al. (2020). *Microplastics pollution as an invisible potential threat to food safety and security, policy challenges and the way forward. International Journal of Environmental Research and Public Health* 17 (24): 1–24. MDPI AG. https://doi.org/10.3390/ijerph17249591.

Verla, A.W., Enyoh, C.E., Verla, E.N., and Nwarnorh, K.O. (2019). *Microplastic–toxic chemical interaction: a review study on quantified levels, mechanism and implication. SN Applied Sciences* 1 (11): 1–30. https://doi.org/10.1007/s42452-019-1352-0.

Walkinshaw, C., Lindeque, P.K., Thompson, R. et al. (2020). *Microplastics and seafood: lower trophic organisms at highest risk of contamination. Ecotoxicology and Environmental Safety* 190: 110066. https://doi.org/10.1016/j.ecoenv.2019.110066.

Watts, A.J.R., Urbina, M.A., Goodhead, R. et al. (2016). *Effect of microplastic on the gills of the shore crab Carcinus maenas. Environmental Science and Technology* 50 (10): 5364–5369. https://doi.org/10.1021/acs.est.6b01187.

Welden, N.A.C. and Cowie, P.R. (2016). *Long-term microplastic retention causes reduced body condition in the langoustine, Nephrops norvegicus. Environmental Pollution* 218: 895–900. https://doi.org/10.1016/j.envpol.2016.08.020.

Wierckx, N., Narancic, T., Eberlein, C. et al. (2018). Plastic biodegradation: challenges and opportunities. In: *Consequences of Microbial Interactions with Hydrocarbons, Oils, and Lipids: Biodegradation and Bioremediation* (ed. R. Steffan), 1–29. Springer International Publishing, Cape Coral, US https://doi.org/10.1007/978-3-319-44535-9_23-1.

Wright, S.L. and Kelly, F.J. (2017). *Plastic and human health: a micro issue? Environmental Science and Technology* 51 (12): 6634–6647. https://doi.org/10.1021/acs.est.7b00423.

Wright, S.L., Thompson, R.C., and Galloway, T.S. (2013). *The physical impacts of microplastics on marine organisms: a review. Environmental Pollution (Barking, Essex: 1987)* 178: 483–492. https://doi.org/10.1016/j.envpol.2013.02.031.

Zhang, Y., Kang, S., Allen, S. et al. (2020). *Atmospheric microplastics: a review on current status and perspectives. Earth-Science Reviews* 203: 103118. Elsevier B.V. https://doi.org/10.1016/j.earscirev.2020.103118.

Ziccardi, L.M., Edgington, A., Hentz, K. et al. (2016). *Microplastics as vectors for bioaccumulation of hydrophobic organic chemicals in the marine environment: a state-of-the-science review. Environmental Toxicology and Chemistry* 35 (7): 1667–1676. https://doi.org/10.1002/etc.3461.

4

Microplastic in the Aquatic Ecosystem and Human Health Implications

Ankita Ojha[1], Ankitendran Mishra[2], and Dhanesh Tiwary[3]

[1] *Department of Chemistry, Maharaja College, Arrah, Bihar, India*
[2] *Department of Metallurgical Engineering, IIT, Banaras Hindu University, Varanasi, India*
[3] *Department of Chemistry, IIT, Banaras Hindu University, Varanasi, India*

4.1 Introduction

The past 50 years have revealed severe threats from extraneous human activities such as waste from industries, agricultural activities, and domestic rubbish, along with urban activities such as construction and mining (Amoatey and Baawain 2019). With the advent of modern times, plastics have become an intricate part of our world. Their extraordinary physical and mechanical properties (lighter weight, higher durability, cost-effectiveness, and versatility) make them excellent materials for use in daily life (Hammer et al. 2012). The excessive use and production synthesis of plastics and lack of management have led to their disproportionate exposure to the water bodies and aquatic systems associated with them (Li et al. 2018). There has been an estimated growth of the plastic industry since 1950, and its production was expected to reach 33 billion tons by 2020 (Rochman 2018). The threatening impact of plastic waste has undoubtedly raised global concerns due to its wider distribution and related environmental issues. A wide range of plastic litter has been found in the environment, especially the bottom of aquatic systems. From tiny beads to large fragments, these materials have been reported to be present in the oceans (Carbery et al. 2018; Yu et al. 2020). As per Europe Reports (2016), there was an estimated 322 million tons of plastic generation in 2016 alone. Plastics, which are well known for their higher durability, non-degradability, raised sustainability, and poor waste management, are challenging to break down into their mineralized forms (Cole et al. 2011; Prata et al. 2019). Plastics contain various types of polymers such as Polypropylene (PP), polyethylene (PE), Polyvinyl Chloride (PVC), Polyamides (PA), Polystyrene (PS), Polyethylene Terephthalates (PET), and others, which are derived from petroleum products and natural gases. They have a wide range of applications, including the medical sector, transportation, packaging, and sports, and continuously enter the marine systems through various channels. The dependence of humans on plastics (especially single-use plastic) has resulted in the unbroken chain of their entrance into the aquatic environments. These plastics break down into smaller sizes due to continuous erosion activities occurring in the environment due to mechanical and chemical forces by the surroundings. These MPs were been discovered around 2004 and have become a topic of significant concern since 2014.

Many researches have focused on marine systems due to the misconception of the removal of plastics in sewage treatment plants. Experiments have shown that MPs lead to toxicity in immunosuppression, growth inhibiting of cells, and causing oxidative stress on the cellular structures (Zhang et al. 2020). Microplastics are micron-sized, water-insoluble particles up to 5 mm in diameter, and have increased significantly in surface water (both freshwater and marine). The presence of these MPs in drinking water sources has raised some serious concerns in relation to human health. They are present in rivers, lakes, ponds, and even tap and bottled water (Koelmans et al. 2019). Significant sources of microplastic (MP) pollution found in the freshwater and marine ecosystem are synthetic textiles, medical and personal care products, raw materials from various industries, and untreated plastic wastes. Out of all these, PP, PET, and PS constitute nearly 70% of the total MP pollutions (Li et al. 2020). There is no such fixed definition for MPs, as they represent a wide range of materials in terms of type, shape, size, and color.

The National Oceanic and Atmospheric Administration (NOAA) has defined these MPs as very small (<5 mm) particles that are ubiquitous. Plastic wastes are added into the oceans every year, and it is estimated around 7.5 mg/m^3/s (average of 1553 tons) of plastic debris is dumped in the Black Sea via the Danube (Xu et al. 2020). As per Wright and Kelly (2017), 250 million tons of plastics are assumed to be dumped and settle on the ocean bed. Microplastics are quite susceptible to other pollutants present in water and their surface hydrophobicity, and the large surface-to-volume ratio makes them an effective carrier of these contaminants. They may be in the form of fragments, granules, spheres, or sometimes even as fibers (Elizalde-Velázquez and Gómez-Oliván 2021). Exposure to the wind, microorganisms, and ultraviolet radiation breaks down the larger plastic fragments into smaller macroplastics, mesoplastics, and microplastics (Alimi et al. 2018). Aqueous metal ions, endocrine disruptor chemicals (EDCs), pharmaceuticals and personal care products (PPCPs), persistent organic pollutants (POPs), and some of the hydrophobic organic compounds are attached to these microplastics. They become concentrated over the surface along with large quantities of low-density plastics (LDPs), and transfer toxins from aquatic ecosystems to human beings (Cole et al. 2011). There has been a ubiquitous presence of MPs detected in the freshwater ecosystems of China; most prominently, they have been found in the inland water of urban areas and estuaries. Microplastic pollution in China, especially in freshwater ecosystems, has raised alarms due to mismanagement and lack of proper rules and regulations.

The fate of MPs and their quantification in human intestinal systems has become a significant concern in recent times (Fu and Wang 2019). Undoubtedly, these pollutants have slowly entered our daily lives at various strata and created a new set of troubles for humans. The progressing years show that these MPs will continue to increase quantitatively in the water bodies, and therefore, their presence can never be underestimated in the future (Rainieri and Barranco 2019). Microplastics contain two types of chemical constituents, which are classified as either additives/polymeric raw materials that originate from the plastics, and chemicals that are adsorbed/absorbed by the MPs from their surroundings. Additives include the class of chemicals added to the plastic materials during their manufacture to give them their desired properties, enhance their performance, and make them resistant to various environmental factors such as light, humidity, microbial attack, and to improve their thermal, mechanical, and electrical properties. Such additives include bisphenol A (BPA), terephthalates, and plasticizers added during their processing (Campanale et al. 2020). The order of the presence of MPs in water is PE = PP > PS > PVC > PET, which is the actual global plastic demand, and the higher affinity for the PVC lies in the fact that it has a higher density as compared to PET (Koelmans et al. 2019). Microplastics are creating some significant challenges when it comes to human health and its risk assessment. Various approaches have been taken to monitor the impact of these contaminants on health and cellular activity, but we lack real-time data and observations in our investigations. The potential threats these contaminants precipitate in human health come in three forms: physical hazards due to the direct interaction with particles; chemical hazards due to monomers and sorbed substances on their surfaces; and most potently, formation of thin biofilms (WHO 2019). Population density and closeness of industries near these water bodies largely control the circulation of these microplastics in the human food chain as shown in Figure 4.1.

This chapter highlights the impact of the hazardous effects of these MPs on human health on various organ systems (gastrointestinal, neurological, cardiovascular, pulmonary, and reproductive) along with their sources and future perspectives for their underrated health impacts.

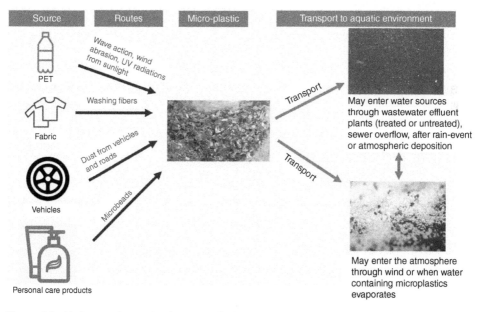

Figure 4.1 Various pathways for the entry of microplastics in the ecosystem. *Source:* From Vivekanand et al. 2021 [48].

4.2 Sources and Food-Chain Entry

Microplastics are homogenously mixed particles, varying in shape, size, and color, and polymers of various constituents. They originate from sources such as manufacturing units and consumer products. Tracking and tracing MPs in the environment is a complicated task in itself. The significant factors that govern the transport and distribution of these contaminants are the material's density, size, and shape. They are classified into two categories; primary microplastics (manufacturing units of microplastics), and secondary MPs (breakdown, disintegration, and defragmentation of larger polymers). Another MP source comes from the wear of plastic products over times, such as fiber from textile industries or wear of tires (Eerkes-Medrano and Thompson 2018). Primary MPs include the class of plastics produced in the micro range in industries, and domestic products manufacturing units such as cleansing agents (face, clothes, etc.). Industrial and domestic waste has become an essential source of MP pollution to marine and freshwater ecosystems. Even though multiple treatment units are available for MP treatment, MP removal efficiency is up to 97% due to the extensive processing volume (Murphy et al. 2016). Primary sources of MP pollution in human systems are sewage treatment plants, weathering and breakdown of the plastic materials discharged into the aquatic systems and runoff from the surface, and soil erosion (Browne 2015). Secondary MPs come from the breakdown of larger plastic by the photodegradation, mechanical breakdown, and biodegradation of these fragments. These MP fragments also adsorb/absorb many chemical pollutants such as heavy metal ions, polychlorinated biphenyls (PCBs), pesticides, polycyclic aromatic hydrocarbons (PAHs), polybrominated diphenyl ethers (PBDEs), among others, which have already been established as significant pollutants of concern (De-la-Torre 2020). Freshwater (streams, ditches, ponds, and lakes) are the most complicated systems in MP pollution because of their more accessible transport and higher retention capacity. They are easily exposed to the MPs through the terrestrial environment, provide easy conduits for the MPs in the marine environment, and act as a sink to deposition MPs in the sediments (Horton and Dixon 2018). Waste disposal allows the entry of more extensive plastic materials into the water bodies through littering or surface runoff. Rivers contribute to around 70–80% of plastic pollution in the ocean and lead to their extensive deposition on the ocean beds (Xu et al. 2019).

Agricultural runoff from farmland adds to the agrarian plastics, or sludge-based microbeads. Urban runoff is mostly unfiltered and contains particles of road paints and vehicle wear (Cole et al. 2014). "City dust" is a term given collectively to the soles of footwear and turf are also potential sources of MPs. In general, it has been observed that MPs have a higher tendency to be present in water bodies near densely populated urban areas and cities. Synthetic textile fibers from washing clothes and cosmetic microbeads constitute a significant MP input in domestic sewage (SAPEA [Science Advice for Policy by European Academies] 2019). Effluents produced from wastewater treatment are produced in large amounts, and even though they are treated effectively, they still contain high and absolute numbers of MPs. As per the study, around 65 million MP particles are released each day in an effluent being discharged from a wastewater treatment plant (WWTP). This means approximately100 particles/population–equivalent/day, which is a very high amount of waste discharged. It has also been estimated that around 1140L of water are released from a secondary WWTP per microplastic particle into the environment from wastewater treatment units (WHO 2019).

The human body is exposed to these contaminants through the direct ingestion of edibles that contain these MPs or through the skin. Ingestion is the major route of entrance of these MPs into human systems. Based on food items consumed, the total estimated MP intake is calculated to be between 39 000–52 000 particles per person^{-1} per year^{-1} (Cox et al. 2019). Food items such as mussels, commercial fishes, sugar, and bottled water are substances that allow a significant amount of MPs to enter into human systems. Particles larger than 50 µm can easily be removed from drinking water during the treatment processes. Reusable water bottles have been reported to have a larger concentration of MPs than single-use plastic bottles, sometimes even more than freshwater bodies (Zhang et al. 2020). Some studies showed that the concentration of MPs/g in sugar, salt, alcohol, and bottled water is around 0.44, 0.11, 0.03, and 0.09, respectively. Microplastics also enter the human system through inhalation as they are airborne and may enter the air through various sources. Another entry point of these MPs is through the skin, where they penetrate the human skin through water while washing or showering (Campanale et al. 2020; De-la-Torre 2020).

In 2016, a comprehensive study was done on the data collected for the MPs present in food items. This study was carried out by European Food Safety Authority (EFSA, 2016). Out of the 13 studies conducted, around 10 showed a significant concentration of MPs in both seawater and freshwater fishes. Not only that, commercially extracted honey has even been shown to contain MPs at substantial concentrations (0.166 fibers/g). As we have shown, MPs have multiple ways to enter human systems. Additionally, ingested food and water containing MPs pose a severe threat due to leaching and weathering of the additives used in plastic production, the presence of residual monomers, and their prolonged interaction with the other pollutants (metal ions or organic molecules) (De-la-Torre 2020). There is still limited data available for the transport mechanisms of these microplastics and their concentrations in air, as shown in Figure 4.2.

4.3 Human Health Implications

The emergence of MPs as contaminants has raised alarms because of their entrance into the human body through the food chain due to ingestion, and causing severe health impacts. Microplastics' effects on human health depend on certain factors such as exposure, concentration, geological areas, and the type of additives used (Rochman 2018). The majority of MPs and their human health impact-based research has been conducted concerning marine ecosystems. Only 4% of the total research has focused on freshwater systems and their consequences (Li et al. 2018). A high amount of polystyrene and polyvinyl chloride are known to accumulate metal ions in seawater. It is assumed that there are 800 times more MPs on the surface than in the neighboring seawater (Xu et al. 2019). A study on the zebrafish (*Danio rerio*) showed that MPs tend to increase the concentration of silver in the gut compared to the other organs of the animal (Xu et al. 2019). However, the impact of MPs on human health is not well estimated due to the lack of regulatory acts by EFSA. Even though the consumption limit for these contaminants has been set at a value of five pieces of MPs for a normal human being, the higher value of these recorded in

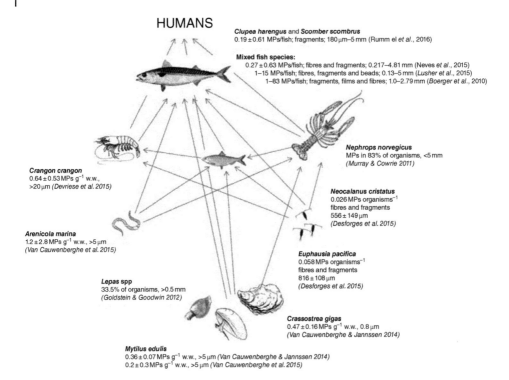

Figure 4.2 A Schematic representation of microplastic translocation from seafood to human beings through food chain. *Source:* Picture taken from Carbery et al. (2018). (Reprinted with permission.)

human feces raised another area of concern (Abbasi et al. 2018; Macmillan 2018). Studies have shown the cellular uptake and direct effect of these contaminants on cellular functions. They are cytotoxic, inflammatory, and lead to the formation of reactive oxygen species (ROS), creating cellular distress (De-la-Torre 2020). These MPs enter the circulatory system via the gut through a trophic transfer at the food chain level and inhibit the efflux pumps. They further induce cytotoxicity in the intestinal cells and create oxidative stress by generating free radicals (due to ROS) (Sharma and Chatterjee 2017). Microplastics alter cell homeostasis through their antioxidant systems (due to overproduced ROS in the cell). These ROS can lead to genetic alterations, physiological damage, carcinogenesis, and instability of cellular structures due to the suppression of antioxidants. At the cellular level, these MPs interfere with the influx pumps, inhibit their activities, cause mitochondrial depolarisations, and induce uncontrolled cellular reactions by ROS. The MPs incessantly attack the signaling pathways of cells, leading to cellular fibrosis, cell autophagy, and sometimes even DNA mutations (Wang et al. 2020).

4.3.1 Digestive System

Microplastics can easily cross cellular barriers through phagocytosis (a process mediated via active vesicles) and persorption (paracellular mechanically-driven approach) and become a part of cellular functions. Cationic particles easily adhere to their surfaces compared to the neutral or anionic particles, and hence enter into active transport mechanisms (Triebskorn et al. 2019). Studies have shown that even the physical adherence of these MPs can lead to tissue necrosis, cell inflammation, and cell proliferation. WHO Reports (2019) stated that MPs may sometimes cross the gut walls and quickly move to other tissues located far away from the mucosal layers. Microplastics can be readily adsorbed in the intestinal lining by M cells, which line intestinal lymphoid tissues (Peyer's patches). The microfolds in the cell are the primary site of MP absorption in the Peyer's patches, where they enter the circulatory system and reach the lymphatic systems (Van Cauwenberghe and Janssen 2014). As per reports, a major part of the ingested MPs (>90%) do not absorb and are directly excreted from the human systems. The M cells of the gastrointestinal tract translocate these MPs by a paracellular transport mechanism, which is mainly responsible for deciding the fate and effects of MPs in the human body (Elizalde-Velázquez and Gómez-Oliván 2021).

Even though a significant portion of MPs pass through without being absorbed, the gut becomes the first organ to be affected by the presence of these contaminants, and hence shows minor effects. Oral intake of MPs at higher concentrations can cause mouth ulcers and inflammation. These MPs can also accumulate in the phagocytes of gut tissues, potentially harm cellular activity and response to infections, and impair local immunity. There are also reports of the microbiome of the gut becoming disturbed by MPs (WHO 2019). Some smaller particles can easily cross the gut wall and translocate to the other tissues, causing further damage to the organ systems. The smaller size of these particles enhances the surface-to-volume ratio. MP molecules hence become an adsorption center for biomolecules such as proteins and lipids, which play a significant role in cellular activity. Some evidence showed that particles having dimensions <150 μm are likely to penetrate through the epithelial lining of the gastrointestinal tract. Only 0.3% is absorbed, and 0.1% of particles more significant than 10 μm in size are likely to reach organ systems and cellular membranes (Barboza et al. 2018).

Human exposure to MPs can lead to severe cases of neurotoxicity and neurogenic disorders. This has been studied *in vivo* by the exposure of cells to the particulates. The neural damages are mainly oxidative stress and the microglia activation in the brain (immune cells) because of the direct contact for the particles that are translocated or migrating to the brain cells from other inflammation sites, damaging the neurons. The prolonged exposure of these particulate matters may result in Alzheimer's disease or incidences of dementia (Chen et al. 2017). Some of the plastic additives, such as phthalates, Pb or Cd containing colorants, and others, are a serious concern because of their more severe damage to human bodies. *In vitro* studies of 3 μm MPs with a controlled intake of 0.2% poly(lactic-co-glycolic acid) in human colon mucosal tissue have shown inflammatory bowel disease (IBD) in patients. Several cases have been reported of intestinal damage after just 90 days of exposure to these pollutants (Sana et al. 2020). The studies conducted on gastric adenocarcinoma cells done by Forte et al. (2016) showed that around 44–100 nm of non-modified PS nanoparticles are engulfed into the tissue through Clathrin-mediated endocytosis. It is estimated that around 10 μ/L of PS concentrations regulated the gene expression of IL-6 and IL-8. This has been shown to induce inflammatory reactions and morphological altercations, which were higher with the reducing size. The co-incubating of cell-lining with positively charged PS nanoparticles reduced the cell lining of Caco2, LS174T, and HT29. These results have been established through optical microscopy and showed completely distorted cellular morphology. Hence, it is well established that a high concentration of PS in the gut lining can lead to cell destruction through apoptotic-mediated cytotoxicity via caspase 3-, 7-, and 8-mediated cytotoxicity (Bexiga et al. 2011; Inkielewicz-Stepniak et al. 2018).

4.3.2 Respiratory System

Microplastics containing PS were studied for human cell linings at various concentrations. These cell linings were Calu-3 epithelial cells, and THP-1 macrophages where LC50 has been reached at 31 and 75 µ/ml for the macrophages above, respectively (Paget et al. 2015). The studies based on polystyrene microplastics (PS-MPs) on human lung epithelial BEAS-2B cells have revealed an inflammatory and toxic impact of materials over the cells by forming ROS. PS-MPs deplete the zonula occludens proteins, and hence reduce the transepithelial electrical resistances. Thus, a high concentration of PS-MPs can trigger such adverse responses, and low concentrations only disturb the protective pulmonary barrier. This opens up an increased risk for lung diseases and damage to the human respiratory system (Dong et al. 2019). They are also known to increase Acetylcholinesterase (AChE) activity and its related neurotransmitters such as threonine, aspartate, and taurine (Deng et al. 2017). Reports are available for lung infections such as coughing and dyspnea, and reduction in the capacity of the lungs have been observed in the workers who work at the nylon flock plants in the US and Canada. Toxic chemicals and additives such as BPA, Triclosan, phthalates, and flame retardants have entered our systems through microplastics. Their disruptive effects have already been discussed and investigated in many research articles. Polystyrene and polyvinyl chlorides used in foam packaging, plates, crockery, etc., are well known for their carcinogenic effects and reproductive abnormalities in humans. Polystyrene nanoparticles can also lead to reduced cell viability, inflammatory response to the gene expressions, and adenocarcinoma of epithelial cells of human gastric cells (Karbalaei et al. 2018). Xu et al. (2019) studied A549 cells for the effects of PS and found that high concentrations inhibited the cellular activity of these cells. They also inhibit the cellular activities of DR5 (both gene expressions and apoptotic proteins) as caspase 8-, 3-, and 9-mediated cytotoxicity. This surely indicated that PS-MP could lead to functional damage of the respiratory system.

4.3.3 Nervous System

Schirinzi et al. (2017) studied the cytotoxic effect of MPs (PE and PS) on the human cerebral cells (T98G) after exposure to various concentrations in terms of oxidative stress and cell viability. The results showed that while there was not much impact of these MPs on cell viability, there was, however, a significant increase in the level of ROS in these cells, which leads to high oxidative stress. This is an essential mechanism by which these MPs exert on the cell on toxicity level.

4.3.4 Placental Barrier

Even though MP presence in the uterus is not a significant concern, it may cross the human placental barrier; an *ex vivo* study of the human placenta model has been performed to study the mechanism of transport of these MPs. It has been found that MPs have been accumulated in the syncytiotrophoblast of the placenta, and the syncytio-trophoblast has been found to play a significant role in nutrient transport across the placental barrier. However, it also becomes a key player in the transportation of MPs across the placental tissues. There is a need for a comprehensive study to understand the transport mechanism of these MPs in the placental barrier that may lead to embryotoxicity (Grafmueller et al. 2015). A high concentration of these MPs may increase ROS on various human-derived cells, leading to inflammatory reactions and may cause apoptosis. However, the chemical composition of MPs may lead to variations in the results. A comprehensive analysis of the hazardous effect of these plastic polymers was done considering health risk assessment, physical attributes, and environmental factors. As per the reports, (PUs), epoxy resins, (PVC), and (PS) were found to be highly hazardous and toxic.

On the other hand, (PE), polyvinyl acetate (PVA), and (PP) were found to be comparatively less harmful than the previously mentioned microplastics. The toxicity of these MPs is affected not only by their chemical constitution but also by their local environment (Lithner et al. 2011). During the polymerization process and processing of plastic materials free radicals are formed, which act as a common factor that induces ROS formation. Further, these free radicals increase in their concentration over time, due to the breakdown of the carbon–hydrogen bond catalyzed by the exposure to light or the effect of transition metals in the weathering of these plastics (Gewert et al. 2015; Wright and Kelly 2017).

4.3.5 Other Health Impacts

Flame retardants, stabilizers, inorganic pigments, and other biocides used as plasticizers in materials are responsible for some breast cancers. Flame retardants (PET, PE, and PVC) are also known for their genotoxicity effects on humans. PVC used as heat stabilizers in many plastic products leads to metabolic changes for calcium, phosphorous, and bone, and may lead to osteomalacia or bone fractures in postmenopausal women. PVC is also well known for lipid peroxidation and it promotes carcinogenesis in some cases. Bisphenol A (BPA), one of the common plasticizers used in water bottles, has been a significant antagonist on endocrinal health. It has an estrogenic impact on human health, predominantly on females, and is widely known for disrupting the hormonal system of the body, causing breast cancer in women and prostate cancer in men (Campanale et al. 2020). It is also well known for causing reproductive disorders in human females and is a significant cause of childhood obesity and heart disease in newborn babies. BPA seriously interferes with the alpha- and beta-receptors of fat tissues and is emerging as the primary cause of obesity in humans by constricting liposomal enzyme lipase, aromatase, and some of the other lipogenesis regulators enzymes. Human males who have a higher level of BPA in their urine also showed a lowered amount of inhibin B and other steroidal hormones (De-la-Torre 2020).

BPA is well known to be an antagonist to estrogens and leads to insulin resistance, inhibits the transcription mediated via thyroid hormone, and hence affects the hormonal balances. The following significant chemicals of concern are phthalate esters which are used as plasticizers, they lead to abnormal sexual behaviors and birth-related defects in newborn babies (Karbalaei et al. 2018). Other plasticizers and chemicals present in microplastics, such as PSs, PCBs, pharmaceuticals, and other metabolites, have been mutagenic, carcinogenic, teratogenic, and damaging to the endocrinal systems. Furthermore, two more chemicals, butyl benzyl phthalate and di-2-Ethylhexyl phthalates have been classified under categories of threatening carcinogens by USEPA. Direct exposure to POPs and other toxic compounds that are adhered to the surfaces of microplastics can create a severe threat to both children and adults (Smith et al. 2018).

Another major concern of these MPs includes the formation of biofilms on their surface, which act as a potent carrier of microbial contamination when they enter the food chain. Even though there is limited evidence for this phenomenon and its impact, the longer transports of these biological entities over the long range can never be ignored. These highly diverse microbial communities formed over the surface of microplastics are termed "plastispheres." They are fairly close to the human pathogen, especially from the genus *Vibrio*, and act as a vector for the pathogens. They seriously threaten the biological quality of water and form a new habitat around the MP environments (Wagner et al. 2014). Studies conducted with lake water indicated that such anti-microbial resistant strains of microbes attached to the surface of MPs are easily carried. They may facilitate plasmid transfers of cells of biofilms to the resistant genes. Some studies also showed that biofilms could ensure bacterial survival in treatment plants, and these pathogens can easily invade new areas which are non-pathogenic (WHO 2019).

4.4 Future Directions and Plausible Solutions

There is not much evidence for the widespread impact of these MPs on human health; however, there is heavy demand for understanding their human exposures (SAPEA 2019). The role of MP penetration into the skin and its effect is a matter still under investigation. Some of the serious points that need to be addressed while dealing with these MPs are their hydrophobic interactions, effects of pH variations, aging of MP particles with time, and changing composition. These factors should be studied for the long-term impact of these MPs on human health. The primary or direct contamination of MP is also a topic of interest to understand their intrinsic, extrinsic, and mixed behaviors, and hence their chemical behaviors due to exposure to weathering. A comprehensive knowledge gap still resides in the fact that there is not enough literature that relates the impact of these MPs on human health. Most of the investigations have been done *in vivo* and still lack groundwork for understanding the real-time exposure results. So there is a lack of a definitive conclusion due to some contradictory results achieved while studying their impact on human beings (Mercogliano et al. 2020).

Despite all these considerations, chemical science combined with geomorphology and hydrology has become the focus when we consider the sources and impact of MPs. Most of the data available for these pollutants are fragmented, especially when we talk about freshwater resources. This creates a challenge in ascertaining the environmental risk assessment of these pollutants and their hazardous human health effects. Now is the time to give attention to the solution of these micro-pollutants, as their concentration is increasing at various trophic levels. Some serious prevention and clean-up processes are needed for their treatment and inhibiting their entrance into the water bodies. We know that single-use plastics are one of the significant contributors to these pollutants, so we need a better replacement of these potent hazardous materials. Over-use of plastic materials has to be decreased, and eco-friendly alternatives to these materials need to be applied in the future for reducing their generation. Better and deeper insight have to be made regarding the risk assessment of these materials, and more case studies are needed for calculating their effect on human beings (Eerkes-Medrano and Thompson 2018). Tracing and tracking these MPs at different trophic levels are still an area of significant concern which demands serious attention. Microbial degradation techniques will undoubtedly prove to be of great help while talking about the combatting of these pollutants, and they have been a topic of investigation for some years. Supplementary data are needed to identify the primary sources of MP contaminations, and standard sampling techniques need to be devised to analyze these MPs in drinking water and fresh water sources.

Some of the ways to avoid exposure to these microplastics consumption include:

- Using ceramic or glassware for heating food items in the microwave. Avoid using plastic containers.
- Try to cool down food before packing it into plastic containers.
- Avoid reusable plastic with codes 3, 6, and 7, which stand for phthalates, styrene, and bisphenol, as they are a major source of microplastics.
- Avoid using a dishwasher, as it becomes a major source of plastics and chemicals leaching into the containers.

4.5 Conclusion

Since we know that the distribution and abundance of MPs in water will increase with time due to the excessive use of plastic, there is a need to curb these materials from the aquatic systems. Freshwater ecosystems containing aquatic biota ranging from microbes to vertebrates along with a number of trophic levels are easily affected by MP pollution. MP pollution has become widespread through the land, fresh water, and other aquatic systems, it is easily penetrated into the environment from the breakdown of larger plastics, and very easily migrates from their sources. MPs are continuously recycled into the environment through the weather cycles. The increasing consumption of plastics and their allies lead to higher exposure to MPs. The ingestive route of these MPs into the human body is mainly through aquatic foods and bottled water. There are many pieces of evidence for the entry of these MPs into the human systems. They are widely known for their neurotoxic, immunotoxic, renal, hepatic, and endocrinal impacts, and sometimes are known to cause cancer. There is a need for an exhaustive study of these MPs for their health impacts, and there are varieties of high uncertainties considering their precautionary principles in applications. The increasing introduction of MPs needs extensive studies for the monitoring and tracking for understanding their health impacts, pathogenesis, and exposure. Even though there are many investigations available, the data is not enough to assess the role of MP and its toxic effect on humans.

References

Abbasi, S., Soltani, N., Keshavarzi, B. et al. (2018). *Microplastics in different tissues of fish and prawn from the Musa Estuary, Persian Gulf. Chemosphere* 205: 80–87.

Alimi, O.S., Budarz, J.F., Hernandez, L.M. et al. (2018). *Microplastics and nanoplastics in aquatic environments: aggregation, deposition, and enhanced contaminant transport. Environmental Science & Technology* 52 (4): 1704–1724.

Amoatey, P. and Baawain, M.S. (2019). *Effects of pollution on freshwater aquatic organisms. Water Environment Research* 91 (10): 1272–1287.

Barboza, L.G.A., Vethaak, A.D., Lavorante, B.R. et al. (2018). *Marine microplastic debris: an emerging issue for food security, food safety, and human health. Marine Pollution Bulletin* 133: 336–348.

Bexiga, M.G., Varela, J.A., Wang, F. et al. (2011). *Cationic nanoparticles induce caspase 3-, 7-, and 9-mediated cytotoxicity in a human astrocytoma cell line. Nanotoxicology* 5 (4): 557–567.

Browne, M.A. (2015). Sources and pathways of microplastics to habitats. In: *Marine Anthropogenic Litter* (eds. M. Bergmann, L. Gutow and M. Klages), 229–244. Cham: Springer International Publishing.

Campanale, C., Massarelli, C., Savino, I. et al. (2020). *A detailed review study on potential effects of microplastics and additives of concern on human health. International Journal of Environmental Research and Public Health* 17 (4): 1212–1238.

Carbery, M., O'Connor, W., and Palanisami, T. (2018). *Trophic transfer of microplastics and mixed contaminants in the marine food web and implications for human health. Environment International* 115 (March): 400–409.

Chen, H., Kwong, J.C., Copes, R. et al. (2017). *Living near major roads and the incidence of dementia, Parkinson's disease, and multiple sclerosis: a population-based cohort study. The Lancet* 389 (10070): 718–726.

Cole, M., Lindeque, P., Halsband, C., and Galloway, T.S. (2011). *Microplastics as contaminants in the marine environment: a review. Marine Pollution Bulletin* 62 (12): 2588–2597.

Cole, M., Webb, H., Lindeque, P.K. et al. (2014). *Isolation of microplastics in biota-rich seawater samples and marine organisms. Scientific Reports* 4: 4528.

Cox, K.D., Covernton, G.A., Davies, H.L. et al. (2019). *Human consumption of microplastics. Environmental Science and Technology* 53 (12): 7068–7074.

De-la-Torre, G.E. (2020). *Microplastics: an emerging threat to food security and human health. Journal of Food Science and Technology* 57 (5): 1601–1608.

Deng, Y., Zhang, Y., Lemos, B. et al. (2017). *Tissue accumulation of microplastics in mice and biomarker responses suggest widespread health risks of exposure. Scientific Reports* 7: 46687.

Dong, C.-D., Chen, C.-W., Chen, Y.-C. et al. (2019). *Polystyrene microplastic particles: in vitro pulmonary toxicity assessment. Journal of Hazardous Materials* 385: 121575.

Eerkes-Medrano, D. and Thompson, R. (2018). Occurrence, fate, and effect of microplastics in freshwater systems. In: *Microplastic Contamination in Aquatic Environments* (ed. E.Y. Zeng), 95–132. Cambridge.

Elizalde-Velázquez, G.A. and Gómez-Oliván, L.M. (2021). *Microplastics in aquatic environments: a review on occurrence, distribution, toxic effects, and implications for human health. Science of the Total Environment* 780: 146551.

Europe, R. (2016). *Plastics – the Facts 2016: An Analysis of European Plastics Production, Demand, and Waste Data*. Brussels: Association of Plastic Manufacturers.

European Food Safety Authority (EFSA) (2016). *Presence of microplastics and nanoplastics in food, with particular focus on seafood. EFSA Journal* 14 (6) https://doi.org/10.2903/j.efsa.2016.4501.

Forte, M., Iachetta, G., Tussellino, M. et al. (2016). *Polystyrene nanoparticles internalization in human gastric adenocarcinoma cells. Toxicology in vitro* 31: 126–136.

Fu, Z. and Wang, J. (2019). *Current practices and future perspectives of microplastic pollution in freshwater ecosystems in China. Science of the Total Environment*, Elsevier BV 691: 697–712.

Gewert, B., Plassmann, M.M., and MacLeod, M. (2015). *Pathways for degradation of plastic polymers floating in the marine environment. Environmental Science. Processes & Impacts* 17 (9): 1513–1521.

Grafmueller, S., Manser, P., Diener, L. et al. (2015). *Bidirectional transfer study of polystyrene nanoparticles across the placental barrier in an ex vivo human placental perfusion model. Environmental Health Perspectives* 123 (12): 1280–1286.

Hammer, J., Kraak, M.H.S., and Parsons, J.R. (2012). Plastics in the marine environment: the dark side of a modern gift. In: *Reviews of Environmental Contamination and Toxicology* (ed. D.M. Whitacre), 1–44. New York, NY: Springer.

Horton, A.A. and Dixon, S.J. (2018). *Microplastics: an introduction to environmental transport processes. WIREs Water* 5 (2): 1–10.

Inkielewicz-Stepniak, I., Tajber, L., Behan, G. et al. (2018). *The role of mucin in the toxicological impact of polystyrene nanoparticles. Materials* 11 (5): 724.

Karbalaei, S., Hanachi, P., Walker, T.R. et al. (2018). *Occurrence, sources, human health impacts and mitigation of microplastic pollution. Environmental Science and Pollution Research* 25 (36): 36046–36063.

Koelmans, A.A., Nor, N.H.M., Hermsen, E. et al. (2019). *Microplastics in freshwaters and drinking water: critical review and assessment of data quality*. In: *Water Research*, vol. 155, 410–422. Elsevier Ltd.

Li, J., Liu, H., and Chen, P, J. (2018). *Microplastics in freshwater systems: a review on occurrence, environmental effects, and methods for microplastics detection. Water Research* 137: 362–374.

Li, C., Busquets, R., and Campos, L.C. (2020). *Assessment of microplastics in freshwater systems: a review*. In: *Science of the Total Environment*, vol. 707, 135578. Elsevier BV.

Lithner, D., Larsson, Å., and Dave, G. (2011). *Environmental and health hazard ranking and assessment of plastic polymers based on chemical composition. Science of the Total Environment* 409 (18): 3309–3333.

MacMillan, A. (2018). *There's plastic in your poop: Study warns microplastics may be hurting us all*; yahoo! online.

Mercogliano, R., Avio, C.G., Regoli, F. et al. (2020). *Occurrence of microplastics in commercial seafood under the perspective of the human food chain. A review. Journal of Agricultural and Food Chemistry* 68 (19): 5296–5301.

Murphy, F., Ewins, C., Carbonnier, F., and Quinn, B. (2016). *Wastewater treatment works (WwTW) as a source of microplastics in the aquatic environment. Environmental Science & Technology* 50 (11): 5800–5808.

Paget, V., Dekali, S., Kortulewski, T. et al. (2015). *Specific uptake and genotoxicity induced by polystyrene nanobeads with distinct surface chemistry on human lung epithelial cells and macrophages. PLoS One* 10 (4): e0123297.

Prata, J.C., da Costa, J.P., Lopes, I. et al. (2019). *Environmental exposure to microplastics: an overview on possible human health effects. Science of the Total Environment* 702: 134455.

Rainieri, S. and Barranco, A. (2019). *Microplastics, a food safety issue? Trends in Food Science and Technology.* Elsevier Ltd. 84: 55–57.

Rochman, C.M. (2018). *Microplastics research – from sink to source in freshwater systems. Science* 360 (6384): 28–29.

Sana, S.S., Dogiparthi, L.K., Gangadhar, L. et al. (2020). *Effects of microplastics and nanoplastics on the marine environment and human health. Environmental Science and Pollution Research* 27 (36): 44743–44756.

SAPEA (Science Advice for Policy by European Academies) (2019). *A Scientific Perspective on Microplastics in Nature and Society.* Berlin: SAPEA.

Schirinzi, G.F., Pérez-Pomeda, I., Sanchís, J. et al. (2017). *Cytotoxic effects of commonly used nanomaterials and microplastics on cerebral and epithelial human cells. Environmental Research* 159: 579–587.

Sharma, S. and Chatterjee, S. (2017). *Microplastic pollution, a threat to the marine ecosystem and human health: a short review. Environmental Science and Pollution Research* 24 (27): 21530–21547.

Smith, M., Love, D.C., Rochman, C.M. et al. (2018). *Microplastics in seafood and the implications for human health. Current Environmental Health Reports* 5 (3): 375–386.

Triebskorn, R., Braunbeck, T., Grummt, T. et al. (2019). *Relevance of nano- and microplastics for freshwater ecosystems: a critical review. TrAC – Trends in Analytical Chemistry* 110: 375–392.

Van Cauwenberghe, L. and Janssen, C.R. (2014). *Microplastics in bivalves cultured for human consumption. Environmental Pollution* 193: 65–70.

Vivekanand, A.C., Mohapatra, S., and Tyagi, V.K. (2021). *Microplastics in aquatic environment: challenges and perspectives. Chemosphere* 282 (June): 131151.

Wagner, M., Scherer, C., Alvarez-Muñoz, D. et al. (2014). *Microplastics in freshwater ecosystems: what we know and what we need to know. Environmental Sciences Europe* 26: 12.

Wang, Y.-L., Lee, Y.-H., Chu, I.-J. et al. (2020). *Potent impact of plastic nanomaterials and micromaterials on the food chain and human health. International Journal of Molecular Sciences* 21 (5): 1727–1740.

WHO (2019). *Microplastics in drinking-water – Key messages,* 1–3. https://www.who.int/water_sanitation_health/water-quality/guidelines/microplastics-in-dw-information-sheet/en (accessed 26 October 2021).

Wright, S.L. and Kelly, F.J. (2017). *Plastic and human health: a micro issue? Environmental Science & Technology* 51 (12): 6634–6647.

Xu, M., Halimu, G., Zhang, Q. et al. (2019). *Internalization and toxicity: a preliminary study of effects of nano plastic particles on human lung epithelial cells. Science of the Total Environment* 694: 133794.

Xu, S., Jie, M., Rong, J. et al. (2020). *Microplastics in aquatic environments: occurrence, accumulation, and biological effects. Science of the Total Environment* 703: 134699.

Yu, Q., Hu, X., Yang, B. et al. (2020). *Distribution, abundance, and risks of microplastics in the environment. Chemosphere,* Elsevier Ltd 249: 126059.

Zhang, Q., Xu, E.G., Li, J. et al. (2020). *A review of microplastics in table salt, drinking water, and air: direct human exposure. Environmental Science and Technology* 54 (7): 3740–3751.

5

Interactions of Microplastics Toward an Ecological Risk in Soil Diversity: An Appraisal

Iqbal Ansari[1], Marlia Mohd Hanafiah[1,2], Maha M. El-Kady[3], Charu Arora[4], and Sumbul Jahan[5]

[1] Department of Earth Sciences and Environment, Faculty of Science and Technology, Universiti Kebangsaan Malaysia (UKM), Bangi, Selangor, Malaysia
[2] Centre for Tropical Climate Change System, Institute of Climate Change, Universiti Kebangsaan Malaysia (UKM), Bangi, Selangor, Malaysia
[3] Department of Self-pollinated Vegetables Crops, Horticulture Research Institute (HRI), Agricultural Research Centre (ARC), Giza, Egypt
[4] Department of Chemistry, Guru Ghasidas University, Bilaspur, Chhattisgarh, India
[5] PG Department of Biotechnology, Vinoba Bhave University, Hazaribag, Jharkhand, India

5.1 Introduction

Plastics are a synthetic material which provides an immense societal benefit. In the world, its production exceeds about 320 million tons (Mt) per year; in which 40% has a one-time use for packaging purposes, and as a result, it is the primary source of plastic waste (PlasticsEurope 2017). da Costa et al. (2016) have opined plastics are polymers which are petrochemical products and contain of polypropylene (PP), propylene oxide and polyethylene (PE). Due to its high durability, inertness, cost effective, high strength, and resistance toward water make it very useful in various wide range of products (Andrady and Neal 2009; Cauwenberghe et al. 2015).

Plastic is versatile substance, it is a cluster of polymers having repeating units of carbon and hydrogen and may contain O, N, S, Cl, Si, and F. It delivers social benefits with a broad range of applications in industries such as construction, medicine, and food preservation (Clark et al. 2016; Singh and Devi 2019). Plastic manufacturing has flourished in the past 70 years as it has become applicable in every sphere of human life. Many reports over the last 10 years state that more than 2.6 billion metric tons (MT) of plastic have been produced globally (Jambeck et al. 2015). Due to sluggish degradation, plastic persists for long durations and is now recognized as a pollutant of the international environment. The MacArthur Foundation 2016 (MacArthur, 2016) reported that only 60% of produced plastic persists in the environment and only 9% of total plastic wastes are recycled. Wilson et al. (2015) reported that average per-capita consumption of developed countries is much higher than in developing countries, but the poor waste management of plastic leads to choking the environment. The United Nations estimated that single use plastic such as bottles, spoons, forks, straws, bags, etc., are about 50% of the total plastic items produced. If the consumption continues with this rate, it has been estimated that by 2050 the quantity of plastic will be larger than fish, by weight (Singh and Devi 2019). The indiscriminate use and inefficient waste management practices lead to threats to terrestrial, marine, industrial, and food security (Bonanno and Orlando-Bonaca 2018; Ryan 2015).

Meanwhile, the attention from large plastic items is shifted toward microscopic plastic particles and fibers. These microscopic plastic particles and fibers are derived from routine plastic waste and come directly from daily use of personal products (Mason et al. 2016). The presence of MPs in coastal waters was first reported in the 1970s (Carpenter and Smith 1972; Carpenter et al. 1972). Such microscopic plastic particles termed as microplastic (MP), are <5 mm in diameter; very similar to plankton, and are an ample and extensive hazardous pollutant to the environment (Clark et al. 2016; Mason et al. 2016; Pirc et al. 2016). Direct use of MP or breakdown of larger plastic results in accumulation of MPs in the environment. Small particles and fragmentation of larger plastic debris are the main source of MP. Use and washing of synthetic polymeric textiles result in accumulation of microfibers in seawater (Pirc et al. 2016). Washing of single garments was identified as a source of pollution. A significant amount of fibers are released from laundries and household washing (Sundt et al. 2014). Synthetic fibers from textile industries in municipal sewage sludge are one of the spreading routes in soil and water. The primary sources include plastic powders used in molding, industrial scrubbers used to clean surfaces, and a variety of industrial processes. MPs are added to various personal care products, including cosmetics. As microbeads wash down the drain, the concern about direct emission of MPs into terrestrial and aquatic environments is growing (McCormick et al. 2014; Mason et al. 2016). MPs can be easily ingested by organisms as it is plankton-like in size, the leaching of adsorbed contaminants and additives could be a source of lethal substances, which may impact the organisms by entering into the food chain (Mason et al. 2016; Rochman et al. 2015; Wright et al. 2013).

In the soil ecosystem, MP pollution was reported by Rillig (2012), their existence and harmful impacts in such complex soil ecosystems have concerned world researchers (Corradini et al. 2019; Huerta Lwanga et al. 2017; Yu et al. 2020). Moreover, studies on MP distribution and data in the soil ecosystems are inadequate because of the lack of advanced investigative analytical methods matrix (Moller et al. 2020). Presently, the studies on terrestrial ecosystems account for approximately 5% of MP-related data, which obviously lags behind that of aquatic systems. MPs from numerous sources may endure multifaceted pathways and environmental cycles in the complex heterogeneous system of soil, which may responsible for ecological stress in various fields (Benckiser 2019; Helmberger et al. 2019; Prata et al. 2020). The ingestion behavior of MPs by soil microbes and animals affects the health and function of soil ecosystems, including negative effects on some soil detritus feeders and habitants (Kim and An 2019; Zhou et al. 2020). Preliminary data revealed the presence of MPs in soil, which alters the soil properties, plant performance, and microbial activities (de Souza Machado et al. 2019; Fei et al. 2020).

Thus, there is a need to carry out investigation on sources of microplastic contamination; therefore, investigations are required to efficiently evaluate MPs effects on soil health and ecosystems.

5.2 Microplastic-Types and Properties

The abundance of MP is found in a diversified range of shapes and densities. This divergence causes scatter in various levels in the aquatic system, and show their presence to organisms at different trophic levels in different habitats (Betts 2008; Cole et al. 2011; Thompson et al. 2009). According to Plastic Europe (2017), various types of plastic have been utilized to describe MPs, viz. PE; polyester (PES), polyvinyl chloride (PVC), acrylic polymers (AC), polyamide (PA), polyether (PT), PP, polystyrene (PS), polyurethane (PU), cellophane (CP), and many more. The families of PE possess both high and low density. MPs have been classified as primary and secondary on the basis of its production and mannerisms. Primary MPs are very tiny plastic particles which are poured into the environment directly through various domestic and industrial effluents, spills, and sewage discharge, or indirectly. MP particles and their types are found in the form of fragments, fibers, pellets, film, and spheres (Kang et al. 2015; Li et al. 2016; Lusher et al. 2015; Rummel et al. 2016). Secondary MPs are produced by steady decaying and the gradual degradation and breakdown of larger plastic debris, which occur in the environment by the action of UV radiation (photo-oxidation), mechanical transformation (e.g. waves abrasion), and degradation by biological agents (Andrady and Neal 2009; Browne et al. 2007; Cole et al. 2011). Finally, these MPs degraded into nanoplastics having 1–100 nm in size.

5.3 Microplastic Sources and Accumulation in Soil and Sediments

MP pollution draws quite a bit of attention as it has been widespread in aquatic environments, sediments, and wetlands fields, and possesses a negative biological effect (Kallie et al. 2019). Maxwell et al. (2020) have opined that MPs are exceptional among the pollutants as these may contaminate soil in different ways. Some organisms may assist in soil contamination and distribution and arbitrate their effects in the soil food web. From various literature it has been found that accumulation of MPs depends on environmental abiotic factors and varies geographically with locations, environmental pressure, time, and hydrodynamic conditions (Hamid et al. 2018). It has been observed that environmental factors play a dynamic role in the dispersal of MPs as compared to anthropogenic factors (Huerta et al. 2017; Zhang et al. 2016). The abiotic factors such as wave currents (Kim et al. 2015), tides, cyclones, wind directions (Browne et al. 2010; Kukulka et al. 2012; Liubartseva et al. 2016; Sadri and Thompson 2014; Thiel et al. 2013), and river hydrodynamics (Besseling et al. 2017), determine the distribution of MPs. First time MP pollution in urban wetlands has been reported by Kallie et al. (2019) which describes particle size <1 mm. Nearly 46 items/kg of dry sediment consisting of 68.5% of MPs were observed.

Similarly, Sajimol et al. (2020) have observed a large quantity of MPs with particle size <5 mm in the beach sediments of Kanaykumari (India). They claimed about 343 particles (67% fiber and 33% fragment) by evaluating 50 g dry sediments (d.s.) from each location.

In addition, a large quantity of MPs in the Beijing River were found in the range of 178 ± 69 to 544 ± 107 items/kg sediment (Wang et al. 2017). Scheurer and Bigalke (2018) have reported the presence of MPs composed of 0.002% of soil dry weight in rural soil reserves, as in remote high mountain areas. The nature of soil can be altered and contaminated with MPs, and are responsible for changes in structure and function of soil, as well in microbial diversity (He et al. 2019; Rillig 2012).

5.4 Migration of Microplastics' Fate in Environment

Numerous tons of plastics had been produced yearly since the 50s. Bläsing and Amelung (2018) found that the yearly plastics production is about 322 million tons. Recently, initiatives such as a gradual increase in plastic recycling and a global trend for reducing the usage of plastic products have been taken to tackle the problem of microplastic pollution. After releasing plastic wastes into the environment, they start a degradation cycle and cause severe environmental problems through splitting and decomposing by microbial organisms, sunlight such as ultraviolet (UV), weather conditions. Gradually, the previous conditions cause the breakdown of plastics into smaller fragments known as MPs (Andrady 2011; Auta et al. 2017). The first observation of MPs was in the marine environment, many studies had been performed to find the MP in marine ecosystem and its effect on aquatic ecosystem (Zhang et al. 2020). A small number of studies focused on occurrence, sources, and effects of MPs in soil ecosystems, because the estimation and extraction of MPs pollution in soils is relatively difficult than its estimation in marine ecosystem. This is because it is mixed with soil aggregates and soil organic matter (SOM). Van Sebille et al. (2015) reported that, there are 93–236 thousand tons of MPs floating on the ocean surface, which equals about 51 trillion particles, based on the latest global estimation of MPs. Some earlier studies revealed that MP pollution in the oceans is huge, but later Zhang and Liu (2018) reported that MP concentrations in soils were found to be greater than in aquatic environments. The accumulation and increase of MPs in soil tend to be a result of numerous environmental origins and human activities such as rainfall, contaminated water sources, plastic mulching, irrigation, floods, sewage sludge, street runoff, and agricultural compost (de Souza Machado et al. 2018; Liu et al. 2017). Earthworms and other soil organisms such as gophers, mites, and moles can share in secondary MPs formation because they uptake plastic debris and convert them into MPs, and transfer these plastic particles from place to place in different soil (Rodríguez-Seijo et al. 2018). Moreover, MPs can work as vectors for the concentration of chemical pollutants by adsorption on their surface, like polycyclic organochlorine pesticides, polychlorinated biphenyls, and aromatic hydrocarbons, as well as heavy metals such as zinc, nickel, lead, and cadmium (Li et al. 2018).

Furthermore, due to the small sizes of MP, fauna can ingest it, which can lead to MPs transferring into the food chains (Liu et al. 2017). When MPs are introduced into soils, they are integrated into a complex consisting of SOM and mineral substituents. The previous mixture can affect the soil biota, change the physical properties of soils, and may reach levels far higher than that by threatening biodiversity and effecting global food production (Scheurer and Bigalke 2018).

5.5 Migration of Microplastics through Soil

There is horizontal and vertical distribution of MPs in soil and this distribution can be determined by many factors related to soil structure and organisms living inside and above soil; some of these soil characteristics are aggregations, soil cracking, macropores, and agricultural practices such as harvesting and plowing (Rillig et al. 2017). Gabet et al. (2003) suggested that bioturbation can share in MPs migration in soil by plant processes (for example, roots elongation, uprooting, and harvesting). Moreover, the biota which live in soil (for example, larvae, earthworms, cocoons, rodents, etc.) swallow and then excrete MPs and thus share in its movement (Cao et al. 2017; Huerta Lwanga et al. 2016; Rillig et al. 2017). MPs can transport vertically from superficial to various depths of soil by earthworms' sanctuaries, and spread horizontally across wide zones by the earthworms and through mosquito larvae feeding (Huerta Lwanga et al. 2016; Hurley and Nizzetto 2018; Ziajahromi et al. 2018). In the same ways, collembola and mites can redistribute and spread MPs by abrasion or chewing MPs, and rodents digging also contribute in MPs' migration in soil by analogous mechanisms (Maaß et al. 2017; Rillig 2012).

5.6 Soil Analysis Methodology

Most of the declared extracting and analysis methods for MPs was originally developed for use in aquatic samples in water and sediments, and are likely not harmonious with the complex nature of soil patterns which need more challenging methods for estimation, quantification, and analysis of MPs in soil (Thomas et al. 2020). Precise consideration for soil type and components such as particles and aggregate sizes, soil dissolved matter, silicates, minerals, and SOM (which is a heterogeneous, lusty mixture consisting of animal and plant remains at various decomposition stages have easily-degradable molecules of lipids, peptides, carbohydrates, and more complex polymeric macromolecules called humic) (Bronick & Lal 2005), and should be required for exact soil analysis (Blume et al. 2016).

5.7 Collection of Samples

For agricultural fields, the Federal Soil Protection and Contaminated Sites Ordinance (FSPCSO) of Germany determined the minimum depth of taking samples from soil as 30 cm (BBodSchV 1999). However, the predominance of agricultural studies use a 5 cm depth of soil for collecting MP contaminated samples (Piehl et al. 2018), while, the U.S. Department of Agriculture recommended collecting control samples from similar soil types of unaffected nearby areas for contaminated sites of concern (Schoeneberger et al. 2012). The majority of studies determining soils contaminated with MPs work on single samples, but for increasing samples representatively and homogeneity, the FSPCSO recommends subdividing each field into at least three subplots, and each subplot should

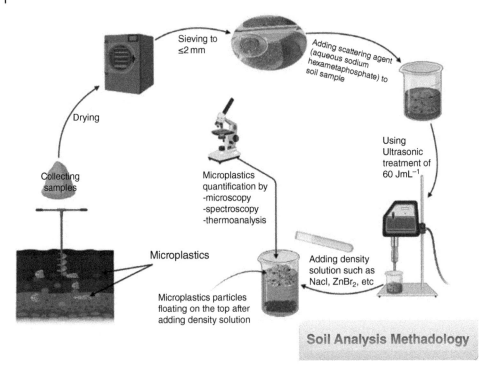

Figure 5.1   The various stages and their details of soil analysis methodology.

contain one combined sample consisting of 15–50 subsamples (BBodSchV 1999). The sample amount for soils with particles smaller than 2 mm should be at least 500 g (ISO 18400-102 2017). In more recent studies, MP screening of agricultural lands included sample quantities of 300 g to various kilograms per plot (Liu et al. 2018; Piehl et al. 2018; Scheurer & Bigalke 2018). The various stages and their details of soil analysis of contaminated land with MPs have been depicted in Figure 5.1.

5.8   Sample Preparation

5.8.1   Drying

Drying soil is a pretreatment before analysis needed to obtain water-free samples, ISO 11464 (2006) recommends to dry soil at 40 °C until weight stability. Other researchers have dried samples at 70 °C for 24 hours (Liu et al. 2018). Furthermore, in a previous study related to thermal decomposition of polymers, Beyler & Hirschler 2002 reported that temperatures above 40 °C may affect the structural and physical properties of polymers by melting, glass transition, or degradation. Results from some paradigms of the study showed that the glass transition temperatures of polybutylene terephthalate, polymethyl methacrylate (PMMA), and PA are 40, 50, and 50–75 °C, respectively. Ethylene-vinyl acetate and natural rubber start melting at 30–65 °C.

5.8.2 Sieving

For preparatory classification of plastic sizes, soil sample sieving is recommended (Bläsing et al. 2017). Soils sieving is usually done by <2 mm (according to the maximum size limits of sand; Schlichting et al. 1995). All dependent soil analyses usually refer to sieved fine soil, so Bläsing et al. (2017) suggested that MP soil sample analysis should be sieved with <5 mm and <1 mm sizes.

5.8.3 Soil Aggregates Dismantling and Density Separation

Sometimes MPs are integrated with soil aggregates and require additional pretreatments to promote soil aggregates degeneration and particle dispersion (Zhang & Liu 2018). But pretreatment should not include soil grinding because that can induce MPs melting by frictional heat and increase soil particles fragmentation (ISO 11464 2006). Dismantling includes elementary shaking for contaminated soil samples in scattering agents (Vermaire et al. 2017; Zhou et al. 2020). Subsequently, using an ultrasonic treatment of $60 \, Jm/L^{-1}$ is found to be sufficient for dismantling soil aggregates because at higher ultrasonic inputs, plant materials can become ruptured (Kaiser & Berhe 2014).

For density separation, the contaminated soil sample is mixed with a density solution which causes plastic particles to float. They are collected after certain amount of time in higher density solutions than that of plastics ($r = 0.9$–$1.6 \, g/cm^3$), for soil minerals such as silica which are higher than $r > 2.0 \, g/cm^3$, MP precipitates at the bottom (Enders et al. 2020; Liu et al. 2020). There are numerous density solutions with varied density numbers (0.8–$1.8 \, g/cm^3$) used to extract different types of MPs from soil samples, these solutions include sodium tungstate dihydrate and sodium polytungstate (SPT), which are suitable for extracting a wide range of MPs such as PP, PE, PS, PVC, PET, PU, PC, PA, EVA, and PMMA from sediments; but the disadvantage is that they are expensive and not commercial chemicals. Furthermore, deionized water and NaCl solution are considered easily available, not expensive, and not harmful on the environmental solutions suitable to extract PE, PS, and PP from soil mineral matrix. There are many more solutions, such as $NaBr$, $CaCl_2$, $ZnBr_2$, $ZnCl_2$, potassium formate, and NaI, which have high density (1.4–1.8) and can extract a wide range of MPs from different types of soils (Campanale et al. 2020).

5.8.4 Removing Soil Organic Matter (SOM)

SOM has similar density as MPs (0.9–$1.6 \, g/cm^3$) and it can only be slightly removed by density separation (Enders et al. 2020). Several studies have used H_2O_2 oxidants for SOM removal, because it was applied in previous studies to extract SOM from sediments (Imhof et al. 2012; Nuelle et al. 2014). Similarly, 96–108% of SOM has been removed by Hurley et al. (2018) from a loamy sand with 30% H_2O_2 at 70 °C. In the same study, PS particles partly deteriorated, while PA particles were destroyed. In addition, other digestion agents have been used to remove SOM from soil samples so as not to interfere with subsequent steps for MP analysis, such as alkaline solutions KOH or NaOH, and acids including H_2SO_4 and HNO_3 (Dehaut et al. 2016).

5.8.5 Microplastics Quantification

After collecting floated particles of MPs and following the reported process, MPs analyses were finally carried out by using the instruments listed, which include:

5.8.5.1 Microscopy

Microscopy can determine shapes, sizes, and particle numbers of MPs at low cost, and is used mainly to analyze large plastic particles (mostly >500 µm) (Lusher et al. 2020; Zhang et al. 2020). Adding fluorescent staining dyes such as Evans blue, Calcofluor white, and Nile red can help in differentiating MPs from the surrounding matrix (Maxwell et al. 2020; Nel et al. 2021). The disadvantage of this method is that visual classification for MPs identification in soil samples, which are solid heterogeneous matrix, is expected to have high error rates around 20–70% (Bläsing & Amelung 2018). Therefore, identification of MPs should be completed with spectroscopic methods such as thermoanalytical approaches, or spectroscopic methods like FTIR (He et al. 2018).

5.8.5.2 Spectroscopy

It helps to analyze chemical and physical properties of MPs concurrently, such as particle sizes and shapes and the polymer types. Both types of microspectroscopic techniques (FTIR and Raman) are used commonly for particles <500 µm, and samples should be placed on a flat filter disc with a bright background that is about 13–47 mm in diameter (Anger et al. 2018; Xu et al. 2019). The disadvantage is that transparent and white items are easily missed in manual selection of suspicious MP particles, which raises error rates because of the bright filter background (Lares et al. 2019), even using automated methods may reduce or increase number of particles when MPs are not proportionately distributed after sample preparation on the filter discs (Anger et al. 2018). Using a FT-IR microspectrometer with a focal plane array (FPA) may reduce measurement times and error rates. Moreover, such detector elements allow for chemical mapping of larger areas on the flat filter and provide multiple measurements for the same particle (Simon et al. 2018).

5.8.5.3 Thermoanalysis

The polymer thermal decomposition at temperatures higher than 500 °C and quantification by distinctive pyrolysis outputs, particles size, and shape should be determined in advance by microspectrometers because it is impossible to do that after thermoanalysis stage (Dümichen et al. 2017; Nguyen et al. 2019). Posteriorly, some improvements were added to use thermoanalytical with time-of-flight (TOF) or FTIR detectors for additional accuracy in analyzing samples (Dierkes et al. 2019; Steinmetz et al. 2020). In addition, instrumental analyzing (thermoanalytical with microspectrometers) becomes more sensitive for MPs quantification in the low mg/kg to µg/kg range (Sullivan et al. 2020). One study showed that thermoanalytical methods including chromatographical separation were more reliable in identifying and quantifying PP, PE, PET, and PS in an organic matter (Becker et al. 2020).

5.9 Interactions and Impacts on Soil Diversity

When MPs are present in soil as contaminants for a long period, it alters the physicochemical properties of the soil. The microbes and animals existing in such conditions also suffer from its impacts.

5.9.1 Soil Properties

MPs can be merged with soil aggregates and lumps. Water holding capacity may be notably increased by PES fibers, water-stable aggregation, and bulk density decreases. MPs may also modify water retention and the permeability of soil, which affects evaporation of water (de Souza Machado et al. 2018). A study by Wan et al. (2019) investigated the water evaporation and drought cracking in two clay soils after the addition of MPs, they reported that both clay soils are prominent and increase more with the increasing MP content. Based on the previous results, MPs can increase water shortages in soil, modify the water cycle in soils, and exacerbate the MPs migration into deep soil layers along rifts (Rillig et al. 2017). Soil enzymes are closely related with soil biochemical processes; these enzymes act as soil fertility indicators and have a fundamental role in the soil nutrient regulation cycle for nutrients such as P, C, and N (Trasar-Cepeda et al. 2008). Studies proved that MPs have significant effects on soil enzyme activity such as catalase, urease, phenol oxidase, and fluorescein diacetate hydrolase (FDAse) (Huang et al. 2019; Liu et al. 2017), which can affect soil quality (Muscolo et al. 2015). Furthermore, the presence of MPs may cause miscalculation of soil carbon storage, since the soil bulk density is an important tool for extrapolating soil carbon storage (Rillig et al. 2018). In addition, some of the MPs carbon content may be masked in the soil as an anthropogenic component of the soil organic carbon pools, since MPs contain high carbon polymers (Rillig et al. 2018).

5.9.2 Soil Microbial Activity

The existence of MPs in soil can affect the living microbes in it, change in physical soil properties such as modification in soil moisture, and soil porosity caused by MPs may change the natural distribution of aerobic and anaerobic microorganisms (Rubol et al. 2013). MPs can also alter pore spaces, and that may lead to loss of natural habitat and the disappearance of domestic microorganisms (Veresoglou et al. 2015). Likewise, the addition of MPs changed the microbial community structure and caused notable reduction in the substrate-induced respiration (SIR) rates, as an indication to changes in the microbial activity in soil as a result for MPs existence (Judy et al. 2019). Moreover, MPs found in soils can change the natural diversity of microorganisms in those soils. Huang et al. (2019) noticed significant numerical increases in some Actinobacteria group members in soils enriched with PE such as *Streptomyces humidus*, *Streptomyces misionensis*, and *Streptomyces iakyrus*, which has the ability to degrade synthetic polymers.

5.9.3 Microplastics Entered Via Food Chains

Lwanga et al. (2017) found that MP particles found in chickens bowels and excrement, when fed with MP-free crops, may have two speculative explanations; (i) macroplastics, when passed through the digestive tract of chickens, are converted to MPs; and (ii) the source of MPs in chickens may come from the feeding on earthworms containing MPs. More evidence about the trophic transport of MPs in terrestrial food chains are reported by Deng et al. (2017). They found MPs in mice tissues such as kidney, liver, and gut; this study shows that mice can digest and store MPs in their organs that will be eaten by other animals in higher trophic systems. Therefore, the bioaccumulation of MPs negatively affects terrestrial food webs and human health. The way of transporting MPs through food chains and their effects in soil animals and floral diversity has been shown in Figure 5.2.

5.9.4 The Effect of MPs on Soil Animals

MPs may stick on the outer body of organisms and directly impede their mobility (Kim & An 2019). The MPs ingestion in most cases is accidental, as organisms sometimes mistake MPs for food (Cole et al. 2013). This MPs ingestion can cause false repletion, which leads to less food intake and energy consumption, decreasing growth, and may ultimately lead to death (da Costa et al. 2016; Setälä et al. 2016). In addition, MPs also can cause mechanical damage to the organisms' organs such as intestinal obstruction, esophagus, metabolism disorders, decreasing reproduction, and biochemical responses as decreasing immune

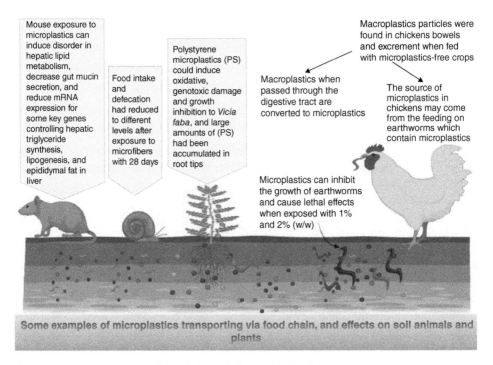

Mouse exposure to microplastics can induce disorder in hepatic lipid metabolism, decrease gut mucin secretion, and reduce mRNA expression for some key genes controlling hepatic triglyceride synthesis, lipogenesis, and epididymal fat in liver

Food intake and defecation had reduced to different levels after exposure to microfibers with 28 days

Polystyrene microplastics (PS) could induce oxidative, genotoxic damage and growth inhibition to *Vicia faba*, and large amounts of (PS) had been accumulated in root tips

Macroplastics when passed through the digestive tract are converted to microplastics

Macroplastics particles were found in chickens bowels and excrement when fed with microplastics-free crops

The source of microplastics in chickens may come from the feeding on earthworms which contain microplastics

Microplastics can inhibit the growth of earthworms and cause lethal effects when exposed with 1% and 2% (w/w)

Some examples of microplastics transporting via food chain, and effects on soil animals and plants

Figure 5.2 The entry route of MPs via food chains, and their effects.

response, etc. (Lahive et al. 2019; Lönnstedt & Eklöv 2016; Wang et al. 2019). Similarly, a study made by Cao et al. (2017) on earthworms suggested that MPs can significantly inhibit their growth and cause a lethal effect when exposed with 1 and 2% (w/w) concentrations. Song et al. (2019) reported that snail (*Achatina fulica*) food intake and defecation had reduced to different levels after exposure to microfibers within 28-day, also they noticed that microfibers induced villous injury to the gastrointestinal wall of snails. Moreover, another study by Lu et al. (2018) reported that mice exposed to MPs can induce disorder in hepatic lipid metabolism, decrease gut mucin secretion, and reduce the expressions of the mRNA for some key genes controlling hepatic triglyceride synthesis, lipogenesis and epididymal fat in the liver. Besides, to the previous hazards, there is the toxicity of additives in plastics manufacturing, toxic contaminants with chemicals, antibiotics and heavy metals all of that can be adsorbed on MPs, which increases soil pollution and amplifies the hazards exposure to biota and humans (Groh et al. 2019; Hahladakis et al. 2018; Hodson et al. 2017; Rodríguez-Seijo et al. 2018).

5.9.5 The Effect of MPs on Plants

Very few studies were carried out to measure the effects of MPs on plants which grow in contaminated soil. Those studies reported a significant effect of MPs on plants, including perennial ryegrass (Lolium perenne), wheat (*Triticum aestivum*), cress (*Lepidium sativum*), beans (*Vicia faba*), and spring onion (*Allium fistulosum*) (Boots et al. 2019; Bosker et al. 2019; Jiang et al. 2019; Qi et al. 2018; de Souza Machado et al. 2019). Similarly, Jiang et al. (2019) reported that, PS MPs could induce oxidative and genetic damage, and growth inhibition to *V. faba* grown in hydroponic systems, and screening with the laser confocal microscopy proved that the large amounts of PS (100 nm) had been accumulated in root tips. In addition, MPs with 4.8 µm can accumulate in *L. sativum* pores and cause significant reduction in germination rate after eight hours of exposure (Bosker et al. 2019). However, piling of MPs in plants may cause a siege effect between the cell connections or between cell wall pores, and thus undermine the transport and absorption of nutrients by the plant (Asli & Neumann 2009; Ma et al. 2010). The effects of three types of MP (fibers, fragments, and beads) on spring onion (*A. fistulosum*) growth had been investigated by de Souza Machado et al. (2019), and they found that MP causes significant changes dependent on particle types in elemental tissue composition, such as water content, C—N ratio, and leaf nitrogen content; biomass; root measurements such as root length, total root area, root average diameter, and the density of root tissues; and root symbiosis relationships with microorganisms.

5.10 Ecotoxicology of Microplastic

MPs can alter the physical properties of soil, which include bulk density and water dynamics (Liu et al. 2017; de Souza Machado et al. 2018, 2019), hence attractively declining overall bulk density of soil; additionally an increased in density of the rhizosphere has been observed (de Souza Machado et al. 2019). Machado et al. (2018) have shown that MPs altered the bulk density, water holding capacity, and the functional relationship between the microbial activity and water-stable aggregates in soil and all were badly affected.

The ecotoxicological effects of MPs on soil animals have been broadly reviewed (Chae & An 2018; Ng et al. 2018; Zhu et al. 2019) and investigated by several researchers. Many researchers have focused on earthworms (Cao et al. 2017; Huerta Lwanga et al. 2016; Judy et al. 2019; Rodríguez-Seijo et al. 2018); however, Lei et al. (2018) have studied the *Caenorhabditis elegans*, a nematode capable of ingesting MPs which accrue in the gut and cause a harmful physiological effect. *C. elegans* usully lives in decaying plant litter slightly larger than bulk soil (Schulenburg & Félix 2017). A recent report by Elizabeth (2021) revealed that MPs were found in human placentas, which possibly affect fetal health and development. Such MPs possibly entered into female bodies through the process of ingestion and inhalation, and then translocated into the placentas. Another similar study by Ragusa et al. (2020) reports the presence of MPs in the placenta, which may possibly alter several cellular regulating pathways such as immunity mechanisms. Moreover, another study in rats reveals that by inhalation of MPs in pregnant rats,, these MPs entered into internal organs, and their presence has been observed in placentas, as well as the fetal liver, lungs, heart, kidney, and brain (Fournier et al. 2020). There are several studies which have shown the alterations and, in some cases, severe ill effects of MPs in animals.

5.11 Mitigation Process of Microplastics

The elimination process of MPs became very challenging due to their tiny size, although some studies have shown good results with the removal process of MPs, especially in wastewater treatment plants and sewage sludge, since MPs are considered the primary source of contamination. Meanwhile, MPs which are concentrated in the sludge have been removed from sewage in WWTPs, and used as fertilizer in agricultural soil (Corradini et al. 2019). There are several conventional and innovative techniques developed by many countries thus far, and used for the removal of MPs in its diversified sources (Hidayaturrahman & Lee 2019; Imhof et al. 2012; Lares et al. 2018; Talvitie et al. 2017; Ziajahromi et al. 2017).

5.11.1 Biological Methods

Various microbes, soil, and compost consortia have been successfully harnessed for the biodegradation of plastic polymers in biological methods of remedial process of MPs. However, in the case of PE, the carbon mineralization rates are very low, even after prolonged periods (Orhan & Buyukgungor 2000; Restrepo-Florez et al. 2014; Starnecker & Menner 1996; Volke-Sepúlveda et al. 2002; Watanabe et al. 2009). The bacteria present in the soil uptake some quantity of low-density polyethylene (LDPE) and convert it into hydroxyl and carbonyl groups (Mukherjee et al. 2017; Watanabe et al. 2009). Novotný et al. (2018) have investigated the virgin and γ-irradiation/high temperature-pretreated linear low-density polyethylene (LLDPE) film biodegradation by using a *Bacillus amyloliquefaciens* strain, which were isolated from composted plastics and identified using microbial identification system BIOLOG and 16S rDNA sequences. The FTIR spectra result reveals that before and after a 60-day treatment, there was a decrease in carbonyl band and flattening of the 1300–1100/cm zone due to bacterial action. Gel Permeation Chromatography (GPC) showed an increase of Mn and Mw of 2300–3700 and 32 200–35 500 g/mol, respectively, resulting in a decrease of polydispersity index, which further suggesting the presence of low molar weight LLDPE oligomers in pretreated LLDPE. The PET plastics can be degraded by *Ideonella sakaiensis* (Yoshida et al. 2016) and PE with the help of marine fungus *Zalerion maritimum* (Paço et al. 2017). An engineered bacterial strain was found suitable for biodegradation of micro-size PET particles as MPs. The study revealed that the combining method of alkali with organisms gives more efficient results in the biodegradation process based on whole-cell-biocatalysis (Jixian et al. 2018). Hence, such an approach will be a potential asset in the bioremediation process of MPs. This will provide a sustainable development for the human race by decaying and hopefully removing MPs from the environment.

5.12 Conclusion and Future Perspectives

MPs are small, heterogeneously mixed plastics, which permeate soil ecosystems through different sources. Various factors affect the flow of MPs in soil ecosystems. Interaction of MPs with soil leads to change in physical, chemical, and biological properties. There are reports on increased adsorption capacity of soil due to interaction with MPs, which ultimately results in increased levels of harmful pollutants in soil. As a result, MP and pollutants adsorbed on soil surfaces transport and accumulate in plats as well as in soil organisms. We recommend research on microorganisms which can degrade MPs as an environmentally friendly way for bioremediation of MP contaminated soil.

References

Andrady, A.L. (2011). *Microplastics in the marine environment. Marine Pollution Bulletin* 62 (8): 1596–1605.

Andrady, A.L. and Neal, M.A. (2009). *Applications and societal benefits of plastics. Philosophical Transactions of the Royal Society of London. Series B, Biological Sciences* 364: 1977–1984.

Anger, P.M., von der Esch, E., Baumann, T. et al. (2018). *Raman microspectroscopy as a tool for microplastic particle analysis. TrAC Trends in Analytical Chemistry* 109: 214–226.

Asli, S. and Neumann, P.M. (2009). *Colloidal suspensions of clay or titanium dioxide nanoparticles can inhibit leaf growth and transpiration via physical effects on root water transport. Plant, Cell & Environment* 32 (5): 577–584.

Auta, H.S., Emenike, C.U., and Hamid, F. S. (2017). *Distribution and importance of microplastics in the marine environment: a review of the sources, fate, effects, and potential solutions. Environment International* 102: 165–176.

BBodSchV. (1999). Federal Soil Protection and Contaminated Sites Ordinance; Number On the Basis of §§ 6, 8, Paragraphs 1 and 2; and § 13, Paragraph 1, Sentence 2; Federal Soil Protection Law of 17 March 1998 (Federal Law Gazette I, p. 502); Bundesgesetzblatt: Berlin, Germany.

Becker, R., Altmann, K., Sommerfeld, T., and Braun, U. (2020). *Quantification of microplastics in a freshwater suspended organic matter using different thermoanalytical methods–outcome of an interlaboratory comparison. Journal of Analytical and Applied Pyrolysis* 148: 104829.

Benckiser, G. (2019). Plastics, micro- and nanomaterials, and virus-soil microbe-plant interactions in the environment. In: *Plant Nanobionics, Nanotechnology in the Life Sciences* (ed. R. Prasad), 83–110. Springer.

Betts, K. (2008). *Why small plastic particles may pose a big problem in the oceans. Environmental Science & Technology* 42 (24): 8995–8995.

Beyler, C.L. and Hirschler, M.M. (2002). Thermal decomposition of polymers. In: *SFPE Handbook of Fire Protection Engineering, Scientific Research*, 2. An Academic Publisher.

Bläsing, M. and Amelung, W. (2018). *Plastics in soil: analytical methods and possible sources. Science of the Total Environment* 612: 422–435.

Bläsing, M., Amelung, W., Schwark, L., and Lehndorff, E. (2017). *Inland navigation: PAH inventories in soil and vegetation after EU fuel regulation 2009/30/EC. Science of the Total Environment* 584: 19–28.

Blume, H.P., Brümmer, G.W., Fleige, H. et al. (2016). *Scheffer/Schachtschabel Soil Science*, 1e. Berlin, Germany: Springer.

Bonanno, G. and Orlando-Bonaca, M. (2018). *Ten inconvenient questions about plastics in the sea. Environmental Science & Policy* 85: 146–154.

Boots, B., Russell, C.W., and Green, D.S. (2019). *Effects of microplastics in soil ecosystems: above and below ground. Environmental Science & Technology* 53 (19): 11496–11506.

Bosker, T., Bouwman, L.J., Brun, N.R. et al. (2019). *Microplastics accumulate on pores in seed capsule and delay germination and root growth of the terrestrial vascular plant* Lepidium sativum. *Chemosphere* 226: 774–781.

Bronick, C.J. and Lal, R. (2005). *Soil structure and management: a review. Geoderma* 124 (1–2): 3–22.

Browne, M.A., Galloway, T., and Thompson, R. (2007). *Microplastic – an emerging contaminant of potential concern? Integrated Environmental Assessment and Management* 3: 559–561.

Campanale, C., Savino, I., Pojar, I. et al. (2020). *A practical overview of methodologies for sampling and analysis of microplastics in riverine environments. Sustainability* 12 (17): 6755.

Cao, D., Wang, X., Luo, X., Liu, G. et al. (2017). Effects of polystyrene microplastics on the fitness of earthworms in an agricultural soil. In: *IOP Conference Series: Earth and Environmental Science*, 61(1), 012148. IOP Publishing.

Carpenter, E.J. and Smith, K.L. Jr. (1972). *Plastics on the Sargasso Sea surface. Science* 175: 1240–1241. https://doi.org/10.1126/SCIENCE.175.4027.1240.

Carpenter, E.J., Anderson, S.J., Harvey, G.R. et al. (1972). *Polystyrene spherules in coastal waters. Science* 178: 749–750. https://doi.org/10.1126/SCIENCE.178.4062.749.

Cauwenberghe, V.L., Devriese, L., Galgani, F. et al. (2015). *Microplastics in sediments: a review of techniques, occurrence and effects. Marine Environmental Research* 111: 5–17.

Chae, Y. and An, Y.J. (2018). *Current research trends on plastic pollution and ecological impacts on the soil ecosystem: a review. Environmental Pollution* 240: 387–395. https://doi.org/10.1016/j.envpol.2018.05.008.

Clark, J.R., Cole, M., Lindeque, P.K. et al. (2016). *Marine microplastic debris: a targeted plan for understanding and quantifying interactions with marine life. Frontiers in Ecology and the Environment* 14 (6): 317–324.

Cole, M., Lindeque, P., Halsband, C. et al. (2011). *Microplastics as contaminants in the marine environment: a review. Marine Pollution Bulletin* 62: 2588–2597. https://doi.org/10.1016/j.marpolbul.2011.09.025.

Cole, M., Lindeque, P., Fileman, E. et al. (2013). *Microplastic ingestion by zooplankton. Environmental Science & Technology* 47 (12): 6646–6655.

Corradini, F., Meza, P., Eguiluz, R. et al. (2019). *Evidence of microplastic accumulation in agricultural soils from sewage sludge disposal. Science of the Total Environment* 671: 411–420.

da Costa, J.P., Santos, P.S., Duarte, A.C., and Rocha-Santos, T. (2016). *(Nano) plastics in the environment–sources, fates and effects. Science of the Total Environment* 566: 15–26.

Dehaut, A., Cassone, A.L., Frère, L. et al. (2016). *Microplastics in seafood: benchmark protocol for their extraction and characterization. Environmental Pollution* 215: 223–233.

Deng, Y., Zhang, Y., Lemos, B., and Ren, H. (2017). *Tissue accumulation of microplastics in mice and biomarker responses suggest widespread health risks of exposure. Scientific Reports* 7 (1): 1–10.

Dierkes, G., Lauschke, T., Becher, S. et al. (2019). *Quantification of microplastics in environmental samples via pressurized liquid extraction and pyrolysis-gas chromatography. Analytical and Bioanalytical Chemistry* 411 (26): 6959–6968.

Dümichen, E., Eisentraut, P., Bannick, C.G. et al. (2017). *Fast identification of microplastics in complex environmental samples by a thermal degradation method. Chemosphere* 174: 572–584.

Elizabeth, C.A. (2021). *Great concern as study finds microplastics in human placentas.* https://news.mongabay.com/2021/01/great-concern-as-study-finds-microplastics-in-human-placentas/ (accessed 26 October 2021).

Enders, K., Tagg, A.S., and Labrenz, M. (2020). *Evaluation of electrostatic separation of microplastics from mineral-rich environmental samples. Frontiers in Environmental Science* 8: 112.

Fei, Y., Huang, S., Zhang, H. et al. (2020). *Response of soil enzyme activities and bacterial communities to the accumulation of microplastics in an acid cropped soil. Science of the Total Environment* 707: 135634.

Fournier, S.B., D'Errico, J.N., Adler, D.S. et al. (2020). *Nanopolystyrene translocation and fetal deposition after acute lung exposure during late-stage pregnancy. Particle Fibre Toxicology* 17 (55) https://doi.org/10.21203/rs.3.rs-39676/v1.

Gabet, E.J., Reichman, O.J., and Seabloom, E.W. (2003). *The effects of bioturbation on soil processes and sediment transport. Annual Review of Earth and Planetary Sciences* 31 (1): 249–273.

Groh, K.J., Backhaus, T., Carney-Almroth, B. et al. (2019). *Overview of known plastic packaging-associated chemicals and their hazards. Science of the Total Environment* 651: 3253–3268.

Hahladakis, J.N., Purnell, P., Iacovidou, E. et al. (2018). *Post-consumer plastic packaging waste in England: assessing the yield of multiple collection-recycling schemes. Waste Management* 75: 149–159.

Hamid, F.S., Mehran, S.B., Norkhairiyah, A. et al. (2018). *Worldwide distribution and abundance of microplastic: How dire is the situation? Waste Management & Research* 36 (10): 873–897.

He, D., Luo, Y., Lu, S. et al. (2018). *Microplastics in soils: analytical methods, pollution characteristics and ecological risks. TrAC Trends in Analytical Chemistry* 109: 163–172.

He, P., Chen, L., Shao, L. et al. (2019). *Municipal solid waste (MSW) landfill: a source of microplastics?-evidence of microplastics in landfill leachate. Water Research* 159: 38–45. https://doi.org/10.1016/j.watres.2019.04.060.

Helmberger, M.S., Tiemann, L.K., and Grieshop, M.J. (2019). *Towards an ecology of soil microplastics. Functional Ecology* 3 (3): 550–560. https://doi.org/10.1111/1365-2435.13495.

Hidayaturrahman, H. and Lee, T.-G. (2019). A study on characteristics of microplastic in wastewater of South Korea: identification, quantification, and fate of microplastics during treatment process. *Marine Pollution Bulletin* 146: 696–702. https://doi.org/10.1016/j.marpolbul.2019.06.071.

Hodson, M.E., Duffus-Hodson, C.A., Clark, A. et al. (2017). *Plastic bag-derived microplastics as a vector for metal exposure in terrestrial invertebrates. Environmental Science & Technology* 51 (8): 4714–4721.

Huang, Y., Zhao, Y., Wang, J. et al. (2019). *LDPE microplastic films alter microbial community composition and enzymatic activities in soil. Environmental Pollution* 254: 112983.

Huerta Lwanga, E., Gertsen, H., Gooren, H. et al. (2016). *Microplastics in the terrestrial ecosystem: implications for* Lumbricusterrestris (Oligochaeta, Lumbricidae). *Environmental Science & Technology* 50 (5): 2685–2691.

Huerta Lwanga, E., Gertsen, H., Gooren, H. et al. (2017). *Incorporation of microplastics from litter into burrows of* Lumbricusterrestris. *Environmental Pollution* 220: 523–531.

Huerta, L.E., Vega, J.M., Quej, V.K. et al. (2017). *Field evidence for transfer of plastic debris along a terrestrial food chain. Scientific Reports* 7: 1–7.

Hurley, R.R. and Nizzetto, L. (2018). *Fate and occurrence of micro (nano) plastics in soils: knowledge gaps and possible risks. Current Opinion in Environmental Science & Health* 1: 6–11.

Hurley, R.R., Lusher, A.L., Olsen, M., and Nizzetto, L. (2018). *Validation of a method for extracting microplastics from complex, organic-rich, environmental matrices. Environmental Science & Technology* 52 (13): 7409–7417.

Imhof, H.K., Schmid, J., Niessner, R. et al. (2012). *A novel, highly efficient method for the separation and quantification of plastic particles in sediments of aquatic environments. Limnology and Oceanography: Methods* 10 (7): 524–537.

ISO 11464 (2006). *Soil Quality – Pretreatment of Samples for Physico-Chemical Analysis.* Technical report; International Organization for Standardization: Geneva, Switzerland.

Jambeck, J.R., Geyer, R., Wilcox, C. et al. (2015). *Plastic waste inputs from land into the ocean. Science* 347 (6223): 768–771.

Jiang, X., Chen, H., Liao, Y. et al. (2019). *Ecotoxicity and genotoxicity of polystyrene microplastics on higher plant* Vicia faba. *Environmental Pollution* 250: 831–838.

Jixian, G., Tongtong, K., Yuqiang, L. et al. (2018). *Biodegradation of microplastic derived from poly (ethylene terephthalate) with bacterial whole-cell biocatalysts. Polymers* 10: 1326. https://doi.org/10.3390/polym10121326.

Judy, J.D., Williams, M., Gregg, A. et al. (2019). *Microplastics in municipal mixed-waste organic outputs induce minimal short to long-term toxicity in key terrestrial biota. Environmental Pollution* 252: 522–531.

Kaiser, M. and Berhe, A.A. (2014). *How does sonication affect the mineral and organic constituents of soil aggregates? – A review. Journal of Plant Nutrition and Soil Science* 177 (4): 479–495.

Kallie, R.T., Hsuan-Cheng, L., David, J.S. et al. (2019). *Associations between microplastic pollution and land use in urban wetland sediments. Environmental Science and Pollution Research Intl.* 26: 22551–22561. https://doi.org/10.1007/s11356-019-04885-w.

Kang, J.K., Kwon, O.Y., Lee, K.W. et al. (2015). *Marine neustonic microplastics around the southeastern coast of Korea. Marine Pollution Bulletin* 96 (1–2): 304–312.

Kim, S.W. and An, Y.J. (2019). *Soil microplastics inhibit the movement of springtail species. Environment International* 126: 699–706.

Kim, I.S., Chae, D.H., Kim, S.K. et al. (eds.) (2015). Factors Influencing the Spatial Variation of Microplastics on High-Tidal Coastal Beaches in Korea. *Archives of Environmental Contamination & Toxicology* 69 (3): 299–309. https://doi.org/10.1007/s00244-015-0155-6. Epub 2015 Apr 12. PMID: 25864179.

Lahive, E., Walton, A., Horton, A.A. et al. (2019). *Microplastic particles reduce reproduction in the terrestrial worm Enchytraeuscrypticus in a soil exposure. Environmental Pollution* 255: 113174.

Lares, M., Ncibi, M.C., Sillanpää, M. et al. (2018). *Occurrence, identification and removal of microplastic particles and fibers in conventional activated sludge process and advanced MBR technology. Water Research* 133: 236–246.

Lares, M., Ncibi, M.C., Sillanpää, M., and Sillanpää, M. (2019). *Intercomparison study on commonly used methods to determine microplastics in wastewater and sludge samples. Environmental Science and Pollution Research* 26 (12): 12109–12122.

Li, H.X., Getzinger, G.J., Fergunson, P.L. et al. (2016). *Effects of toxic leachate from commercial plastics on larval survival and settlement of the barnacle* Amphibalanus Amphitrite. *Environmental Science & Technology* 50 (2): 924–931.

Li, X., Chen, L., Mei, Q. et al. (2018). *Microplastics in sewage sludge from the wastewater treatment plants in China. Water Research* 142: 75–85.

Liu, H., Yang, X., Liu, G. et al. (2017). *Response of soil dissolved organic matter to microplastic addition in Chinese loess soil. Chemosphere* 185: 907–917.

Liu, M., Lu, S., Song, Y. et al. (2018). *Microplastic and mesoplastic pollution in farmland soils in suburbs of Shanghai, China. Environmental Pollution* 242: 855–862.

Liu, M., Lu, S., Chen, Y. et al. (2020). Analytical methods for microplastics in environments: current advances and challenges. In: *Microplastics in Terrestrial Environments* (eds. D. He and Y. Luo), 3–24. Springer.

Lönnstedt, O.M. and Eklöv, P. (2016). *Environmentally relevant concentrations of microplastic particles influence larval fish ecology. Science* 352 (6290): 1213–1216.

Lu, L., Wan, Z., Luo, T. et al. (2018). *Polystyrene microplastics induce gut microbiota dysbiosis and hepatic lipid metabolism disorder in mice. Science of the Total Environment* 631: 449–458.

Lusher, A.L., Hernandez-Milian, G., O'Brien, J. et al. (2015). *Microplastic and macroplastic ingestion by deep diving oceanic cetacean: the True's beaked whale* Mesoplodon mirus. *Environmental Pollution* 199: 185–191.

Lusher, A.L., Bråte, I.L.N., Munno, K. et al. (2020). *Is it or isn't it: the importance of visual classification in microplastic characterization. Applied Spectroscopy* 74 (9): 1139–1153.

Ma, X., Geiser-Lee, J., Deng, Y., and Kolmakov, A. (2010). *Interactions between engineered nanoparticles (ENPs) and plants: phytotoxicity, uptake and accumulation. Science of the Total Environment* 408 (16): 3053–3061.

Maaß, S., Daphi, D., Lehmann, A., and Rillig, M.C. (2017). *Transport of microplastics by two collembolan species. Environmental Pollution* 225: 456–459.

MacArthur, E., Waughray, D., and Stuchtey, M.R. (2016). *The New Plastics Economy, Rethinking the Future of Plastics*. World Economic Forum.

Mason, S.A., Garneau, D., Sutton, R. et al. (2016). *Microplastic pollution is widely detected in us municipal wastewater treatment plant effluent. Environmental Pollution* 218: 1045–1054.

Maxwell, S.H., Melinda, K.F., and Matthew, G. (2020). *Counterstaining to separate Nile Red-stained microplastic particles from terrestrial invertebrate biomass. Environmental Science & Technology* 54 (9): 5580–5588.

McCormick, A., Hoellein, T.J., Mason, S.A. et al. (2014). *Microplastic is an abundant and distinct microbial habitat in an urban river. Environmental Science & Technology* 48 (20): 11863–11871.

Moller, J.N., Loder, M.G.J., and Laforsch, C. (2020). *Finding microplastics in soils: a review of analytical methods. Environmental Science & Technology* 54 (4): 2078–2090.

Mukherjee, S., Roy Chaudhuri, U., and Kundu, P.P. (2017). *Anionic surfactant induced oxidation of low-density polyethylene followed by its microbial bio-degradation. International Biodeterioration and Biodegradation* 117: 255–268.

Muscolo, A., Settineri, G., and Attinà, E. (2015). *Early warning indicators of changes in soil ecosystem functioning. Ecological Indicators* 48: 542–549.

Nel, H.A., Chetwynd, A.J., Kelleher, L. et al. (2021). *Detection limits are central to improve reporting standards when using Nile red for microplastic quantification. Chemosphere* 263: 127953.

Ng, E.L., Huerta, L.E., Eldridge, S.M. et al. (2018). *An overview of microplastic and nanoplastic pollution in agroecosystems. Science of the Total Environment* 627: 1377–1388. https://doi.org/10.1016/j.scito tenv.2018.01.341.

Nguyen, B., Claveau-Mallet, D., Hernandez, L.M. et al. (2019). *Separation and analysis of microplastics and nanoplastics in complex environmental samples. Accounts of Chemical Research* 52 (4): 858–866.

Novotný, Č., Malachová, K., Adamus, G. et al. (2018). *Deterioration of irradiation/high-temperature pretreated, linear low-density polyethylene (LLDPE) by* Bacillus amyloliquefaciens. *International Biodeterioration & Biodegradation* 132: 259–267. https://doi.org/10.1016/j.ibiod.2018.04.014.

Nuelle, M.T., Dekiff, J.H., Remy, D., and Fries, E. (2014). *A new analytical approach for monitoring microplastics in marine sediments. Environmental Pollution* 184: 161–169.

Orhan, Y. and Buyukgungor, H. (2000). *Enhancement of biodegradability of disposable polyethylene in controlled biological soil. International Biodeterioration and Biodegradation* 45: 49–55.

Paço, A., Duarte, K., da Costa, J.P. et al. (2017). *Biodegradation of polyethylene microplastics by the marine fungus* Zalerion maritimum. *Science of the Total Environment* 586: 10–15. https://doi.org/10.1016/j.scitotenv.2017.02.017.

Piehl, S., Leibner, A., Löder, M.G. et al. (2018). *Identification and quantification of macro-and microplastics on an agricultural farmland. Scientific Reports* 8 (1): 1–9.

Pirc, U., Vidmar, M., Mozer, A., and Kržan, A. (2016). *Emissions of microplastic fibers from microfiber fleece during domestic washing. Environmental Science and Pollution Research* 23 (21): 22206–22211.

PlasticsEurope (2017). *Plastics – The Facts 2017.* Brussels: PlasticsEurope (44 pp).

Prata, J.C., da Costa, J.P., Lopes, I. et al. (2020). *Environmental exposure to microplastics: An overview on possible human health effects. Science of the Total Environment* 702: 134455.

Qi, Y., Yang, X., Pelaez, A.M. et al. (2018). *Macro-and micro-plastics in soil-plant system: effects of plastic mulch film residues on wheat* (Triticum aestivumi) *growth. Science of the Total Environment* 645: 1048–1056.

Ragusa, A., Svelato, A., Santacroce, C. et al. (2021). *Plasticenta: first evidence of microplastics in human placenta. Environment International* 146: 106274. https://doi.org/10.1016/j.envint.2020.106274.

Restrepo-Florez, J.-M., Bassi, A.&., and Thompson, M.R. (2014). *Microbial degradation and deterioration of polyethylene – a review. International Biodeterioration and Biodegradation* 88: 83–90.

Rillig, M.C. (2012). *Microplastic in terrestrial ecosystems and the soil? Environmental Science & Technology* 46: 6453–6454.

Rillig, M.C., Ingraffia, R., and de Souza Machado, A.A. (2017). *Microplastic incorporation into soil in agroecosystems. Frontiers in Plant Science* 8: 1805.

Rillig, M.C., de Souza Machado, A.A., Lehmann, A., and Klümper, U. (2018). *Evolutionary implications of microplastics for soil biota. Environmental Chemistry* 16 (1): 3–7.

Rillig, M.C., Lehmann, A., de Souza Machado, A.A., and Yang, G. (2019). Microplastic effects on plants. *New Phytologist* 223: 1066–1070. https://doi.org/10.1111/nph.15794.

Rochman, C.M., Tahir, A., Williams, S.L., and Baxa, D.V.V. (2015). *Anthropogenic debris in seafood: plastic debris and fibers from textiles in fish and bivalves sold for human consumption. Scientific Reports* 5: 14340.

Rodríguez-Seijo, A., Santos, B., da Silva, E.F. et al. (2018). Low-density polyethylene microplastics as a source and carriers of agrochemicals to soil and earthworms. *Environmental Chemistry* 16 (1): 8–17.

Rubol, S., Manzoni, S., Bellin, A., and Porporato, A. (2013). *Modeling soil moisture and oxygen effects on soil biogeochemical cycles including dissimilatory nitrate reduction to ammonium (DNRA). Advances in Water Resources* 62: 106–124.

Rummel, C.D., Löder, M.G.J., Fricke, N.F. et al. (2016). *Plastic ingestion by pelagic and demersal fish from the North Sea and Baltic Sea. Marine Pollution Bulletin* 102 (1): 134–141.

Ryan, P.G. (2015). *A brief history of marine litter research.* In: *Marine Anthropogenic Litter* (eds. M. Bergmann, L. Gutow and M. Klages), 1–25. Cham: Springer.

Scheurer, M. and Bigalke, M. (2018). *Microplastics in Swiss floodplain soils. Environmental Science & Technology* 52 (6): 3591–3598.

Schlichting, E., Blume, H.-P., and Stahr, K. (1995). *BodenkundlichesPraktikum-Eine Einführung in pedologischesArbeitenfürÖkologen, insbesondere Land- und Forstwirte, und fürGeowissenschaftlicher*, 2e., neubearbeiteteAufl, 295. Berlin, Boston: Blackwell Wissenschafts-Verlag.

Schoeneberger, P., Wysocki, D., Benham, E., and Staff, S.S. (2012). *Field Book for Describing and Sampling Soils, Version 3.0.* Lincoln, NE, USA: Natural Resources Conservation Service, National Soil Survey Center.

Schulenburg, H. and Félix, M.-A. (2017). *The natural biotic environment of Caenorhabditis elegans. Genetics* 206: 55–86. https://doi.org/10.1534/genet ics.116.195511.

Setälä, O., Norkko, J., and Lehtiniemi, M. (2016). *Feeding type affects microplastic ingestion in a coastal invertebrate community. Marine Pollution Bulletin* 102 (1): 95–101.

Simon, M., van Alst, N., and Vollertsen, J. (2018). *Quantification of microplastic mass and removal rates at wastewater treatment plants applying Focal Plane Array (FPA)-based Fourier Transform Infrared (FT-IR) imaging. Water Research* 142: 1–9.

Singh, A.P. and Devi, A.S. (2019). Plastic waste: a review. *International Journal of Advanced Scientific Research and Management* 4 (3): 47–51.

Song, Y., Cao, C., Qiu, R. et al. (2019). *Uptake and adverse effects of polyethylene terephthalate microplastics fibers on terrestrial snails (Achatina fulica) after soil exposure. Environmental Pollution* 250: 447–455.

de Souza Machado, A.A., Lau, C.W., Kloas, W. et al. (2019). *Microplastics can change soil properties and affect plant performance. Environmental Science & Technology* 53 (10): 6044–6052.

Starnecker, A. and Menner, M. (1996). *Assessment of biodegradability of plastics under simulated composting conditions in a laboratory test system. International Biodeterioration and Biodegradation* 37: 85–92.

Steinmetz, Z., Kintzi, A., Muñoz, K., and Schaumann, G.E. (2020). *A simple method for the selective quantification of polyethylene, polypropylene, and polystyrene plastic debris in soil by pyrolysis-gas chromatography/mass spectrometry. Journal of Analytical and Applied Pyrolysis* 147: 104803.

Sullivan, G.L., Gallardo, J.D., Jones, E.W. et al. (2020). *Detection of trace sub-micron (nano) plastics in water samples using pyrolysis-gas chromatography time of flight mass spectrometry (PY-GCToF). Chemosphere* 249: 126179.

Sundt, P., Schulze, P. E. & Syversen, F. (2014). *Sources of Microplastic Pollution to the Marine Environment.* Report No. M-321/2015, Mepex for the Norwegian Environment Agency, Miljodirektoratet, 86.

Talvite, J., Mikola, A., Koistinen, A. et al. (2017). Solutions to microplastic pollution - Removal of microplastics from wastewater effluent with advanced wastewater treatment technologies. *Water Research* 123: 401–407. https://doi.org/10.1016/j.watres.2017.07.005.

Thomas, D., Schütze, B., Heinze, W.M., and Steinmetz, Z. (2020). *Sample preparation techniques for the analysis of microplastics in soil – a review. Sustainability* 12 (21): 9074.

Trasar-Cepeda, C., Leirós, M.C., and Gil-Sotres, F. (2008). *Hydrolytic enzyme activities in agricultural and forest soils. Some implications for their use as indicators of soil quality. Soil Biology and Biochemistry* 40 (9): 2146–2155.

Van Sebille, E., Wilcox, C., Lebreton, L. et al. (2015). *A global inventory of small floating plastic debris. Environmental Research Letters* 10 (12): 124006.

Veresoglou, S.D., Halley, J.M., and Rillig, M.C. (2015). *Extinction risk of soil biota. Nature Communications* 6 (1): 1–10.

Vermaire, J.C., Pomeroy, C., Herczegh, S.M. et al. (2017). *Microplastic abundance and distribution in the open water and sediment of the Ottawa River, Canada, and its tributaries. Facets* 2 (1): 301–314.

Volke-Sepúlveda, T., Saucedo-Castañeda, G., Gutiérrez-Rojas, M. et al. (2002). *Thermally treated low density polyethylene biodegradation by Penicillium pinophilum and Aspergillus Niger. Journal of Applied Polymer Science* 83: 305–314.

Wan, Y., Wu, C., Xue, Q., and Hui, X. (2019). *Effects of plastic contamination on water evaporation and desiccation cracking in soil. Science of the Total Environment* 654: 576–582.

Wang, J., Liu, X., Li, Y. et al. (2019). *Microplastics as contaminants in the soil environment: a mini-review. Science of the Total Environment* 691: 848–857.

Watanabe, T., Ohtake, Y., Asabe, H. et al. (2009). *Biodegradability and degrading microbes of low-density polyethylene. Journal of Applied Polymer Science* 111: 551–559.

Wilson, D.C., Rodic, L., Modak, P. et al. (2015). *Global Waste Management Outlook.* UNEP.

Wright, S.L., Thompson, R.C., and Galloway, T.S. (2013). *The physical impacts of microplastics on marine organisms: a review. Environmental Pollution* 178: 483–492.

Xu, J.L., Thomas, K.V., Luo, Z., and Gowen, A.A. (2019). *FTIR and Raman imaging for microplastics analysis: state of the art, challenges and prospects. TrAC Trends in Analytical Chemistry* 119: 115629.

Yu, H., Hou, J., Dang, Q. et al. (2020). *Decrease in bioavailability of soil heavy metals caused by the presence of microplastics varies across aggregate levels. Journal of Hazardous Materials* 395: 122690.

Zhang, G.S. and Liu, Y.F. (2018). *The distribution of microplastics in soil aggregate fractions in southwestern China. Science of the Total Environment* 642: 12–20.

Zhang, S., Liu, X., Hao, X. et al. (2020). *Distribution of low-density microplastics in the mollisol farmlands of Northeast China. Science of the Total Environment* 708: 135091.

Zhou, Y., Liu, X., and Wang, J. (2020). *Ecotoxicological effects of microplastics and cadmium on the earthworm* Eisenia foetida. *Journal of Hazardous Materials* 392: 122273.

Ziajahromi, S., Kumar, A., Neale, P.A., and Leusch, F.D. (2018). *Environmentally relevant concentrations of polyethylene microplastics negatively impact the survival, growth and emergence of sediment-dwelling invertebrates. Environmental Pollution* 236: 425–431.

Yoshida, S., Hiraga, K., Takehana, T. et al. (2016). *A bacterium that degrades and assimilates poly (ethylene terephthalate). Science* 351: 1196–1199.

Zhang, K., Su, J., Xiong, X. et al. (2016). *Microplastic pollution of lakeshore sediments from remote lakes in Tibet plateau, China. Environonental Pollution* 219: 450–455.

Zhu, F., Zhu, C., Wang, C. et al. (2019). *Occurrence and ecological impacts of microplastics in soil systems: a review. Bulletin of Environmental Contamination & Toxicology* 102 (6): 741–749. https://doi.org/10.1007/s00128-019-02623-z.

Ziajahromi, S., Neale, P.A., Rintoul, L. et al. (2017). *Wastewater treatment plants as a pathway for microplastics: development of a new approach to sample wastewater-based microplastics. Water Research* 112: 93–99.

6

Microplastics in the Air and Their Associated Health Impacts

Akanksha Rajput, Rakesh Kumar, Antima Gupta, and Shivali Gupta

Department of Environmental Sciences, University of Jammu, Jammu, India

6.1 Introduction

With the innovation of plastic in the 1950s, it became a large part of our modern lives. Due to their versatile nature, functionality, and cost-effectiveness, plastics have been in high demand all around the world. But their extensive use over the years has been tainted with problems such as plastic waste disposal and the introduction of nano-, micro-, and macro-plastics into the environment. Currently, managing plastic waste is a huge managerial task for town planners as well as for the scientific community. Rough estimates suggest that since 1950, approximately 8300 million metric tons (MMT) of different types of plastics have been manufactured worldwide (Geyer et al. 2017), with an annual increment of around 3%. In the health sector, plastics have found their immense usefulness; the global market for plastic-based medical supplies in 2020 was \$25.1 billion, and is projected to increase to \$29.4 billion in 2021 (Research and Markets 2020).

A greater consequence of plastic pollution is the emergence of MPs as environmental contaminants. MPs have been widely distributed in aquatic and terrestrial ecosystems far from their primary sources of origin. These MPs are present even in the remotest regions on the earth such as the deep oceans, mountains, and polar regions; endangering the pristine ecosystems of such regions. This raises an urgent need to identify their sources and understand the transport, fate, and impact of microplastics. Moreover, studies have reported the presence of MPs in Arctic biota, like benthic organisms (Fang et al. 2018), cetaceans (Moore et al. 2020), fish, and seabirds (Baak et al. 2020). Several authors have suggested that seabirds could be used as potential bioindicators of ecosystem health (Mallory et al. 2010), as well as to monitor the distribution of MPs in the marine environments (Van Franeker & Law 2015).

Plastic debris is commonly classified based on the size, shape, pigment, types of polymer used, place of origin, and original usage. Classification based on the size of the plastic, given by Barnes et al. (2009) has been provided in Table 6.1. MPs are typically defined as plastics having a diameter ranging from 1 μm to 5 mm (Thompson et al. 2004; Hartmann et al. 2019; Zhang et al. 2020). They are extremely heterogeneous in their physical and

Plastic and Microplastic in the Environment: Management and Health Risks, First Edition.
Edited by Arif Ahamad, Pardeep Singh, and Dhanesh Tiwary.
© 2022 John Wiley & Sons Ltd. Published 2022 by John Wiley & Sons Ltd.

Table 6.1 Categories of plastics based on their size.

Type of plastic	Size	Example
Macro plastics	>20 mm	Plastic debris like plastic bottles
Meso plastics	5–20 mm	Large plastic particles like resin pellets
Microplastics	<5 mm	Small plastic derived from the breakdown of macroplastics
Nano plastics	0.2–2 mm	Smaller MP particles

Note: The term mega-debris (>100 mm) is also used and can be applied to large debris items such as derelict fishing nets.
Source: Barnes et al. (2009)

chemical composition. Based on their origin, MPs are categorized as (i) primary and (ii) secondary microplastics. Primary MPs are the tiny particles designed for industrial and other commercial uses, such as microbeads in cosmetic and health care products, synthetic microfibers that shed from clothing, furnishings, fishing nets, resin pellets used as raw material in plastic manufacturing, etc. (Cole et al. 2011); while secondary MPs are formed from the weathering or degradation of primary plastics (Abbasi et al. 2019) and indirectly from the photocatalytic breakdown (oxidation process and/or mechanical weathering) of larger plastic particles from textile industries, agriculture, plastic material, electronics, construction, etc. (Galloway 2015; Patchaiyappan et al. 2020). Plastic fibers are one of the biggest contributors to microplastics.

Worldwide production of plastics reached 368 million tons in 2019, which is an increase of 47.2% since 2009 (250 million tons) (Tiseo 2021). China alone accounts for one-quarter of the global plastic production with considerably growing exports from $14.4 billion in 2009 to $48.3 billion in 2019. As per the annual report of Central Pollution Control Board (CPCB), the estimated production of plastics in India was 3 360 043 tons/annum for the year 2018–2019. In 2018, about 3.1 MMT of MPs were lost in the environment worldwide. Out of which, 1.41 MMT was contributed by the loss of rubber from tire abrasion, followed by urban dust (0.65 MMT), road markings (0.59 MMT), washing of clothing (0.26 MMT), weathering of marine coatings (0.05 MMT), through upstream plastic production (0.03 MMT), and microbeads contained in cosmetics (0.01 MMT) (Tiseo 2021). The regional distribution of microplastics lost into the environment by various means has been shown in Table 6.2. As plastic production grows worldwide, the concern of its disposal and degradation has increased.

The presence of MPs has been reported from every segment of the environment, and there is a plethora of scientific studies available on MPs in the marine environment (Alimba & Faggio 2019; Prata et al. 2019). Research conducted on the distribution of MPs in marine environments has played a significant role in drawing attention to policymakers and the common public to the ubiquitous presence of MPs and their deleterious effects. Several studies have also reported the presence of MP contaminants in the freshwater ecosystems (Eriksen et al. 2013; Rodrigues et al. 2018), soils and sediments (Abidli et al. 2018; Reed et al. 2018; Zhang et al. 2018), atmosphere (Abbasi et al. 2019; Dris et al. 2016), and food items (Kosuth et al. 2018). However, very limited information on atmospheric MP is available (Prata 2018; Wright et al. 2020; Zhang et al. 2020). One major emerging health issue is due to the inhalation of MPs particles suspended in the air, also known as atmospheric MPs.

Table 6.2 Regional loss of microplastics (%) to the environment as of 2018.

Sources	Western Europe	Africa	Latin America and the Caribbean	Middle East	Japan	NAFTA (including the rest of North America)	Oceania	Asia (excluding India, Japan and China)	India	China
Microbeads from cosmetics and personal care products	3	16	8	6	1	10	1	17	9	22
Loss of rubber from tire abrasion	13	3	6	5	2	20	1	14	6	18
Loss through weathering of marine coatings	18	6	10	5	4	22	0	6	4	19
Loss via washing of textiles	3	3	5	6	1	13	1	20	12	27
Road markings	18	6	10	5	4	22	0	6	4	19
City dust	1	22	8	6	0	3	0	21	14	20
Loss of plastic during upstream plastic production[a]	15	6	5	2	4	17	0	9	11	28

[a] From virgin plastic pellets.
Source: Tiseo (2019).

6.2 Microplastics in the Atmosphere

Microplastics accumulate in urban dust and soils; owing to their low densities, they tend to be readily blown up by winds. Recent studies have reported the presence of MPs in the ambient atmospheres of urban and suburban environments. The presence of MPs in the air enhances their chance of human exposure through the inhalation of fine, airborne materials and unintended ingestion of contaminated geo-solids. Because of low density and environmental persistence, MPs can be transported to remote locations far from their sources, and deposited through wet or dry deposition. Despite their increasing health concerns, there is limited information available on atmospheric MPs, their sources of origin, transport, and associated health risks. Some of the physical and chemical characteristics of atmospheric MPs are been discussed below.

6.2.1 Physical Characteristics

Microplastics appear in the atmosphere in different shapes; generally described as beads, spheres, foam, pellets, fibers, flakes, films, fragments, etc. Shapes of MPs depend upon factors such as the source of primary MP origin, degradation, transportation, and residence time; e.g. sharp-edged MP particulates depict they have recently been introduced in the environment, while the smooth-edged MPs suggest a prolonged introduction to the atmosphere (Hidalgo-Ruz et al. 2012). MP particulates of variable shapes have been reported in the literature; for example, in China, fiber-shaped atmospheric MPs form a large bulk constituting 67–95% of the atmospheric MP load (Cai et al. 2017; Liu et al. 2019a; Zhou et al. 2017). Similarly, Dris et al. (2016, 2017), reported fiber-shaped MPs constituted more than 90% of the load in France. In another study, Liu et al. (2019b) reported that 60% of MP particulates were composed of fiber-shaped fragments in the West Pacific region. The fragment-shaped MPs were found to constitute 68%, and more than 90% of the MPs at the Pyrenees Mountains and in Hamburg (Germany), respectively (Allen et al. 2019; Klein & Fischer 2019). Likewise, Abbasi et al. (2019) reported that the granular-shaped MPs (65.9%) were found dominating in Iran. The introduction of fiber-shaped MPs into the atmosphere is generally linked with the textile industry (Cai et al. 2017), whereas the weathering and tearing of larger plastic items (Zhang et al. 2020) and disposable plastics (Zeng 2018) produce fragment-shaped microplastics.

Similarly in the atmosphere, variable sizes of MP particulates have been reported. In Shanghai and Yantai, MPs having a size of <500 μm were reported (Liu et al. 2019a). In Hamburg, approximately 60% of the MP fragments were of size <63 μm and 30% had a size range of 63–300 μm, fibers having a length between 300–5000 μm were also observed (Klein & Fischer 2019). Likewise, in the Pyrenees Mountains, fiber-shaped MPs having a size of <300 μm were found (Allen et al. 2019). In the work of Bergmann et al. (2019), 80% of MPs were reported to have a size ≤25 μm in the European region, and 98% of MPs had a size of <100 μm at the Arctic region. The results of various studies suggest that the abundance of MPs are negatively correlated with their size (Bergmann et al. 2019; Zhang et al. 2020). The smaller particles tend to have a greater atmospheric residence time in comparison to the larger particles.

MPs of various colors like red, yellow, brown, orange, tan, off-white, gray, and blue, (Bergmann et al. 2019) have been observed in the atmosphere, although they are most commonly found in red and blue colors (Hidalgo-Ruz et al. 2012) since light colors may be difficult to detect during visual interpretation (Dris et al. 2015). Blue and black were the dominant colors found from studies conducted in Shanghai city (Liu et al. 2019a) and superglacial debris (Ambrosini et al. 2019). However, discoloration of MPs can happen during the preparation of samples for analysis, and should be taken into consideration while data reporting and interpreting (Zhang et al. 2020)

6.2.2 Chemical Characteristics

Plastics are formed by polymerization of different monomers into macromolecular chains using additive chemicals such as initiators, catalysts, and solvents. These additives are not bound with the matrix of polymers; hence, they leach out into the atmosphere. Their leaching rates depend upon factors such as volatility of molecular weight of chemicals, the permeability of polymer, pH, and temperature of the surrounding. Sometimes monomers are also leached out along with the additives (Galloway 2015). Lithner et al. (2011) reported that polyurethane (PU), polyacrylonitrile (PAN), polyvinyl chloride (PVC), and acrylonitrile-butadiene-styrene (ABS) fall under the category of highest hazard ranking polymers. while high-density polyethylene (HDPE), low-density polyethylene (LDPE), polyethylene terephthalate (PET), and polypropylene (PP) possess the lowest relative hazard score. Most hazardous polymers are produced from monomers that are classified as carcinogenic and/ or mutagenic in nature. As MPs are formed from plastics, so their composition depends on the properties of the monomers used to a larger extent (thermoplastics and thermosets). The most common types of polymers used are PP (19.3%), LDPE (17.5%), HDPE (12.3%), PVC (10.2%), PUR (7.7%), PET (7.4%), and PS (6.6%) (Plastics Europe 2018). Seawater composition of MPs suggests the presence of polyethylene as the predominant polymer, followed by PP and polystyrene; whereas on beaches and subtidal regions, polyethylene, polystyrene, PP, and polyester are major polymers (Zeng 2018). In Shanghai, PE, PET, PES (polyester), PAN, PAA [poly (N-methyl acrylamide)], and rayon comprised 91% of chemical constituents of MPs (Liu et al. 2019a); in Dongguan city, PE, PS, and PP were found (Cai et al. 2017); in Arctic snow, plasticized rubber, polyamides, and varnish were found; while polyimide, rubber, varnish, ethylene vinyl acetate (EVA), and PE were observed in European snow (Bergmann et al. 2019).

6.2.3 Sources and Generation

Plastic itself is an inert substance, but chemicals are often added to it for color, flexibility, rigidity, heat resistance, UV resistance, etc. It may be photo degraded by sunlight, the wind shear effect, and collisions with other suspended particulates, all of which breaks it into smaller fragments and releases different chemicals in the environment. These fragments become MPs, leading to the high bioavailability of fibers in the ambient atmospheres as well as other environmental components. The sources of MPs are closely linked to the high population density regions (Browne et al. 2011) and its distribution into various

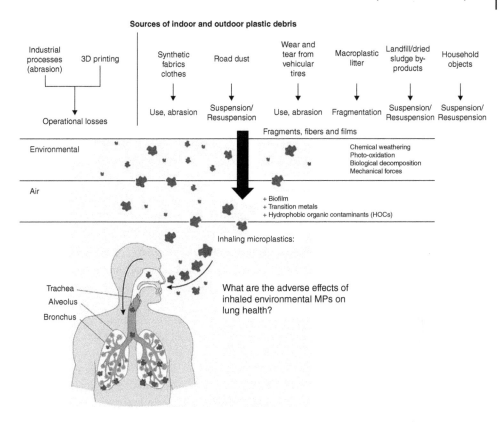

Figure 6.1 Major sources of atmospheric microplastics in indoor and outdoor environments and subject to human inhalation (Amato-Lourenço et al. 2020).

environmental components is strongly influenced by prevailing eolian currents and particle densities (Prata 2018). Common sources of MPs in the urban atmosphere include synthetic fibers from clothing and furnishing, abrasion of tires, building materials, incineration, municipal solid waste, cosmetics, industrial emissions, (Amato-Lourenço et al. 2020; Catarino et al. 2018; Dris et al. 2016), sewage sludge used as fertilizer, degradation of agricultural PE sheets (Wright & Kelly 2017), synthetic particles used in horticulture, and tumble dryer exhaust (Prata 2018).

Urban road dust is a sink for many of the MPs, owing to low density MPs being blown into the air by eolian activity and vehicular movements. In industrial environments, cutting and grinding processes contribute significantly to the atmospheric MPs (Wright & Kelly 2017). Identification of polymers, of which MPs are made, may also serve as a useful tool in the identification of the primary sources and characteristic transport pathways of MPs. For example, PET is the most common form of polymer found in indoor MPs, which derives its origin from textile industries (Kuczenski & Geyer 2010); whereas in the outdoor and aquatic environments, it may be transferred through the wind action and water runoff, respectively (Zhang et al. 2020). The morphology of MPs is highly variable if based on their origin from variable sources. Many studies have reported that fibers dominate atmospheric MPs, with a length >5 μm and width 3 μm (Cai et al. 2017). Fragment-shaped MPs have also been found dominating along with fibers in aquatic ecosystems such as seawater, freshwater, and beaches (Fu & Wang 2019; Zeng 2018), and terrestrial environments. In the urban and suburban regions in Paris, Dris et al. (2016) reported that 50% of the fibers in the atmospheric fallout were natural, being cotton or wool, and 21% were transformed from natural polymers like rayon, and 17% were pure synthetic fibers, mainly polyethylene-terephthalate. Moreover, owing to low density than the seawater, aerosolization of sea-surface MPs could happen from wind action and sea spray (Wright & Kelly 2017).

6.2.4 Fate and Dispersion

Transportation of MPs from their source of production and utilization through air and water are predominantly responsible for their dispersal in air and contamination in water bodies. However, the amount of MPs being transported through the atmosphere has not yet been estimated. Like other suspended particulates in the atmosphere, MPs' behavior and transport is affected generally by the following factors: (i) vertical pollution concentration gradient; (ii) wind speed; (iii) wind direction; (iv) atmospheric precipitation (affecting particles >2.5 mm); and (v) temperature (Prata 2018). The residence time of atmospheric MPs depends upon their size, density, and shape. Suspended MPs are eventually transported to the nearest water bodies depending on the above factors, and ultimately dumped into the oceans, acting as the sink.

To a larger extent, marine contamination occurs directly from the terrestrial areas (Gasperi et al. 2018). Slow degradation of the plastic waste by microbes like *Bacillus cereus*, Micrococcus sp., or Corynebacterium, oxidation, heat, light, or hydrolysis have also been reported (Smith et al. 2018). At higher altitudes, suspended atmospheric MPs are brought down by precipitation (Bergmann et al. 2019). The role of MPs as cloud condensation nuclei cannot be ruled out; however, it needs to be investigated. Assessments of atmospheric deposition on the glaciers have resulted in successfully calculating the total flux of the region. Due to the region's low temperature and unaltered conditions, deposition on the ice cores also provides an insight into the temporal variation (Turner et al. 2019). Few studies explained the potential sources of MPs in such far-flung areas by computing possible trajectories involved using the HYSPLIT model (Allen et al. 2019). However, still more research is required to figure out the possible transport efficiency, ice nucleation, precipitation scavenging, and source-pathway-sink of the MPs in the atmosphere.

6.3 Measurement of Atmospheric Microplastics

6.3.1 Sampling and Analysis

Concerns related to the health effects of atmospheric MP necessitated the quantification of the MPs in the atmosphere. As the research field is still in the preliminary phase, there is no standardized universal method for characterization, occurrence, and quantification of MPs and their impacts. Different investigators have applied several analytical approaches and interpretations. Most commonly, two types of sampling techniques have been employed, i.e. (i) passive sampling and (ii) active sampling techniques.

In the passive sampling technique, the collection of total atmospheric fallout is carried out via various methods such as sweeping, vacuum, active pumped sampling, and through a funnel into a glass bottle. Several researchers have examined MPs in road dust samples collected manually using brushes and a collecting pan (Abbasi et al. 2017; Dehghani et al. 2017). However, in the active type of sampling techniques, pumped samplers are used for a sample collection from a known volume of air over a specified time duration (Dris et al. 2017; Hayward et al. 2010; Liu et al. 2019a,b). Meteorological conditions must also be taken into consideration for the interpretation of results; therefore, passive sampling considers the deposited MPs, whereas the active samplers provide the mass of MPs suspended in the air instead. For analytical purposes, common techniques used for MP examination are visual microscopic examination, Raman spectroscopy (Zhao et al. 2017), Fourier transform infrared spectroscopy (FTIR) (Gaston et al. 2020), and pyrolysis-gas chromatography-mass chromatography (Fries et al. 2013), etc.

6.3.2 Atmospheric Abundance of Microplastics

Only a few studies have been conducted for quantitative and qualitative analysis of MPs in different study areas. Studies conducted in Paris, China, and London using passive sampling technique showed deposition of MPs to be 29–280 particles/m^2/day, 175–313 particles/m^2/day, and 575–1008 particles/m^2/day, respectively (Cai et al. 2017; Wright et al. 2020). In addition to a dry deposition, wet deposition can be a possible driver in the deposition of MPs from the atmosphere (Dris et al. 2016). The correlation of snow events with MPs may be explained by a study showing MPs as efficient cloud ice nuclei (Ganguly & Ariya 2019; Allen et al. 2019). Melted snow samples collected from Europe and the Artic showed MPs deposition range from 190 to 154×10^3 particles/L and 0–14.4×10^3 particles/L, respectively (Bergmann et al. 2019), the abundance of which was found to be four to seven fold higher than the previously conducted studies in Dongguan and Paris (Cai et al. 2017; Dris et al. 2017). Also, when compared to outdoor concentration, indoor concentration was higher, which may be a possible important source contributing to overall MPs. Indoor concentration ranged from 1.0 to 60.0 fibers/m^3, whereas outdoor concentration was found to be in the range of 0.3–1.5 fibers/m^3 (Dris et al. 2017). Studies conducted by Abbasi et al. (2018, 2019) with industrialized and urbanized street dust revealed the presence of about 900 MPs and 250 microrubbers (MRs)/15g sample. Liu et al. (2019a) quantified that fibers comprised 67% of all MPs, whereas fragments and granules comprised 30%, and 54% of the observed particles were composed of synthetic compounds. Similarly, in London, Wright et al. (2020) reported 15 different types of polymers in the air samples, the majority (92%) of which was fibers. A similar study by Stanton et al. (2019) reported a high prevalence of natural textile fibers (97.7%) at the campus of the University of Nottingham (UK), compared to extruded textile fibers.

6.4 Health Impacts of Microplastics

6.4.1 Routes of Exposure and Interaction with Body Tissues

In recent decades, MPs have emerged as a serious environmental health hazard. The direct risk is due to their presence in food, water, and air (Katyal et al. 2020) which in turn depends on their exposure, concentrations (Smith et al. 2018), degree of uptake, translocation, redistribution, and retention within the body tissue (Galloway 2015). MPs may also enter the human body through the intake of pharmaceutical drugs or infusion of small particles into packed food, beverages, etc. Indirectly, MPs may also find their way into the human body via inhalation as they fall from the atmosphere, by dermal contact with dust, or by degradation of plastic products from different industries such as textiles (Revel et al. 2018). For instance, crawling infants can easily take up fallout due to their frequent hand-to-mouth contact if fibrous MPs are settled on the floor (Gasperi et al. 2018). Atmospheric deposition of MPs into the reservoirs that supply tap water may contaminate our water supplies. Moreover, plastic in the water bodies attracts other contaminants like organic pollutants and toxic metals, and colonizes microbes. The pollutants like DDT, hexachlorobenzene, polycyclic aromatic hydrocarbons (PAH), mercury, etc. are sorbed by the MPs, which may result in genotoxicity, reproductive toxicity, mutagenicity, and carcinogenicity when ingested (Wright & Kelly 2017; Gasperi et al. 2018). Atmospheric particles and fibers that enter the nose and mouth and deposit in the upper respiratory tract are inhalable, whereas those reaching the deeper lung are categorized as respirable (Gasperi et al. 2018). Particles with larger aerodynamic diameter settle down easily and do not remain suspended for a long time. The uptake efficiency of particles depends upon surface charge and hydrophilicity in addition to shape and size (Galloway 2015; Wright & Kelly 2017), and the long, thin fibers are incompletely phagocytosed. They are biologically more active than short fibers, and the other persistent particles may translocate into the epithelial layers and induce acute or chronic inflammatory processes (Amato-Lourenço et al. 2020). Additionally, owing to their larger surface area, small MP particulates may induce an intense release of chemotactic factors in the respiratory system, which prevents the macrophage migration and increases permeability leading to chronic inflammation, known as dust overload (Donaldson et al. 2000).

Another route of uptake of MP into the gastrointestinal tract (GIT) is via persorption; which is the mechanical kneading of solid particles through the gaps of the single-layered epithelium of GIT and into the circulatory system (Wright & Kelly 2017). Volkheimer (1975) used PVC as a non-biodegradable MP along with starch to study this process in animals (rat, guinea pigs, rabbit, chicken, dogs, and pigs). Persorption of the starch molecule with a diameter of 150 μm in humans and its elimination through urine confirms that large and inert particles may translate from the gut to other parts of the body (Volkheimer 1977).

6.4.2 Health Impacts

Airborne MPs can enter the human body directly through inhalation, posing serious health risks. Recent investigations have detected natural and synthetic fibers from items of clothing and furnishings, and fragments of large plastics in the urban and suburban dust. Additionally, MPs can potentially enter the human food chain through terrestrial and sea foods (Wright & Kelly 2017). Their presence in some marine animals such as fish and crustaceans is well documented and they have been found to be present in sea salts (Karami et al. 2017; Yang et al. 2015), honey and sugar (Liebezeit & Liebezeit 2013), tapped water (Kosuth et al. 2017), and bottled water (Schymanski et al. 2018). Worldwide, scientific investigations have been conducted to understand the risk associated with humans exposure to plastics and their toxicity mechanisms (Canesi et al. 2015; Deng et al. 2017; Détrée

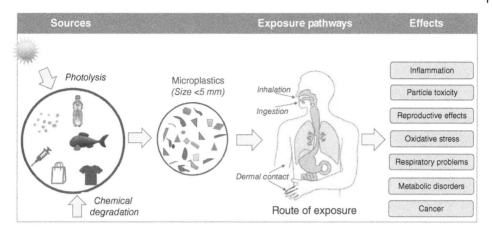

Figure 6.2 Pathways of microplastics exposure and toxicity to humans. *Source for bottle image:* kristiinaros/2 images/Pixabay.

& Gallardo-Escárate 2018; Farhat et al. 2011; Furukuma & Fuji 2016; Geiser et al. 2005; Watts et al. 2015; Wen et al. 2018; Wick et al. 2010; Wright et al. 2013; Xu et al. 2017; Yacobi et al. 2008; Zargorski et al. 2003). The major mechanisms through which MP toxicity happens include (i) MPs as a carrier of microbial contamination and harmful chemicals; (ii) disturbance of the energy homeostasis and metabolism; (iii) oxidative stress and cytotoxicity; (iv) interference with immune functions; (v) neurotoxicity; and (vi) translocation of microplastics to the circulatory system and distant tissues, among others (Prata et al. 2020).

After inhaling, the persistent nature of MPs limits the chance of removal from the body, causing chronic inflammation, which further aggravates the risk of cancer (Prata et al. 2020). Most of the MP fibers entering the respiratory tract are intercepted by the lung lining fluid, while some of them may escape the mucociliary clearance mechanisms of the lung (Wright & Kelly 2017; Gasperi et al. 2018). Another newly discovered phenomenon known as the "Plastisphere" (MP colonizing microbes) has also been cited to inflict indirect impacts on human health (Yang et al. 2020). MPs provide new microbial niches in the aquatic environment, such as bacteria developing biofilms by secretion of proteins, allowing them to stick to each other and the surfaces, thus enabling them to survive in hostile environments (Keswani et al. 2016), facilitating their dispersal and transport through water, air, or land (Revel et al. 2018). Kirstein et al. (2016) found that microbial species like Vibrio spp. can colonize the MPs that may be inhaled into the human body along with particulates. When ingested, these bacteria may cause watery diarrhea, often accompanied by abdominal cramping, nausea, vomiting, fever, and chills. Moreover, owing to their larger surface area, MPs tend to accumulate harmful chemicals including organic chemicals and toxic metals, which may be leached into body tissues (Crawford & Quinn 2017).

Monomers and other additives used in plastic production may leach from the matrix inside the organism, exposing body tissues to chemicals such as bisphenol A (BPA) and phthalates. BPA is a commonly used monomer for making polycarbonate (PC) plastic and epoxy resins, for lining food and beverage cans, which can leach out of the matrix and contaminate food and drinks (Galloway 2015) and affect human health by altering the liver function, brain function, reproductive system, insulin resistance, and may affect the development of offspring in pregnant woman (Srivastava & Godara 2017). MPs have also been reported to be enriched with several harmful contaminants such as PAHs and PCBs (Crawford & Quinn 2017; Frias et al. 2010). Phthalate ester is one of the contaminants that can cause abnormal sexual development and birth defects (Cheng et al. 2013). Many of these chemicals are also known as endocrine disruptors, capable of interfering with endogenous hormones even when present at quite low concentrations (Cole et al. 2011). Additionally, exposure to vehicular pollution has been associated with mild cognitive impairment among seniors, raising the risk of development of Alzheimer's (Ranft et al. 2009), and a higher incidence of dementia (Chen et al. 2017). Studies have found that human exposure to MPs may lead to various cancers, a weakened immune system, oxidative stress, DNA damage, inflammation, and more (Gasperi et al. 2018).

Several investigations have reported the translocation of MPs upon exposure in test animals as well as humans (Eyles et al. 2001; West-Eberhard 2019; Wick et al. 2010; Zauner et al. 2001). As the result of inflammation from MP exposure, the permeability of epithelial barriers increases, thus favoring their translocation to other body tissues (West-Eberhard 2019). Cell culture studies have also confirmed the internalization of PS nanospheres up to 1 μm (Zauner et al. 2001). The perfusion of placenta model-based studies has also suggested that 240 nm PS particles can cross the placental barrier (Wick et al. 2010). Small particles detected in the stomach, intestine, kidney, heart, and fetuses of rats, and the brains of fish suggest that MPs can pass through the gut barrier and reach the bloodstream. Moreover, MPs in the circulatory system may result in inflammation, vascular occlusions, increased coagulability, pulmonary hypertension, etc. (Churg & Brauer 2000; Jones et al. 2003; Zargorski et al. 2003), and reach the liver and kidney, affecting their functioning.

Exposure to MPs is also known to cause oxidative stress and cytotoxicity. Oxidative stress is a state of imbalance in the production and accumulation of oxygen-reactive species in the body, which can lead to cell and tissue damage, and the ability of a biological system to detoxify these reactive products (Pizzino et al. 2017). Oxidative stress may be caused by the release of oxidizing species adsorbed to the MPs surface (e.g. metals) or due to reactive oxygen species released during the inflammatory response (Schirinzi et al. 2017; Valavanidis et al. 2013). Several investigations have been carried out to examine the cytotoxicity of MPs (Furukuma & Fuji 2016; Geiser et al. 2005; Yacobi et al. 2008). Inside the cell, MPs are not membrane-bound, potentially interacting with intercellular structures, as reported by Geiser et al. (2005). Further, *in vitro* studies have also shown that exposure to plastic is known to induce cytotoxicity. For example, workers handling nylon and polyester in the industries are exposed to higher levels of MPs, which can lead to severe health outcomes such as interstitial lung disease, coughing, dyspnea, and reduced lung capacity (Gasperi et al. 2018). Cellulosic and plastic fibers from the lung tissue of patients with different types of lung cancer have been reported (Pauly et al. 1994, 1998). Cytotoxic effects caused due to MPs effects could be worse if the person is suffering from obesity and cardiovascular disease (Cipelli et al. 2014).

Furthermore, MPs have been reported to disrupt the immune function and increase the risk of immune or neurodegenerative disorders (Canesi et al. 2015; Détrée & Gallardo-Escárate 2018). Inhalation of MP particulates contaminated by chemical and biological pathogens (Vethaak & Leslie 2016) may contribute to the increasing incidence of cardiovascular, neurodegenerative diseases, etc. MP inhalation may also trigger autoimmune disorders through oxidative stress, particle translocation, the release of immune modulators, and activation of immune cells, resulting in exposure to self-antigens and production of autoantibodies (Farhat et al. 2011). Still, there are many information gaps and contradictions to the studies related to health impacts, therefore more scientific, systematic, and data-driven studies are required to understand the distribution, causes, and risk factors associated with MP exposure to humans and to further decipher the pathways of their toxicities.

6.5 Conclusions and Future Perspectives

Atmospheric MPs have emerged as a new class of environmental contaminants having heterogeneous characteristics, multiplicity of sources, transport, and fate in the environment. Recent studies have regarded atmospheric MPs as a significant threat to human health, and their presence has been detected in the atmosphere of urban, suburban, and remote areas. Their presence in the far-off pristine areas distant from their primary sources suggests their potential long-distance atmospheric transport, owing to their low densities and long persistence in the environment. Published work suggests that fibers and fragments are the most frequently identified MP shapes in the atmosphere. Exposure to MPs by ingestion, inhalation, and/or dermal contact may cause particle toxicity, oxidative stress, inflammatory lesions, and increased uptake or translocation. Additionally, MPs are carriers of harmful chemicals and colonize pathogenic microbes, which could result in inflammation, vascular occlusions, increased coagulability, pulmonary hypertension, etc. However, research work on atmospheric MPs is in the early stages. Extensive scientific studies are required worldwide for (i) development of standards and protocols for sampling and characterization; (ii) apportionment of their sources; and (iii) understanding of the causes, and risk factors associated with MP exposure to humans.

References

Abbasi, S., Keshavarzi, B., Moore, F. et al. (2017). *Investigation of microrubbers, microplastics and heavy metals in street dust: a study in Bushehr city, Iran. Environmental Earth Sciences* 76 (23) https://doi.org/10.1007/s12665-017-7137-0.

Abbasi, S., Soltani, N., Keshavarzi, B. et al. (2018). *Microplastics in different tissues of fish and prawn from the Musa Estuary, Persian Gulf. Chemosphere* 205: 80–87. https://doi.org/10.1016/j.chemosphere.2018.04.076.

Abbasi, S., Keshavarzi, B., Moore, F. et al. (2019). *Distribution and potential health impacts of microplastics and microrubbers in air and street dusts from Asaluyeh County, Iran. Environmental Pollution* 244: 153–164. https://doi.org/10.1016/j.envpol.2018.10.039.

Abidli, S., Antunes, J.C., Ferreira, J.L. et al. (2018). *Microplastics in sediments from the littoral zone of the north Tunisian coast (Mediterranean Sea). Estuarine, Coastal and Shelf Science* 205: 1–9.

Alimba, C.G. and Faggio, C. (2019). *Microplastics in the marine environment: current trends in environmental pollution and mechanisms of toxicological profile. Environmental Toxicology and Pharmacology* 68: 61–74. https://doi.org/10.1016/j.etap.2019.03.001.

Allen, S., Allen, D., Phoenix, V.R. et al. (2019). Atmospheric transport and deposition of microplastics in a remote mountain catchment. *Nature Geoscience* 12: 339–344. https://doi.org/10.1038/s41561-019-0335-5.

Amato-Lourenço, L.F., dos Santos Galvão, L., de Weger, L.A. et al. (2020). *An emerging class of air pollutants: potential effects of microplastics to respiratory human health? Science of the Total Environment* 749: 141676. https://doi.org/10.1016/j.scitotenv.2020.141676.

Ambrosini, R., Azzoni, R.S., Pittino, F. et al. (2019). *First evidence of microplastic contamination in the supraglacial debris of an alpine glacier. Enviromental Pollution* 253: 297–301. https://doi.org/10.1016/j.envpol.2019.07.005.

Baak, J.E., Provencher, J.F., and Mallory, M.L. (2020). *Plastic ingestion by four seabird species in the Canadian Arctic: comparisons across species and time. Marine Pollution Bulletin* 158: 111386. https://doi.org/10.1016/j.marpolbul.2020.111386.

Barnes, D.K., Galgani, F., Thompson, R.C., and Barlaz, M. (2009). *Accumulation andfragmentation of plastic debris in global environments. Philosophical Transactions of the Royal Society B* 364 (1526): 1985–1998.

Bergmann, M., Mützel, S., Primpke, S. et al. (2019). *White and wonderful? Microplastics prevail in snow from the Alps to the Arctic. Science Advances* 5: eaax1157. https://doi.org/10.1126/sciadv.aax1157.

Browne, M.A., Crump, P., Niven, S.J. et al. (2011). *Accumulation of microplastic on shorelines worldwide: sources and sinks. Environmental Science & Technology* 45: 9175–9179.

Cai, L., Wang, J., Peng, J. et al. (2017). *Characteristic of microplastics in the atmospheric fallout from Dongguan city, China: preliminary research and first evidence. Environmental Science and Pollution Research* 24 (32): 24928–24935. https://doi.org/10.1007/s11356-017-0116-x.

Canesi, L., Ciacci, C., Bergami, E. et al. (2015). *Evidence for immunomodulation and apoptotic processes induced by cationic polystyrene nanoparticles in the hemocytes of marine bivalve* Mytilus. *Marine Environmental Research* 11: 34–40. https://doi.org/10.1016/j.marenvres.2015.06.008.

Catarino, A.I., Macchia, V., Sanderson, W.G. et al. (2018). *Low levels of microplastics (MPs) in wild mussels indicate that MP ingestion by humans is minimal compared to exposure via household fibres fallout during a meal. Environmental Pollution* 237: 675–684. https://doi.org/10.1016/j.envpol.2018.02.069.

Chen, H., Kwong, J.C., Copes, R. et al. (2017). *Living near major roads and the incidence of dementia, Parkinson's disease, and multiple sclerosis: a population-based cohort study. Lancet* 389 (10070): 718–726. https://doi.org/10.1016/S0140-6736(16)32399-6.

Cheng, Z., Nie, X.-P., Wang, H.-S., and Wong, M.-H. (2013). *Risk assessments of human exposure to bioaccessible phthalate esters through market fish consumption. Environment International* 57-58: 75–80.

Churg, A. and Brauer, M. (2000). *Ambient atmospheric particles in the airways of human lungs. Ultrastructural Pathology* 24: 353–361. https://doi.org/10.1080/019131200750060014.

Cipelli, R., Harries, L., Okuda, K. et al. (2014). *Bisphenol A modulates the metabolic regulator oestrogen-related receptor-α in T-cells. Reproduction* 147: 419–426.

Cole, M., Lindeque, P., Halsband, C., and Galloway, T.S. (2011). *Microplastics as contaminants in the marine environment: a review. Marine Pollution Bulletin* 62 (12): 2588–2597. https://doi.org/10.1016/j.marpolbul.2011.09.025.

Crawford, C.B. and Quinn, B. (2017). The interactions of microplastics and chemical pollutants. In: *Microplastic Pollutants* (eds. C.B. Crawford and B. Quinn), 131–157. Elsevier https://doi.org/10.1016/B978-0-12-809406-8.00006-2.

Dehghani, S., Moore, F., and Akhbarizadeh, R. (2017). *Microplastic pollution in deposited urban dust, Tehran metropolis, Iran. Environmental Science and Pollution Research* 24: 20360–20371. https://doi.org/10.1007/s11356-017-9674-1.

Deng, Y., Zhang, Y., Lemos, B., and Ren, H. (2017). *Tissue accumulation of microplastics in mice and biomarker responses suggest widespread health risks of exposure. Scientific Reports* 7: 46687. http://doi.org/10.1038/srep46687.

Détrée, C. and Gallardo-Escárate, C. (2018). *Single and repetitive microplastics exposure induce immune system modulation and homeostasis alteration in the edible mussel* Mytillus galloprovincialis. *Fish & Shellfish Immunology* 83: 52–60. https://doi.org/10.1016/j.fsi.2018.09.018.

Donaldson, K., Stone, V., Gilmour, P.S. et al. (2000). *Ultrafine particles: mechanisms of lung injury. Physical Transactions of Royal Society* 358 (1775): 2741–2749. https://doi.org/10.1098/rsta.2000.0681.

Dris, R., Gasperi, C.J., Rocher, A.V. et al. (2015). *Microplastic contamination in an urban area: a case study in greater Paris. Environmental Chemistry* 12: 592–599.

Dris, R., Gasperi, J., Saad, M. et al. (2016). *Synthetic fibers in atmospheric fallout: a source of microplastics in the environment? Marine Pollution Bulletin* 104 (1–2): 290–293. https://doi.org/10.1016/j.marpolbul.2016.01.006.

Dris, R., Gasperi, J., Mirande, C. et al. (2017). *A first overview of textile fibers, including microplastics, in indoor and outdoor environments. Environmental Pollution* 221: 453–458. https://doi.org/10.1016/j.envpol.2016.12.013.

Eriksen, M., Mason, S., Wilson, S. et al. (2013). *Microplastic pollution in the surface waters of the Laurentian Great Lakes. Marine Pollution Bulletin* 77 (1–2): 177–182. https://doi.org/10.1016/j.marpolbul.2013.10.007.

Eyles, J.E., Bramwell, V.W., Williamson, E.D., and Alpar, H.O. (2001). *Microsphere translocation and immunopotentiation in systemic tissues following intranasal administration. Vaccine* 19: 4732–4742. https://doi.org/10.1016/S0264-410X(01)00220-1.

Fang, C., Zheng, R., Zhang, Y. et al. (2018). *Microplastic contamination in benthic organisms from the Arctic and sub-Arctic regions. Chemosphere* 209: 298–306. https://doi.org/10.1016/j.chemosphere.2018.06.101.

Farhat, S.C.L., Silva, C.A., Orione, M.A.M. et al. (2011). *Air pollution in autoimmune rheumatic diseases: a review. Autoimmunity Reviews* 11: 14–21. https://doi.org/10.1016/j.autrev.2011.06.008.

Frias, J.P.G.L., Sobral, P., and Ferreira, A.M. (2010). *Organic pollutants in microplastics from two beaches of the Portuguese coast. Marine Pollution Bulletin* 60 (11): 1988–1992. https://doi.org/10.1016/j.marpolbul.2010.07.030.

Fries, E., Dekiff, J.H., Willmeyer, J. et al. (2013). *Identification of polymer types and additives in marine microplastic particles using pyrolysis-GC/MS and scanning electron microscopy. Environmental Sciences Process Impacts* 15: 1949–1956. https://doi.org/10.1039/c3em00214d.

Fu, Z. and Wang, J. (2019). *Current practices and future perspectives of microplastic pollution in freshwater ecosystems in China. Science of the Total Environment* 691: 697–712. https://doi.org/10.1016/j.scitotenv.2019.07.167.

Furukuma, S. and Fuji, N. (2016). *in vitro cytotoxicity evaluation of plastic marine debris by colony-forming assay. Japanese Journal of Environmental Toxicology* 19 (2): 71–81. https://doi.org/10.11403/jset.19.71.

Galloway, T.S. (2015). Micro- and nano-plastics and human health. In: *Marine Anthropogenic Litter* (eds. M. Bergmann, L. Gutow and M. Klages), 343–366. Springer https://doi.org/10.1007/978-3-319-16510-3.

Ganguly, M. and Ariya, P.A. (2019). *Ice nucleation of model nano-micro plastics: a novel synthetic protocol and the influence of particle capping at diverse atmospheric environments. ACS Earth and Space Chemistry*, Special Issue: New Advances In Organic Aerosol Chemistry, 3: 1729–1739. https://doi.org/10.1021/acsearthspacechem.9b00132.

Gasperi, J., Wright, S.L., Dris, R. et al. (2018). *Microplastics in air: are we breathing it in? Current Opinion in Environmental Science and Health* 1: 1–5. https://doi.org/10.1016/j.coesh.2017.10.002.

Gaston, E., Woo, M., Steele, C. et al. (2020). *Microplastics differ between indoor and outdoor air masses: insights from multiple microscopy methodologies. Applied Spectroscopy* 74 (9): 1079–1098. https://doi.org/10.1177/0003702820920652.

Geiser, M., Rothen-Rutishauser, B., Kapp, N. et al. (2005). *Ultrafine particles cross cellular membranes by nonphagocytic mechanisms in lung and in cultured cells. Environmental Health Perspectives* 113 (11): 1555–1560. https://doi.org/10.1289/ehp.8006.

Geyer, R., Jambeck, J.R., and Law, K.L. (2017). *Production, use, and fate of all plastics ever made. Science Advances* 3 (7): 25–29. https://doi.org/10.1126/sciadv.1700782.

Hartmann, N.B., Hüffer, T., Thompson, R.C. et al. (2019). *Are we speaking the same language? Recommendations for a definition and categorization framework for plastic debris. Environmental Science and Technology* 53: 1039–1047. https://doi.org/10.1021/acs.est.8b05297.

Hayward, S.J., Gouin, T., and Wania, F. (2010). *Comparison of four active and passive sampling techniques for pesticides in air. Environmental Science & Technology* 44: 3410–3416. https://doi.org/10.1021/es902512h.

Hidalgo-Ruz, V., Gutow, L., Thompson, R.C., and Thiel, M. (2012). *Microplastics in the marine environment: a review of the methods used for identification and quantification. Environmental Science & Technology* 46: 3060–3075. https://doi.org/10.1021/es2031505.

Jones, A.E., Watts, J.A., Debelak, J.P. et al. (2003). *Inhibition of prostaglandin synthesis during polystyrene microsphere – induced pulmonary embolism in the rat. American Journal of Physiology; Lung Cellular and Molecular Physiology* 284: L1072–L1081. https://doi.org/10.1152/ajplung.00283.2002.

Karami, A., Golieskardi, A., Choo, C.K. et al. (2017). The presence of microplastics in commercial salts from different countries. *Scientific Reports* 7: 46173.

Katyal, D., Kong, E., and Villanueva, J. (2020). *Microplastics in the environment: impact on human health and future mitigation strategies. Environmental Health Review* 63 (1): 27–31. https://doi.org/10.5864/d2020-005.

Keswani, A., Oliver, D.M., Gutierrez, T., and Quilliam, R.S. (2016). *Microbial hitchhikers on marine plastic debris: human exposure risks at bathing waters and beach environments. Marine Environmental Research* 118: 10–19.

Kirstein, I.V., Kirmizi, S., Wichels, A. et al. (2016). *Dangerous hichhikers? Evidence for potentially pathogenic Vibrio spp. on microplastic particles. Marine Environmental Research* 120: 1–8. https://doi.org/10.1016/j.marenvres.2016.07.004.

Klein, M. and Fischer, E.K. (2019). *Microplastic abundance in atmospheric deposition within the metropolitan area of Hamburg, Germany. Science of the Total Environment* 685: 96–103. https://doi.org/10.1016/j.scitotenv.2019.05.405.

Kosuth, M., Wattenberg, E.V., Mason, S.A. et al. (2017). *Synthetic Polymer Contamination in Global Drinking Water*. Orb Media.

Kosuth, M., Mason, S.A., and Wattenberg, E.V. (2018). *Anthropogenic contamination of tap water, beer, and sea salt. PLoS One* 13 (4): e0194970. https://doi.org/10.1371/journal.pone.0194970.

Kuczenski, B. and Geyer, R. (2010). *Material flow analysis of polyethylene terephthalate in the US, 1996–2007. Resources, Conservation and Recycling* 54: 1161–1169. https://doi.org/10.1016/j.resconrec.2010.03.013.

Liebezeit, G. and Liebezeit, E. (2013). *Non-pollen particulates in honey and sugar. Food Additives & Contaminants. Part A, Chemistry, Analysis, Control, Exposure & Risk Assessment* 30: 2136–2140.

Lithner, D., Larsson, A., and Dave, G. (2011). *Environmental and health hazard ranking and assessment of plastic polymers based on chemical composition. Science of the Total Environment* 409 (18): 3309–3324. https://doi.org/10.1016/j.scitotenv.2011.04.038.

Liu, C., Li, J., Zhang, Y. et al. (2019a). *Widespread distribution of PET and PC microplastics in dust in urban China and their estimated human exposure. Environment International* 128: 116–124. https://doi.org/10.1016/j.envint.2019.04.024.

Liu, K., Wang, X., Fang, T. et al. (2019b). *Source and potential risk assessment of suspended atmospheric microplastics in Shanghai. Science of the Total Environment* 675: 462–471. https://doi.org/10.1016/j.scitotenv.2019.04.110.

Mallory, M.L., Robinson, S.A., Hebert, C.E., and Forbes, M.R. (2010). *Seabirds as indicators of aquatic ecosystem conditions: a case for gathering multiple proxies of seabird health. Marine Polluion Bulletin* 60: 7–12. https://doi.org/10.1016/j.marpolbul.2009.08.024.

Moore, R.C., Loseto, L., Noel, M. et al. (2020). *Microplastics in beluga whales (Delphinapterus leucas) from the Eastern Beaufort Sea. Marine Pollution Bulletin* 150: 110723. https://doi.org/10.1016/j.marpolbul.2019.110723.

Patchaiyappan, A., Dowarah, K., Zaki Ahmed, S. et al. (2020). *Prevalence and characteristics of microplastics present in the street dust collected from Chennai metropolitan city, India. Chemosphere* 269: 128757. https://doi.org/10.1016/j.chemosphere.2020.128757.

Pauly, J.L., Rodriguez, M.I., Falzone, C.M. et al. (1994). *Methods for viewing, identifying, and enumerating natural and synthetic fibers in human lungs. American Journal of Respiratory and Critical Care Medicine* 149: A8IO.

Pauly, J.L., Stegmeier, S.J., Allaart, H.A. et al. (1998). *Inhaled cellulosic and plastic fibers found in human lung tissue. Cancer Epidemiology, Biomarkers & Prevention* 7: 419–428.

Pizzino, G., Irrera, N., Cucinotta, M. et al. (2017). *Oxidative stress: harms and benefits for human health. Oxidative Medicine and Cellular Longevity* 2017: 8416763. https://doi.org/10.1155/2017/8416763.

PlasticsEurope (2018). *Plastics – The Facts 2018: An Analysis of European Plastics Production, Demand and Waste Data*. Association of Plastics Manufacturers. https://www.plasticseurope.org/en/resources/publications/3-plastics-facts-2016.

Prata, J.C. (2018). *Airborne microplastics: consequences to human health? Environmental Pollution* 234: 115–126. https://doi.org/10.1016/j.envpol.2017.11.043.

Prata, J.C., da Costa, J.P., Girão, A.V. et al. (2019). *Identifying a quick and efficient method of removing organic matter without damaging microplastic samples. Science of the Total Environment* 686: 131–139. https://doi.org/10.1016/j.scitotenv.2019.05.456.

Prata, J.C., da Costa, J.P., Lopes, I. et al. (2020). *Environmental exposure to microplastics: an overview on possible human health effects. Science of the Total Environment* 702: 134–455.

Ranft, U., Schikowski, T., Sugiri, D. et al. (2009). *Long-term exposure to traffic-related particulate matter impairs cognitive function in the elderly. Environmental Research* 109 (8): 1004–1011. https://doi.org/10.1016/j.envres.2009.08.003.

Reed, S., Clark, M., Thompson, R., and Hughes, K.A. (2018). *Microplastics in marine sediments near Rothera research station, Antarctica. Marine Pollution Bulletin* 133: 460–463.

Research and Markets. (2020). Global Medical Plastics Market Outlook 2020–2021: Impact Assessment of the COVID-19 Pandemic. https://www.globenewswire.com/news-release/2020/05/15/2034060/0/en/Global-Medical-Plastics-Market-Outlook-2020-2021-Impact-Assessment-of-the-COVID-19-Pandemic.html (accessed 26 October 2021).

Revel, M., Châtel, A., and Mouneyrac, C. (2018). *Micro(nano)plastics: a threat to human health? Current Opinion in Environmental Science and Health* 1: 17–23. https://doi.org/10.1016/j.coesh.2017.10.003.

Rodrigues, M.O., Abrantes, N., Gonçalves, F.J.M. et al. (2018). Spatial and temporal distribution of microplastics in water and sediments of a freshwater system (Antuã River, Portugal). *Science of the Total Environment* 633: 1549–1559.

Schirinzi, G.F., Pérez-Pomeda, I., Sanchís, J. et al. (2017). *Cytotoxicity effects of commonly used nanomaterials and microplastics on cerebral and epithelial human cells. Environmental Research* 159: 579–587. https://doi.org/10.1016/j.envres.2017.08.043.

Schymanski, D., Goldbeck, C., Humpf, H.-U., and Fürst, P. (2018). *Analysis of microplastics in water by micro-Raman spectroscopy: release of plastic particles from different packaging into mineral water. Water Research* 129: 154–162.

Smith, M., Love, D.C., Rochman, C.M., and Neff, R.A. (2018). *Microplastics in seafood and the implications for human health. Current Environmental Health Reports* 5 (3): 375–386. https://doi.org/10.1007/s40572-018-0206-z.

Srivastava, R.K. and Godara, S. (2017). *Use of polycarbonate plastic products and human health. International Journal of Basic and Clinical Pharmacology* 2 (1): 12–17.

Stanton, T., Johnson, M., Nathanail, P. et al. (2019). *Freshwater and airborne textile fibre populations are dominated by 'natural', not microplastic, fibres. Science of the Total Environment* 666: 377–389. https://doi.org/10.1016/j.scitotenv.2019.02.278.

Thompson, R.C., Olson, Y., Mitchell, R.P. et al. (2004). *Lost at sea: where is all the plastic? Science* 304 (5672): 838. https://doi.org/10.1126/science.1094559.

Tiseo, I. (2019). Global loss of microplastics to the environment by region 2018. https://www.statista.com/statistics/1020286/losses-microplastics-environment-by-region (accessed January 22 2021).

Tiseo, I. (2021). Global plastic production 1950–2019. https://www.statista.com/statistics/282732/global-production-of-plastics-since-1950 (accessed Feburay 16 2021)

Turner, S., Horton, A.A., Rose, N.L., and Hall, C. (2019). *A temporal sediment record of microplastics in an urban lake, London, UK. Journal of Paleolimnology* 61: 449–462. https://doi.org/10.1007/s10933-019-00071-7.

Valavanidis, A., Vlachogianni, T., Fiotakis, K., and Loridas, S. (2013). *Pulmonary oxidative stress, inflammation and cancer: respirable particulate matter, fibrous dusts and ozone as major causes of lung carcinogenesis through reactive oxygen species mechanisms. International Journal of Environmental Research and Public Health* 10 (9): 3886–3907.

Van Franeker, J.A. and Law, K.L. (2015). *Seabirds, gyres and global trends in plastic pollution. Environmental Pollution* 203: 89–96. https://doi.org/10.1016/j.envpol.2015.02.034.

Vethaak, A.D. and Leslie, H.A. (2016). *Plastic debris is a human health issue. Environmental Science & Technology* 50: 6825–6826.

Volkheimer, G. (1975). *Hematogenous dissemination of ingested polyvinyl chloride particles. Annals of the New York Academy of Sciences* 246 (1): 164–171.

Volkheimer, G. (1977). *Passage of particles through the wall of the gastrointenstinal tract. Environmental Health Perspectives* 9: 215–225.

Watts, A.J.R., Urbina, M.A., Corr, S. et al. (2015). *Ingestion of plastic microfibers by the crab* Carcinus maenas *and its effect on food consumption and energy balance. Environmental Science and Technology* 49 (24): 14597–14604. http://doi.org/10.1021/acs.est.5b04026.

Wen, B., Zhang, N., Jin, S.-R. et al. (2018). *Microplastics have a more profound impact than elevated temperatures on the predatory performance, digestion and energy metabolism of an Amazonian cichlid. Aquatic Toxicology* 195: 67–76. https://doi.org/10.1016/j.aquatox.2017.12.010.

West-Eberhard, M.J. (2019). *Nutrition, the visceral immune system, and the evolutionary origins of pathogenic obesity. Proceedings of the National Academy of Sciences of the United States of America* 116 (3): 723–731. https://doi.org/10.1073/pnas.1809046116.

Wick, P., Malek, A., Manser, P. et al. (2010). *Barrier capacity of human placenta for nanozised materials. Environmental Health Perspectives* 118 (3): 432–436. https://doi.org/10.1289/ehp.0901200.

Wright, S.L. and Kelly, F.J. (2017). *Plastic and human health: a micro issue? Environmental Science & Technology* 51: 6634–6647. https://doi.org/10.1021/acs.est.7b00423.

Wright, S.L., Rowe, D., Thompson, R.C., and Galloway, T.S. (2013). *Microplastic ingestion decreases energy reserves in marine worms. Current Biology* 23 (23): R1031–R1033. https://doi.org/10.1016/j.cub.2013.10.068.

Wright, S.L., Ulke, J., Font, A. et al. (2020). *Atmospheric microplastic deposition in an urban environment and an evaluation of transport. Environment International* 136: 105411. https://doi.org/10.1016/j.envint.2019.105411.

Xu, X.-Y., Lee, W.T., Chan, A.K.Y. et al. (2017). *Microplastics ingestion reduces energy intake in the clam* Atactodea striata. *Marine Pollution Bulletin* 124 (2): 798–802. https://doi.org/10.1016/j.marpolbul.2016.12.027.

Yacobi, N.R., DeMaio, L., Xie, L. et al. (2008). *Polystyrene nanoparticles trafficking across alveolar epithelium. Nanomedicine: Nanotechnology, Biology and Medicine* 4 (2): 139–145. https://doi.org/10.1016/j.nano.2008.02.002.

Yang, D., Shi, H., Li, L. et al. (2015). *Microplastic pollution in table salts from China. Environmental Science & Technology* 49: 13622–13627.

Yang, Y., Liu, W., Zhang, Z. et al. (2020). *Microplastics provide new microbial niches in aquatic environments. Applied Microbiology and Biotechnology* 104: 6501–6511.

Zargorski, J., Debelak, J., Gellar, M. et al. (2003). *Chemokines accumulate in the lungs of rats with severe pulmonary embolism induced by polystyrene microspheres. Journal of Immunology* 171: 5529–5536. https://doi.org/10.4049/jimmunol.171.10.5529.

Zauner, W., Farrow, N.A., and Haines, A.M.R. (2001). *in vitro uptake of polystyrene microspheres: effect of particle size, cell line and cell density. Journal of Controlled Release* 71: 39–35. https://doi.org/10.1016/S0168-3659(00)00358-8.

Zeng, E.Y. (ed.) (2018). *Microplastic Contamination in Aquatic Environments – An Emerging Matter of Environmental Urgency.* Elsevier https://doi.org/10.1016/C2016-0-04784-8.

Zhang, S., Yang, X., Gertsen, H. et al. (2018). *A simple method for the extraction and identification of light density microplastics from soil. Science of the Total Environment* 616–617: 1056–1065. https://doi.org/10.1016/j.scitotenv.2017.10.213.

Zhang, Y., Kang, S., Allen, S. et al. (2020). *Atmospheric microplastics: a review on current status and perspectives. Earth-Science Reviews* 203: 103118.

Zhao, S., Danley, M., Ward, J.E., and Mincer, T.J. (2017). *Analytical methods using Raman microscopy. Analytical Methods* 9: 1470–1478. https://doi.org/10.1039/C6AY02302A.

Zhou, Q., Tian, C., and Luo, Y. (2017). *Various forms and deposition fluxes of microplastics identified in the coastal urban atmosphere. Chinese Science Bulletin* 62: 3902–3909.

7

Plastic Marine Litter in the Southern and Eastern Mediterranean Sea: Current Research Trends and Management Strategies

Soha Shabaka

Hydrobiology Lab, Department of Marine Environment, National Institute of Oceanography and Fisheries (NIOF), Cairo, Egypt

7.1 Introduction

Pollution from plastic waste is one of the most important challenges facing humans today, with negative impacts on coastal environments, and consequently, human wellbeing (Bergmann et al. 2015; Eriksen et al. 2014). Plastic waste stems from human activities along with the lack of financial support from governments to find alternative solutions, which further exacerbates the problem. Microplastics (MPs), another form of plastic litter, have gained particular attention due to their small sizes (ranging from 1 μm to 5 mm), thus they have serious implications to aquatic organisms and human health (Conkle et al. 2018; Frias and Nash 2019; Lin 2016; Smith et al. 2018; Tanaka and Takada 2016). Plastic marine litter and MPs exhibit characteristics associated with creeping crises, which emerges from pressures that accumulate gradually and exhibit a slow manifestation pattern (Mæland and Staupe-Delgado 2020).

Higher-income countries have over 10 times more per capita rate of plastic waste generation than many low- and middle-income countries, while LMICs rank top in mismanaged plastic waste (Jagath et al. 2020). Mismanaged plastic waste is dumped and eventually finds its way to aquatic ecosystems (Jambeck et al. 2015), despite this fact, the extent of marine plastic pollution, in particular MPs, is poorly understood in LMICs (Akindele and Alimba 2021; Alimi et al. 2021). This lack of information hampers the development of policies and interventions that can reduce the negative impacts of the current plastic waste burden, and makes it more challenging to develop alternative solutions. Therefore, LMICs in the Mediterranean Sea, namely Lebanon, Palestine, Egypt, Tunisia, Algeria, and Morocco were chosen for this study. Higher-income countries belonging to the eastern Mediterranean Sea, Greece, Turkey, Cyprus, and Israel, were also included. The current research aims to determine if the economy of these countries has affected the marine litter research trends and outcomes, and to highlight the possible collaborations that would benefit LMICs and thus improve and overcome the current knowledge gaps. The eastern Mediterranean was

Plastic and Microplastic in the Environment: Management and Health Risks, First Edition.
Edited by Arif Ahamad, Pardeep Singh, and Dhanesh Tiwary.
© 2022 John Wiley & Sons Ltd. Published 2022 by John Wiley & Sons Ltd.

found to be the most MPs contaminated zone due to the trapping of MPs, combined with high coastal population density and waste generation (Lots et al. 2017). Egypt is the largest southern Mediterranean country in terms of population (~100 M), bounded by both the Mediterranean and the Red Seas with the Nile River running across it. It was speculated that the Nile River contributes significantly to the input of MPs into the Mediterranean Sea (Lebreton et al. 2012). This is a very important aspect, since the Levant Basin is a closed area of the Mediterranean Sea, separated from the western part of the Mediterranean Sea by the Sicily channel, and it is prone to accumulate debris, mainly from the Nile. Lebanon has a relatively small population (~6.7 M) but has a 210 km long shoreline, including several bays and islands. Lebanon has poor solid waste management with adverse influence on the Mediterranean Sea. This situation is not different in other LMICs in the Mediterranean Sea. The aim of this work is to analyze marine litter research trends in the countries of the Levant Basin and the southern Mediterranean Sea and provide an overview of the abundance and distribution of MPs, methods of analysis, and knowledge gaps.

7.2 Analysis of Marine Litter Research Trends in the Southern and Eastern Mediterranean Sea Countries

Marine litter research is a relatively recent topic in the southern and eastern Mediterranean countries, where it started to emerge in 2015 in Greece. By searching the SciVal (searched 21 May 2021) for the topic "Microplastics; Marine Debris; Litter" (Topic T.4380), a total of 1727 publications covering the period between 2015 to 2020 were found in the Mediterranean Sea countries, where the most active countries were Italy, Spain, and France, respectively. Only 11% of these publications were produced by authors from southern and eastern Mediterranean countries, where Greece and Turkey were the most active countries, while Egypt, Algeria, and Palestine have been recently involved (Figures 7.1 and 7.2). Both the

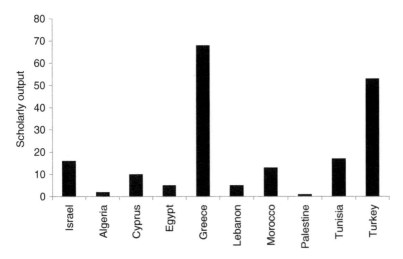

Figure 7.1 Overall scholarly output of the southern and eastern Mediterranean countries in the period between 2015 to >2020. *Source:* data was retrieved from SciVal on 21 May 2021.

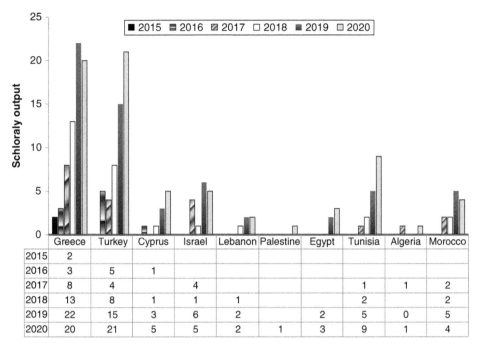

	Greece	Turkey	Cyprus	Israel	Lebanon	Palestine	Egypt	Tunisia	Algeria	Morocco
2015	2									
2016	3	5	1							
2017	8	4		4				1	1	2
2018	13	8	1	1	1			2		2
2019	22	15	3	6	2		2	5	0	5
2020	20	21	5	5	2	1	3	9	1	4

Figure 7.2 Yearly scholarly output of the southern and eastern Mediterranean countries. *Source:* data was retrieved from SciVal on 21 May 2021.

number of scholarly output and the involved countries steadily increased from 2015 to 2021 (Figure 7.2).

Field Weight Citation Impact (FWCI) was applied in order to gain insight into the relative citation performance of countries with relatively poor Scopus coverage in marine litter research. When classifying these publications by their FWCI, international collaboration, and scholarly output, it was observed that Lebanon, Algeria, Tunisia, and Palestine had the highest international collaboration (>80%). In addition, Lebanon, Algeria, and Tunisia had the highest FWCI (>3) regardless of their lower scholarly output in comparison to other countries in the region, which indicates the importance of addressing MP and marine litter research in eastern and southern Mediterranean countries, particularly in LMICs of the region (Figure 7.3). International collaboration is indispensable for technology transfer and capacity building in LMICs.

The analysis of publications showed 23 subject areas under the Topic "MPs; Marine Debris; Litter" with environmental science, agriculture and biological science, and earth and planetary science being the most addressed areas with high FWCI scores (Figure 7.4). Other subject areas such as neuroscience and economics, and econometrics showed high FWCI despite being the least addressed disciplines. Greece and Turkey showed the highest count of the main- and sub-subject areas, which is also a function of their scholarly output (Figure 7.5).

Higher-income countries belonging to the EU, Greece, and Cyprus tend to provide an inclusive approach to marine litter management; for instance, large-scale investigation

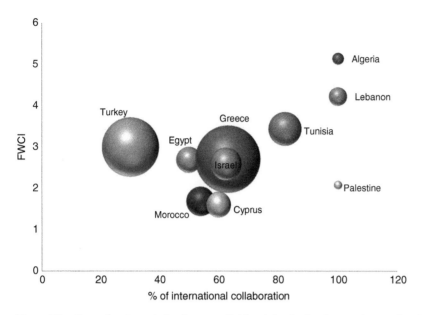

Figure 7.3 Chart showing relation between field weight citation impact, international collaboration (percentage), and scholarly output. Bubble size: Scholarly Output.

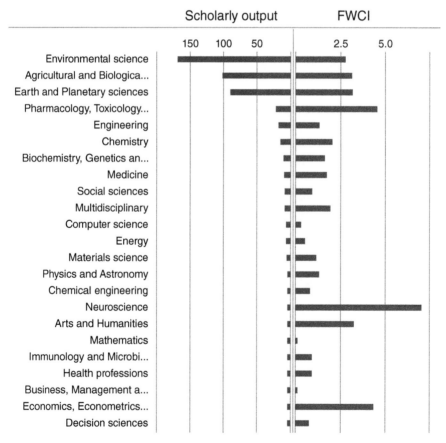

Figure 7.4 Subject areas, scholarly output, and their field weight citation index (FWCI) under the topic "Microplastics; Marine Debris; Litter." *Source:* Data was retrieved from SciVal on 21 May 2021 for Greece, Turkey, Cyprus, Israel, Lebanon, Palestine, Egypt, Tunisia, Algeria, and Morocco.

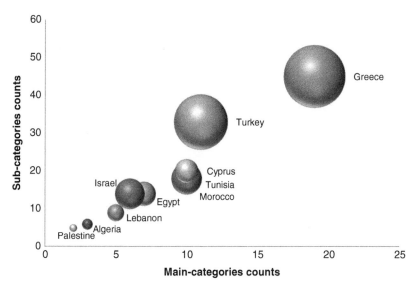

Figure 7.5 Chart showing relation between the counts of sub-categories and main-categories subject areas, and scholarly output. Bubble size: Scholarly Output.

of the implementation of Fishing For Litter, a scheme to remove marine litter from the seafloor (Ronchi et al. 2019), or applying large-scale harmonized protocols for the assessment of marine litter (Fortibuoni et al. 2019). Moreover, higher-income countries tend to apply protocols and bioindicators of marine litter impacts on threatened marine species to respond to the requirements set by the European Union's Marine Strategy Framework Directives (MSFD) and developing thresholds for good environmental status (Claro et al. 2019; Fossi et al. 2018; Matiddi et al. 2019). Applying advanced techniques, such as mapping and modeling (Kikaki et al. 2020; Politikos et al. 2020; Themistocleous et al. 2020; Topouzelis et al. 2020), or artificial intelligence and deep learning techniques (Kylili et al. 2020; Musić et al. 2020) for the assessment of the sources and fate of marine litter is a common research trend in higher-income countries. Other high-income countries such as Turkey focus mainly on the assessment of MPs pollution in the marine environment and wastewater (Güven et al. 2017; Gündoğdu et al. 2018a; Gündoğdu and Çevik 2017), or experiments to assess the impact of MPs on marine organisms (Berber 2019; Tunali et al. 2020). Other important topics, such as the assessment of ghost fishing causes and impacts were also addressed (Dagtekin et al. 2019; Yildiz and Karakulak 2016). Israel focuses mainly on the sources and fate of plastic litter (Jalón-Rojas et al. 2019; Pasternak et al. 2017; Portman and Brennan 2017). Concerning LMICs, Tunisia has an obvious preference to conduct research related to the detection of MPs in different environmental matrices (Tables 7.1–7.3). While Morocco has the tendency to study the spatial distribution of macro marine litter on the seafloor and beaches (Loulad et al. 2019; Mghili et al. 2020; Nachite et al. 2019), in Egypt research focuses on the impact assessment of MPs on different aquatic organisms (Hamed et al. 2019, 2020, 2021; Mohsen et al. 2020, 2021).

Table 7.1 Microplastic detection in seawater in the eastern and southern Mediterranean regions.

Area	Year of sampling	Sampling method	Extraction method	Identification method	Quantification method	Quality control	Concentration	Dominant polymer	Dominant shape	Dominant size	References
Turkey	July and August 2015	Manta net of mesh size 333 μm (Surface water). Sampling net of mesh size 200 μm (Water column).	Filtered on a sieve of mesh size 26 μm. Digestion with 35% H_2O_2	Stereomicroscope FTIR	Counted	Yes	16 339–520 213 items km^{-2}	nd	Fibers and fragments	0.1 and 2.5 mm	Güven et al. (2017)
Turkey (Iskenderun Bay and Mersin Bay)	November and October 2016	Manta net (333 μm mesh size)	Sieving 300 μm Wet peroxide oxidation. Density separation NaCl. Sieving 300 μm.	Microscopy	Counted	No	0.376 items m^{-2}	nd	Fragments, films	2.9 mm	Gündoğdu and Çevik (2017)
Turkey (Mersin Bay)	November 2016 and April 2017 (before and after flood)	Manta net (333 μm mesh size)	Wet peroxide oxidation. Density separation sieves of mesh size 1 mm, 0.5 mm, 0.3 mm, and 0.1 mm	Stereomicroscope. ATR-FTIR	Counted	Yes	539 189 items km^{-2} and 7 699 716 items km^{-2} pre- and post-flood	PE, PP, SAAC, PS	Fragments, films, and fibers	0.1–0.5 mm	Gündoğdu et al. (2018b)

Location	Date	Sampling method	Treatment	Identification			Concentration	Polymer	Shape	Size	Reference
Lebanon	May 2018	Manta net (52 μm mesh size)	Filtered on a sieve mesh size of 850 μm. The recovered water was vacuum filtered on GF/A filters.	Stereomicroscope. Micro-Raman	Counted	Yes	4.3 ± 2.2 particle m^{-3}	PE	Fragments	< 0.2 mm	Kazour et al. (2019)
Israel	July 2013–May 2015	Manta net (333 μm mesh size)	No	Magnifying glass and stereo microscope	counted	No	33.99 ± 84.82– 2.09 ± 2.87 particle m^{-3}	nd	Fragments	0.3–5 mm	van der Hal et al. (2017)
Egypt	December 2017	Phytoplankton net (20 μm mesh size) (Surface water and water column)	Filtered on 20 μm mesh size sieve. Washed with DW. Dried at 40 °C	Stereomicroscope DSC	Counted	Yes	nd	PP	Fragments	< 0.1 mm	Shabaka et al. (2019)
Tunisia (Gulf of Gabes)	October–November 2017	200 μm mesh size trawl net	Filtered on sieve mesh size of 200 μm. Wet peroxide oxidation. Filtered on GF/A filter paper.	Stereomicroscope. ATR-FTIR	Counted	Yes	63 739 items km^{-2}	PE, PP	Fragments and films	0.2–2 mm	Zayen et al. (2020)

(Continued)

Table 7.1 (Continued)

Area	Year of sampling	Sampling method	Extraction method	Identification method	Quantification method	Quality control	Concentration	Dominant polymer	Dominant shape	Dominant size	References
Tunisia (Bizerte lagoon)	August 2019	Pumping (Surface water)	Filtered on a sieve of mesh size 300 μm. Digestion with KOH. Density separation with NaI. Filtered on GF/A	Stereomicroscope. ATR-FTIR	Counted and converted to equivalent weight based on WWF (2019)	Yes	453.0 ± 335.2 Items m^{-3}	PE, PP	Fibers and fragments	>0.3 mm	Wakkaf et al. (2020a)
Tunisia (Bizerte lagoon)	May 2019	Pumping (near Bottom water)	Filtered on a sieve of mesh size 300 μm. Digestion with KOH. Density separation with NaI. Filtered on GF/A	Stereomicroscope. ATR-FTIR	Counted	Yes	400 ± 0.2 particles m^{-3}	PE, PP	Fibers and films		Wakkaf et al. (2020b)

Table 7.2 Microplastic detection in sediments in the eastern and southern Mediterranean regions.

Area	Year of sampling	Sampling method	Extraction method	Identification method	Quantification method	Quality control	Concentration	Dominant polymer	Dominant shape	Dominant size	References
Greece (Salamina Island)	August 2016	Collection of the first 5 cm in 1×1 m quadrate (Beach)	Density separation with NaCl	1) Magnifying lens. 2) ATR-FTIR. 3) SEM	Counted	No	nd	PE, EPS	Pellets	1–5 mm	Tziourrou et al. (2019)
Greece (Northern Crete)	2013–2015	Surface and subsurface samples down to 10 cm were collected in 0.4×0.4 m quadrate	Collected plastics separated by hand using forceps	Visual	Counted and weighted	No	nd	nd	Pellets and fragments	0–4 mm	Karkanorachaki et al. (2018)
Greece (Kea Island)	September 2012	The first 3 cm in 1×1 m quadrate (Beach)	Density separation	Microscopy FTIR		No	1.5–15.7 items m^{-2}	PE-PP	Fragments, pellets	>1 mm	Kaberi et al. (2013)
Turkey	July and August 2015	Van Veen bottom sampler	Density separation (NaCl). Filtered on 26 μm sieve. Digestion with 35% H_2O_2	Stereomicroscope FTIR	Counted	No	nd		Fibers and fragments	0.1 and 2.5 mm	Güven et al. (2017)

(Continued)

Table 7.2 (Continued)

Area	Year of sampling	Sampling method	Extraction method	Identification method	Quantification method	Quality control	Concentration	Dominant polymer	Dominant shape	Dominant size	References
Turkey (Datça Peninsula)	September 2018	Collection of the first 3–4 cm in 0.5×0.5 m quadrate	Digestion with 35% H_2O_2. Density separation (NaCl). Filtered on GF/F. 4. Dried	Stereomicroscope ATR-FTIR	Counted	Yes	1154.4 ± 700.3 particles kg^{-1}	PE, PP-A	Fragments and fibers	nd	Yabanlı et al. (2019)
Cyprus	Jul and Aug 2016	Core (20×60 cm) Beach	Cascade of sieves 5 mm and 1 mm	Visual	Counted and weighed	No	45497 ± 11456 particles m^{-3} $481\pm131\,gm^{-3}$	nd	Pellets and fragments	<5 mm >1 mm	Duncan et al. (2018)
Lebanon	May 2018	2 cm of sediments were sampled using a steel ring (Beach)	Density separation Filtered on GF/A filter	Stereomicroscope. Micro-Raman	Counted	Yes	2433 ± 2000M items kg^{-1} DW	PP	Fragments	>1 mm	Kazour et al. (2019)
Egypt	November 2017	Collection of the first 5 cm in 0.25×0.25 m quadrate (Beach)	Separation with water siphon. Filtered on 20 μm mesh size sieve. Dried at 40°C	Stereomicroscope. DSC	Counted	Yes	242 items kg^{-1} dry sediments (0.554 g)	PET, PP, LDPE, HDPE	Fragments	<0.5 mm	Shabaka et al. (2019)
Tunisia (Gulf of Gabes-Kerkennah archipelago)	March 2018	Collection of the first 3–5 cm in 0.25×0.25 m quadrate	Sieved on cascades of sieves. Density separation using NaI. Filtered on 0.45 μm filter. Washed with DW. Digestion using 10% KOH.	Stereomicroscope. ATR-FTIR	Counted	Yes	595 items m^{-2}	PE, PP	Fragments and fibers	<1 mm	Chouchene et al. (2020)

Location	Date	Sampling	Extraction	Identification		Separation	Concentration	Polymer	Shape	Size	Reference
Tunisia Lagoon of Bizerte	January 2016	Collection of the first 2–3 cm in 0.25×0.25 m quadrate	Density separation by NaCl, Filtered on 10 μm Filter	Stereomicroscope	Counted	No	7960±6.84 Items kg⁻¹ DW	nd	Fiber and fragments	1.39±0.27 −0.51±0.19 mm	Abidli et al. (2017)
Tunisia (Sidi Mansour)	March 2018	Collection of the first 2–3 cm in 0.25×0.25 m quadrate	Sieved on 5 mm Density separation (NaCl). Filtered on 0.45 μm filter. Washed with DW. Digestion (10%KOH)	Fluorescence staining method under microscope. ATR-FTIR	Counted	No	1348±20–2932 ±63 items m⁻²	PE, PP	Fragments and pellets	0.1 to 1.0 mm	Chouchene et al. (2019)
Tunisia lagoon of Bizerte	Nov 2017	Collection of the first 2–3 cm in 0.25×0.25 m quadrate (Fresh water streams)	Dried. Density separation (NaCl). Filtered on 10 μm filters.	Stereomicroscope. ATR-FTIR	Counted	Yes	2340±227.15 and 6920±395.97 items kg⁻¹ DW	PE, PP	Fibers	0.59±0.17 −1.73±1.39 mm	Toumi et al. (2019)
Tunisia (South and north Lake of Tunis, Carthage, Goulette and Menzel Bourguiba)	June 2017	Collection of the first 2–3 cm in 0.25×0.25 m quadrate	Dried Density separation (NaCl). Filtered on GF/C filters	Stereomicroscope. ATR-FTIR	Counted	Yes	141.20±25.98 to 461.25±29.74 items kg⁻¹	PE, PP, PS	Fibers	1.60±0.97 mm −2.18±0.76 mm	Abidli et al. (2018)

(Continued)

Table 7.2 (Continued)

Area	Year of sampling	Sampling method	Extraction method	Identification method	Quantification method	Quality control	Concentration	Dominant polymer	Dominant shape	Dominant size	References
Algeria (Annaba Gulf)	June–July 2018	Collection of the first 2–3 cm in 0.25×0.25 m quadrate	Dried Density separation (NaCl). Filtered on GF/A filters	Stereomicroscope. ATR-FTIR	Counted	Yes	182.66±27.32. – 649, 33±184.02 p kg^{-1}	PE, PP, PET	Fibers and fragments	0.81±0.06 mm –2.16±0.34 mm	Tata et al. (2020)
Morocco (Northeast of Tetouan)	Seasonal of (2015)	Collection of the first 5 cm in 0.5×0.5 m quadrate	Drying Sieving on cascade sieves	Visual	Weighed	No	foam 8.09 g, line nd 2.36 g, and film 0.37 g	foam, line nd and film	Foam	nd	Alshawafi et al. (2017)

Table 7.3 Microplastic detection in marine organisms in the eastern and southern Mediterranean regions.

Area	Year of sampling	Organism	Extraction method	Identification method	Quantification method	Quality control	Concentration	Dominant polymer	Dominant shape	Dominant size	References
Greece (Dodecanese Archipelagos of the southeastern Aegean Sea)	November 2014 to October 2015	Shrimps	Emptying the stomach content	Stereomicroscope. ATR-FTIR	Counted	Yes	5% occurrence	Nylon	Filaments	nd	Bordbar et al. (2018)
Greece	2018–2019	Fish	Digestion of with 15% H_2O_2. Filtered on GF/C	Stereomicroscope. ATR-FTIR	Counted	Yes	0.23 ± 0.05– 0.43 ± 0.13 (21–30%)	PE, PP	Filaments, fragments	1296 ± 169 667 ± 166	Tsangaris et al. (2020)
Turkey	July and August in 2015	Fish	Digestion of the digestive tract with 35% H_2O_2. Filtered using a 26 µm sieve	Stereomicroscope. ATR-FTIR	Counted	Yes	2.36 particles per fish (58%)	nd	nd	656 ± 803 µm	Güven et al. (2017)
Turkey	September 2019 to March 2020	Mussels	Digestion of with 30% H_2O_2. Filtered on GF/C	Stereomicroscope. FTIR	Counted	Yes	0.69 item/ mussel and 0.23 item g^{-1} fresh weight of soft tissue (48%)	PET, PP, PE	Fragments and fibers	1.66 ± 1.45 mm	Gedik and Eryaşar (2020)
Lebanon	May 2018	Fish	Digestion of digestive tract with KOH. Filtered on GF/A	Stereomicroscope. Micro-Raman	Counted	Yes	2.5 ± 0.3 (83.4%)		Fragments and fibers	<200 µm	Kazour et al. (2019)

(Continued)

Table 7.3 (Continued)

Area	Year of sampling	Organism	Extraction method	Identification method	Quantification method	Quality control	Concentration	Dominant polymer	Dominant shape	Dominant size	References
Lebanon	May 2018	Oyster	Digestion of whole flesh with KOH. Filtered on GF/A	Stereomicroscope. Micro-Raman	Counted	Yes	7.2 ± 1.4 Items/ individual and 0.45 ± 0.3 items g^{-1} ww	PS	Fragments and fibers	<200 µm	
Israel	July and November 2017,	Ascidians	Digestion of Oozid with KOH. Followed by acid treatment. Vacuum-filtered on GF/A filter	Microscopy	Counted	Yes	1.37 ± 1.29 particles/ individual (100%)	nd	Fibers	<400 µm	Vered et al. (2019)
Israel	October 2016	Fish	Digestion of digestive tract with KOH. Filtered on 125 µm sieve	ND	Counted	No	68.09 ± 26.91– 1.47 ± 5.21 (n = 88) (92%)	nd	Fragments, fibers, and films	0.3–5 mm	Van der Hal et al. (2018)
Egypt	December 2018 and February, April, and May 2019	Fish	Digestion of digestive tracts (KOH). Filtered on GF/C. Dried	Stereomicroscope. Compound microscope. Thermal analysis.	Counted	Yes	28 ± 21–7527 ± 9551 (100%) $1.67 – 302.09$ mg MP kg^{-1}	PP, HDPE, and PEVA	Filaments and fragments	25–100 µm and 100–300 µm	Shabaka et al. (2020)

Location	Date	Species	Digestion	Detection			Concentration	Polymer	Type	Size	Reference
Tunisia Bizerte lagoon	May 2019	Mussels	Digestion of soft tissue with KOH. Density separation with NaI. Filtered on GF/A.	Stereomicroscope. ATR-FTIR	Counted	Yes	7.7 ± 3.8 items mussel^{-1} (97%)	PE, PP	Fibers and fragments		Wakkaf et al. (2020b)
Tunisia (Bizerte, Sousse, Monastir, Sfax, Gulf of Gabes)	July and September 2018	Fish	Digestion of the digestive tract and part of the muscles with KOH. Density separation (KI). Filtered on GF/C	Stereomicroscope. Micro-Raman	Counted	Yes	3.63 ± 0.35–6.11 ± 0.48 items g^{-1} of digestive tract (100%) 1.78 ± 0.26–6.03 ± 0.47 items g^{-1} of fish muscle (100%)	PEVA, HDPE, PA	nd	1.2–3 µm	Zitouni et al. (2020)
Tunisia Bizerte lagoon	March 2018	Mussels, gastropods, and cephalopods	Digestion of the soft tissue with KOH. Density separation (NaCl). Filtered on GF/A	Stereomicroscope. ATR-FTIR	Counted	Yes	703.95 ± 109.80 to 1482.82 ± 19.20 items kg^{-1} wet weight 1031.10 ± 355.69 100%	PE, PP	Fibers	0.1–1 mm	Abidli et al. (2019)

nd: no data; PP: Polypropylene; PE: Polyethylene; LDPE: Low density polyethylene; HDPE: High density polyethylene; PA: polyamide; PEVA: polyethylene-vinyl acetate; PS: Polystyrene; PET: Polyethylene terephthalate; EPS: Expanded Polystyrene; SAAC: Styrene/Allyl Alcohol Copolymer.

7.3 Microplastics Abundance in the Marine Environment of the Southern and Eastern Mediterranean Countries

A few studies have been conducted to assess marine MPs in the southern and eastern Mediterranean Sea countries, with the highest research output from Tunisia (Tables 7.1–7.3). Although there was a consistency in the sampling of different environmental matrices among different countries—for instance, sampling of surface water involved towing of manta net, sampling of sediments was performed for the first few centimeters in a defined area inshore, and marine organisms were collected from the field and the whole digestive tracts for fish or the soft tissue (in case of other marine invertebrates) were digested—it was difficult to compare MPs abundance among different regions due to the variation of the subsequent extraction techniques, and expression of the results (Tables 7.1 and 7.2). Another important aspect that would greatly affect the comparability of the results is the cut size that determines "MPs." While some studies considered particles of <0.5 mm corresponded to MPs, others included particles <5 mm under MPs. According to Martellini et al. (2018) and Van Cauwenberghe et al. (2015), the lack of homogeneity in the expression of the results and in the sampling and identification techniques makes the comparison between different studies very difficult. It is worthy to mention that quality control and precaution measures to prevent contamination while processing the samples have been followed in most studies (Tables 7.1–7.3).

According to Lots et al. (2017), relatively higher levels of MPs contamination were observed in the sampling stations of Greece and Turkey (242 ± 93 items kg^{-1} dry weight (DW), and 248 ± 47 items kg^{-1} DW, respectively) and in Tel Aviv in Israel (168 ± 93 items kg^{-1} DW), in comparison to an average of 147 ± 14 items kg^{-1} for west Mediterranean shores. Discharges from wastewater treatment plants significantly contributed to MP pollution in the northeast Mediterranean Sea (Akarsu et al. 2020). Regardless of the expression of the final results of MPs contamination from inshore sediments, higher densities were observed in Lebanon, Tunisia, Egypt, and Algeria (Table 7.2) when compared to west Mediterranean Sea countries. MP density recorded in a Bizerte lagoon in Tunisia (>7000 items kg^{-1} DW) was one of the highest MP abundances in sediments reported globally, which was attributed to heavy marine and industrial activities (Table 7.2). High densities of more than 1000 items kg^{-1} were also reported for other regions in Tunisia, Lebanon, and Turkey (Table 7.2).

Comparing MPs recovered from marine organisms was of interest since the extraction and processing technique was more or less similar. According to Shabaka et al. (2020), the fact that the Mediterranean Sea is the region most polluted with MPs contradicted with small quantities of MPs extracted from the wild fish of the Mediterranean countries. Shabaka et al. (2020) reported a large number of MPs from eight fish species in an urban harbor located on the Mediterranean coast of Egypt in comparison to an average range of 1–3 particles reported from other Mediterranean regions. In the current review, the results of MPs from organisms collected in the Levant Basin and Tunisia were high in comparison to different regions in the western Mediterranean Sea (Table 7.3). A substantial amount of MPs was extracted from the teleost *Serranus scriba*, a commercial fish species in an urban harbor in Tunisia, the authors recorded average concentrations that ranged between 1000 to 6000 items kg^{-1} of fish muscle, which indicated the high potential of MPs transferred to the human diet (Zitouni et al. 2020). The latter authors reported that polyethylene vinyl acetate (PEVA) and PE were the dominant polymers in fish muscles and digestive tracts, which were in accordance with the results obtained for fish species from an urban harbor in Egypt (Shabaka et al. 2020).

7.4 Microplastics Characterization and Identification Techniques

The identification and characterization are fundamental steps for the assessment of MPs' effects and fate in the marine environment. Several analytical tools have been used to identify MPs including Fourier Transformed Infra-Red spectrometry (FT-IR), Raman spectrometry, high-temperature gel-permeation chromatography (HT-GPC) with IR detection, SEM-EDS, and Pyrolysis-Gas Chromatography combined with Mass Spectroscopy (Pyr-GC-MS) (Lusher et al. 2017). FT-IR coupled with micro-Raman spectroscopy is commonly used for MP identification (Harrison et al. 2012; Käppler et al. 2016; Löder and Gerdts 2015; Tagg et al. 2015). It was observed that ATR-FTIR coupled with microscopy is a commonly used technique for MPs characterization and identification in most Mediterranean countries (Tables 7.1–7.3), since it provides the possibility to study individual microscopic plastic particles. However, several drawbacks of the latter technique have been observed; for instance, it is difficult to use for routine analysis, time-consuming, expensive, and is not always available (Dümichen et al. 2017; Majewsky et al. 2016; Renner et al. 2018). In addition, external funds and international collaboration with some LMICs such as Lebanon, Algeria, and Tunisia gave the opportunity to use ATR-FTIR. Otherwise, alternative methods that might limit the effectiveness of MPs assessment were used, such as FTIR, microscopy, or even magnifying lens. Thermal analysis using DSC and combustion techniques was adopted in Egypt, where it proved to be efficient for routine analysis (Shabaka et al. 2019, 2020).

Polyethylene (PE) and polypropylene (PP) were the dominant polymers recovered from different environmental matrices in the study regions (Tables 7.1–7.3). PE and PP are commonly found floating in marine surface water, as well as in pelagic and filter-feeder animals (Cole et al. 2013; Khan et al. 2015; Klein et al. 2015). Moreover, fibers and fragments were the dominant types of MPs indicating a secondary source of MP contamination; i.e. fragmented from larger plastic items, which usually results from local anthropogenic impact. Plastic contamination from abandoned fishing nets and ropes, in addition to the high level of sewage input, are common anthropogenic stressors in the MENA region (Middle East and North Africa) (Ouda et al. 2021).

7.5 Microplastics in Coastal Areas Affected by Rivers

Data on MP occurrence within the sediments of tributary rivers are very limited. Mersin Bay, in Turkey, is highly affected by river run-off. The number of MPs in surface waters was assessed before and after a flood event, the authors found that the number of MPs increased from 539 189 items km^{-2} to 7 699 716 items km^{-2}, pre- and post-flood, resp., accompanied by a reduction in the particles sizes (Table 7.1). In addition, 14 plastic polymers were detected, eight of which were not detected before the flood (Gündoğdu et al. 2018b). It was speculated that the Nile River contributes significantly to the input of MPs into the Mediterranean Sea (Lebreton et al. 2012). This is a very important aspect since the Levant Basin is a closed area of the Mediterranean Sea, separated from the western part of the sea by the Sicily channel, and it is prone to accumulate debris mainly from the Nile and major tributaries of the Mediterranean. Main tributaries of the Nile River in the Mediterranean coast of Egypt are currently being investigated under a project funded by the National Institute of Oceanography and Fisheries (NIOF), Egypt. This would provide very important insight into MPs contamination in the region.

7.6 Socioeconomic Impact of Plastic Marine Litter and Reduction Approaches

Marine litter, particularly plastics, has severe socioeconomic implications on tourism, maritime transport, aquaculture, fishing industry, and leisure activities (Zielinski et al. 2019). The socioeconomic impact of marine litter has been investigated in Turkey, Greece, Cyprus, Israel, and Morocco. These studies were mainly conducted to provide a baseline for marine litter management plans (Aydın et al. 2016). According to the socio-economic analysis of Abalansa et al. (2020), the consumption of plastics was linked mainly to multiple economic sectors such as; transportation, agriculture, fisheries, packaging, and construction, where the accompanied changes occurred at the environmental (pollution), ecosystem (ingestion of plastics and ghost fishing), and ecosystem services levels (supply of seafood and cultural benefits), with possible impact on human welfare, such as loss of jobs and income. Wastewater treatment plants were found to contribute largely to MP threats to aquatic ecosystems (Akarsu et al. 2017, 2020; Gündoğdu et al. 2018b; Mourgkogiannis et al. 2018). According to Freeman et al. (2020), there are no policies or regulations requiring the removal of MPs during wastewater treatment, and collaboration among researchers, industry stakeholders, and policy decision-makers is strongly encouraged to introduce innovative technology within the wastewater sector to overcome this problem. Ghost fishing, one of the serious impacts of plastic marine litter on fish mortality, has been addressed in a number of studies (Dagtekin et al. 2019; Ozyurt et al. 2017; Yildiz and Karakulak 2016). Yildiz and Karakulak (2016) recommended setting separate zones for artisanal and industrial fisheries to reduce the impact of ghost fishing. Since the land-based source is also a major contributor to marine litter, several approaches have been developed to reduce and/ or prevent litter from entering the marine environment. As such, the importance of promoting the co-responsibility of the public and the enhancement of public awareness were addressed in some studies (Latinopoulos et al. 2018; Veiga et al. 2016). Data collected by NGOs were found to be essential to fill in the marine litter knowledge gap along the Mediterranean Sea shores (Vlachogianni et al. 2020). A marine litter watch app was developed for citizen education and to raise awareness (Golumbeanu et al. 2017). Citizen science, recreational divers, and enthusiasts of marine conservation have been participating in regular clean-ups and contributed to data collection (Pasternak et al. 2019). In addition, the application of product design methods to prevent littering by beachgoers had a direct influence on the beachgoer's behavior (Portman et al. 2019; Portman and Behar 2020). The polluter pay approach has been suggested, the willingness of beach visitors to pay an entrance fee or increase in local tax to clean up marine litter was proposed in Greece and other European beaches (Brouwer et al. 2017).

7.7   Knowledge Gaps and Recommendation for Future Research

More studies to assess marine litter and MPs in the eastern and southern Mediterranean Sea countries are urgently needed. Common analytical protocols to characterize and quantify marine litter should be adopted in the region in compliance with EU regulations. Developing common and comparable monitoring approaches and data management systems are of importance.

However, the identification and quantification of the small plastic particles are challenging, since adopting a size cut of more than 1 mm, is underestimating these items. It is evident that there is a weakness in the enforcement of policies and interventions that may reduce the negative impacts of the current plastic waste burden, particularly in LMICs of the Mediterranean Sea. Scientific cooperation among Mediterranean countries should be promoted particularly for LMICs; this is indispensable for the capacity development and funding of local and regional projects. Marine litter management requires further research as follows:

1) Identification of the densities and types, and evaluation of sources and accumulation areas of litter, including maritime transport, agricultural, and industrial activities, and the integration of GIS and mapping techniques for that purpose.
2) Evaluation of lost fishing gears and the potential loss of fish stocks due to ghost fishing.
3) Development of modeling programs for the identification of sources and fate of marine litter.
4) Evaluation of degradation rates of different types of plastic litter.
5) Identification of the sources of MPs (primary and secondary).
6) Setting indicators to assess MPs and their effects including fauna, flora, and human health.
7) Evaluation of the direct socio-economic impact.
8) Developing education programs and evaluating of its effectiveness.
9) Developing reduction measures and evaluating their effectiveness in different countries based on their economic class.

References

Abalansa, S., El Mahrad, B., Vondolia, G.K. et al. (2020). *The marine plastic litter issue: a social-economic* analysis. *Sustainability* 12 (20): 8677.

Abidli, S., Toumi, H., Lahbib, Y., and Trigui El Menif, N. (2017). *The first evaluation of microplastics in sediments from the complex lagoon channel of Bizerte (northern Tunisia).* *Water, Air, & Soil Pollution* 228 (7) https://doi.org/10.1007/s11270-017-3439-9.

Abidli, S., Antunes, J.C., Ferreira, J.L. et al. (2018). *Microplastics in sediments from the littoral zone of the north Tunisian coast (Mediterranean Sea). Estuarine, Coastal and Shelf Science* 205: 1–9. https://doi.org/10.1016/j.ecss.2018.03.006.

Abidli, S., Lahbib, Y., and Trigui El Menif, N. (2019). *Microplastics in commercial Molluscs from the lagoon of Bizerte (northern Tunisia). Marine Pollution Bulletin* 142: 243–252. https://doi.org/10.1016/j.marpolbul.2019.03.048.

Akarsu, C., Kideyş, A.E., and Kumbur, H. (2017). *Microplastic threat to aquatic ecosystems of the municipal wastewater treatment plant. Turk Hijyen ve Deneysel Biyoloji Dergisi* 74: 73–78.

Akarsu, C., Kumbur, H., Gökdağ, K. et al. (2020). *Microplastics composition and load from three wastewater treatment plants discharging into Mersin Bay, north eastern Mediterranean Sea. Marine Pollution Bulletin* 150: 1–13. https://doi.org/10.1016/j.marpolbul.2019.110776.

Akindele, E.O. and Alimba, C.G. (2021). *Plastic pollution threat in Africa: current status and implications for aquatic ecosystem health. Environmental Science and Pollution Research* 28: 7636–7651. https://doi.org/10.1007/s11356-020-11736-6.

Alimi, O.S., Fadare, O.O., and Okoffo, E.D. (2021). *Microplastics in African ecosystems: current knowledge, abundance, associated contaminants, techniques, and research needs. Science of the Total Environment* 755: 1–13. https://doi.org/10.1016/j.scitotenv.2020.142422.

Alshawafi, A., Analla, M., Alwashali, E., and Aksissou, M. (2017). *Assessment of marine debris on the coastal wetland of Martil in the north-east of Morocco. Marine Pollution Bulletin* 117 (1–2): 302–310. https://doi.org/10.1016/j.marpolbul.2017.01.079.

Aydın, C., Güven, O., Salihoğlu, B. et al. (2016). *The influence of land use on coastal litter: an approach to identify abundance and sources in the coastal area of Cilician Basin, Turkey. Turkish Journal of Fisheries and Aquatic Sciences* 16 (1): 29–39.

Berber, A.A. (2019). *Polystyrene nanoplastics trigger toxicity on two different aquatic organisms* (Brachionus plicatilis, Daphnia magna). *Fresenius Environmental Bulletin* 28 (8): 6146–6152.

Bergmann, M., Gutow, L., and Klages, M. (2015). *Marine Anthropogenic Litter*. Springer International Publishing https://doi.org/10.1007/978-3-319-16510-3.

Bordbar, L., Kapiris, K., Kalogirou, S., and Anastasopoulou, A. (2018). *First evidence of ingested plastics by a high commercial shrimp species* (Plesionika narval) *in the eastern Mediterranean. Marine Pollution Bulletin* 136: 472–476. https://doi.org/10.1016/j.marpolbul.2018.09.030.

Brouwer, R., Hadzhiyska, D., Ioakeimidis, C. et al. (2017). *The social costs of marine litter along European coasts. Ocean and Coastal Management* 138: 38–49.

Chouchene, K., Da Costa, J.P., Wali, A. et al. (2019). *Microplastic pollution in the sediments of Sidi Mansour Harbor in Southeast Tunisia. Marine Pollution Bulletin* 146: 92–99. https://doi.org/10.1016/j.marpolbul.2019.06.004.

Chouchene, K., Rocha-Santos, T., and Ksibi, M. (2020). *Types, occurrence, and distribution of microplastics and metals contamination in sediments from south west of Kerkennah archipelago, Tunisia. Environmental Science and Pollution Research* 28: 46477–46487. https://doi.org/10.1007/s11356-020-09938-z.

Claro, F., Fossi, M.C., Ioakeimidis, C. et al. (2019). *Tools and constraints in monitoring interactions between marine litter and megafauna: insights from case studies around the world. Marine Pollution Bulletin* 141: 147–160.

Cole, M., Lindeque, P., Fileman, E. et al. (2013). Microplastic ingestion by zooplankton. *Environ. Sci. Technol.* 47: 6646–6655.

Conkle, J.L., Báez Del Valle, C.D., and Turner, J.W. (2018). *Are we underestimating microplastic contamination in aquatic environments? Environmental Management* 61: 1–8. https://doi. org/10.1007/s00267-017-0947-8.

Dagtekin, M., Ozyurt, C.E., Misir, D. et al. (2019). *Rate and causes of lost "gillnets and entangling nets" in the black sea coasts of Turkey. Turkish Journal of Fisheries and Aquatic Sciences* 19 (8): 699–705.

Dümichen, E., Eisentraut, P., Bannick, C.G. et al. (2017). *Fast identification of MPs in complex environmental samples by a thermal degradation method. Chemosphere* 174: 572–584. https:// doi.org/10.1016/j.chemosphere.2017.02.010.

Duncan, E.M., Arrowsmith, J., Bain, C. et al. (2018). *The true depth of the Mediterranean plastic problem: extreme microplastic pollution on marine turtle nesting beaches in Cyprus. Marine Pollution Bulletin* 136: 334–340. https://doi.org/10.1016/j.marpolbul.2018.09.019.

Eriksen, M., Lebreton, L.C.M., Carson, H.S. et al. (2014). *Plastic pollution in the world's oceans: more than 5 trillion plastic pieces weighing over 250,000 tons afloat at sea. PLoS One* 9 https:// doi.org/10.1371/journal.pone.0111913.

Fortibuoni, T., Ronchi, F., Mačić, V. et al. (2019). *A harmonized and coordinated assessment of the abundance and composition of seafloor litter in the Adriatic-Ionian macroregion (Mediterranean Sea). Marine Pollution Bulletin* 1 (39): 412–426.

Fossi, M.C., Pedà, C., Compa, M. et al. (2018). *Bioindicators for monitoring marine litter ingestion and its impacts on Mediterranean biodiversity. Environmental Pollution* 237: 1023–1040.

Freeman, S., Booth, A.M., Sabbah, I. et al. (2020). *Between source and sea: the role of wastewater treatment in reducing marine MPs. Journal of Environmental Management* 266.

Frias, J.P.G.L. and Nash, R. (2019). *Microplastics: finding a consensus on the definition. Marine Pollution Bulletin* 138: 145–147. https://doi.org/10.1016/j.marpolbul.2018.11.022.

Gedik, K. and Eryaşar, A.R. (2020). *Microplastic pollution profile of Mediterranean mussels* (Mytilus galloprovincialis) *collected along the Turkish coasts. Chemosphere* 260: 127570. https://doi.org/10.1016/j.chemosphere.2020.127570.

Golumbeanu, M., Nenciu, M., Galatchi, M. et al. (2017). *Marine litter watch app as a tool for ecological education and awareness raising along the Romanian Black Sea coast. Journal of Environmental Protection and Ecology* 18 (1): 348–362.

Gündoğdu, S. and Çevik, C. (2017). *Micro- and mesoplastics in Northeast Levantine coast of Turkey: the preliminary results from surface samples. Marine Pollution Bulletin* 118 (1–2): 341–347.

Gündoğdu, S., Çevik, C., Güzel, E. et al. (2018a). *Microplastics in municipal wastewater treatment plants in Turkey: a comparison of the influent and secondary effluent concentrations. Environmental Monitoring and Assessment* 190 (11): 1–10.

Gündoğdu, S., Çevik, C., Ayat, B. et al. (2018b). *How microplastics quantities increase with flood events? An example from Mersin Bay NE Levantine coast of Turkey. Environmental Pollution* 239: 342–350. https://doi.org/10.1016/j.envpol.2018.04.042.

Güven, O., Gökdağ, K., Jovanović, B. et al. (2017). *Microplastic litter composition of the Turkish territorial waters of the Mediterranean Sea, and its occurrence in the gastrointestinal tract of fish. Environmental Pollution* 223: 286–294.

Hamed, M., Soliman, H.A.M., Osman, A.G.M. et al. (2019). *Assessment the effect of exposure to microplastics in Nile Tilapia* (Oreochromis niloticus) *early juvenile: I. blood biomarkers. Chemosphere* 228: 345–350.

Hamed, M., Soliman, H.A.M., Osman, A.G.M. et al. (2020). *Antioxidants and molecular damage in Nile Tilapia* (Oreochromis niloticus) *after exposure to microplastics. Environmental Science and Pollution Research* 27 (13): 14581–14588.

Hamed, M., Soliman, H.A.M., Badrey, A.E.A. et al. (2021). *Microplastics induced histopathological lesions in some tissues of tilapia* (Oreochromis niloticus) *early juveniles. Tissue and Cell* 71: 9.

Harrison, J.P., Ojeda, J.J., and Romero-González, M.E. (2012). The applicability of reflectance micro-Fourier-transform infrared spectroscopy for the detection of synthetic microplastics in marine sediments. *Sci. Total Environ.* 416: 455–463. https://doi.org/10.1016/j.scitotenv.2011.11.078.

Jagath, P., Gamaralalage, D. and Onogawa, K. (2020). Strategies to reduce marine plastic pollution from land-based sources in low- and middle-income countries. UNEP report https://www.unep.org/ietc/resources/policy-and-strategy/strategies-reduce-marine-plastic-pollution-land-based-sources-low-and (accessed 26 October 2021).

Jalón-Rojas, I., Wang, X.H., and Fredj, E. (2019). *A 3D numerical model to track marine plastic debris (TrackMPD): sensitivity of microplastic trajectories and fates to particle dynamical properties and physical processes. Marine Pollution Bulletin* 141: 256–272.

Jambeck, J.R., Geyer, R., Wilcox, C. et al. (2015). Plastic waste inputs from land into the ocean. *Science* 347 (6223): 768–771. https://doi.org/10.1126/science.1260352.

Kaberi, H., Tsangaris, C., Zeri, C. et al. (2013, 2013). Microplastics along the shoreline of a Greek island (Kea Island, Aegean Sea): types and densities in relation to beach orientation, characteristics and proximity to sources. In: *Proceedings of the 4th International Conference on Environmental Management.* Engineering, Planning and Economics (CEMEPE) and SECOTOX Conference, Mykonos Island, Greece, 197e202.

Käppler, A., Fischer, D., Oberbeckmann, S. et al. (2016). *Analysis of environmental MPs by vibrational microspectroscopy: FTIR, Raman or both? Analytical and Bioanalytical Chemistry* 408 (29): 8377–8391.

Karkanorachaki, K., Kiparissis, S., Kalogerakis, G.C. et al. (2018). *Plastic pellets, meso- and microplastics on the coastline of Northern Crete: distribution and organic pollution. Marine Pollution Bulletin* 133: 578–589. https://doi.org/10.1016/j.marpolbul.2018.06.011.

Kazour, M., Jemaa, S., Issa, C. et al. (2019). *Microplastics pollution along the Lebanese coast (Eastern Mediterranean Basin): occurrence in surface water, sediments and biota samples. Science of the Total Environment* 696: 133933. https://doi.org/10.1016/j.scitotenv.2019.133933.

Khan, F.R., Syberg, K., Shashoua, Y., and Bury, N.R. (2015). Influence of polyethylene microplastic beads on the uptake and localization of silver in zebrafish (Danio rerio). *Environ. Pollut.* 206: 73–79.

Kikaki, A., Karantzalos, K., Power, C.A. et al. (2020). *Remotely sensing the source and transport of marine plastic debris in Bay Islands of Honduras (Caribbean Sea). Remote Sensing* 12 (11): 1–17.

Klein, S., Worch, E., and Knepper, T.P. (2015). Occurrence and spatial distribution of microplastics in river shore sediments of the Rhine-Main area in Germany. *Environ. Sci. Technol.* 49: 6070–6076.

Kylili, K., Hadjistassou, C., and Artusi, A. (2020). *An intelligent way for discerning plastics at the shorelines and the seas. Environmental Science and Pollution Research* 27 (34): 42631–42643.

Latinopoulos, D., Mentis, C., and Bithas, K. (2018). *The impact of a public information campaign on preferences for marine environmental protection. The case of plastic waste.* Marine Pollution Bulletin 131: 151–162.

Lebreton, L., Greer, S., and Borrero, J. (2012). *Numerical modelling of floating debris in the world's oceans. Marine Pollution Bulletin* 64 (3): 653–661. https://doi.org/10.1016/j.marpolbul.2011.10.027.

Lin, V.S. (2016). *Research highlights: impacts of microplastics on plankton. Environmental Science. Processes & Impacts* 18: 160–163. https://doi.org/10.1039/c6em90004f.

Löder, M.G. and Gerdts, G. (2015). *Methodology used for the detection and identification of microplastics—a critical appraisal. Marine Anthropogenic Litter*: 201–227. https://doi.org/10.1007/978-3-319-16510-3_8.

Lots, F.A., Behrens, P., Vijver, M.G. et al. (2017). *A large-scale investigation of microplastic contamination: abundance and characteristics of microplastics in European beach sediment. Marine Pollution Bulletin* 123 (1–2): 219–226. https://doi.org/10.1016/j.marpolbul.2017.08.057.

Loulad, S., Houssa, R., Ouamari, N.E. et al. (2019). *Quantity and spatial distribution of seafloor marine debris in the Moroccan Mediterranean Sea. Marine Pollution Bulletin* 139: 163–173.

Lusher, A.L., Welden, N.A., Sobral, P., and Cole, M. (2017). *Sampling, isolating and identifying microplastics ingested by fish and invertebrates. Analytical Methods* 9: 1346–1360.

Mæland, C.E. and Staupe-Delgado, R. (2020). Can the global problem of marine litter be considered a crisis? *Risk, Hazards Crisis in Public Policy* 11: 87–104. https://doi.org/10.1002/rhc3.12180.

Majewsky, M., Bitter, H., Eiche, E., and Horn, H. (2016). *Determination of microplastic polyethylene (PE) and polypropylene (PP) in environmental samples using thermal analysis (TGA-DSC). Science of the Total Environment* 568: 507–511. https://doi.org/10.1016/j.scitotenv.2016.06.017.

Martellini, T., Guerranti, C., Scopetani, C. et al. (2018). *A snapshot of microplastics in the coastal areas of the Mediterranean Sea. TrAC Trends in Analytical Chemistry* 109: 173–179. https://doi.org/10.1016/j.trac.2018.09.028.

Matiddi, M., DeLucia, G.A., Silvestri, C. et al. (2019). *Data collection on marine litter ingestion in sea turtles and thresholds for good environmental status. Journal of Visualized Experiments* (147): 1–9.

Mghili, B., Analla, M., Aksissou, M. et al. (2020). *Marine debris in Moroccan Mediterranean beaches: an assessment of their abundance, composition and sources. Marine Pollution Bulletin* 160: 1–7.

Mohsen, M., Zhang, L., Sun, L. et al. (2020). *Microplastic fibers transfer from the water to the internal fluid of the sea cucumber* Apostichopus japonicus. *Environmental Pollution* 257: 1–7.

Mohsen, M., Zhang, L., Sun, L. et al. (2021). *Effect of chronic exposure to microplastic fibre ingestion in the sea cucumber* Apostichopus japonicus. *Ecotoxicology and Environmental Safety* 209: 1–8.

Mourgkogiannis, N., Kalavrouziotis, I.K., and Karapanagioti, H.K. (2018). *Questionnaire-based survey to managers of 101 wastewater treatment plants in Greece confirms their potential as plastic marine litter sources. Marine Pollution Bulletin* 133: 822–827.

Musić, J., Kružic, S., Stančić, I. et al. (2020). Detecting underwater sea litter using deep neural networks: an initial study, 2020. *5th International Conference on Smart and Sustainable Technologies*. SpliTech.

Nachite, D., Maziane, F., Anfuso, G. et al. (2019). *Spatial and temporal variations of litter at the Mediterranean beaches of Morocco mainly due to beach users*. Ocean and Coastal Management 179: 1–5.

Ouda, M., Kadadou, D., Swaidan, B. et al. (2021). *Emerging contaminants in the water bodies of the Middle East and North Africa (MENA): a critical review*. Science of the Total Environment 754: 142177. https://doi.org/10.1016/j.scitotenv.2020.142177.

Ozyurt, C.E., Buyukdeveci, F., and Kiyaga, V.B. (2017). *Ghost fishing effects of lost bottom trammel nets in a storm: a simulation*. Fresenius Environmental Bulletin 25 (12): 8109–8118.

Pasternak, G., Zviely, D., Ribic, C.A. et al. (2017). *Sources, composition and spatial distribution of marine debris along the Mediterranean coast of Israel*. Marine Pollution Bulletin 114 (2): 1036–1045.

Pasternak, G., Ribic, C.A., Spanier, E. et al. (2019). *Nearshore survey and cleanup of benthic marine debris using citizen science divers along the Mediterranean coast of Israel*. Ocean and Coastal Management 175: 17–32.

Politikos, D.V., Tsiaras, K., Papatheodorou, G. et al. (2020). *Modeling of floating marine litter originated from the Eastern Ionian Sea: transport, residence time and connectivity*. Marine Pollution Bulletin 150: 1–12.

Portman, M.E. and Behar, D. (2020). *Influencing beach littering behaviors through infrastructure design: an in situ experimentation case study*. Marine Pollution Bulletin 156: 1–9.

Portman, M.E. and Brennan, R.E. (2017). *Marine litter from beach-based sources: case study of an Eastern Mediterranean coastal town*. Waste Management 69: 535–544.

Portman, M.E., Pasternak, G., Yotam, Y. et al. (2019). *Beachgoer participation in prevention of marine litter: using design for behavior change*. Marine Pollution Bulletin 144: 1–10.

Renner, G., Schmidt, T.C., and Schram, J. (2018). *Analytical methodologies for monitoring micro (nano) plastics: which are fit for purpose?* Current Opinion in Environmental Science and Health 1: 55–61. https://doi.org/10.1016/j.coesh.2017.11.001.

Ronchi, F., Galgani, F., Binda, F. et al. (2019). *Fishing for litter in the Adriatic-Ionian macroregion (Mediterranean Sea): strengths, weaknesses, opportunities and threats*. Marine Policy 100: 226–237. https://doi.org/10.1016/j.marpol.2018.11.041.

Shabaka, S.H., Ghobashy, M., and Marey, R. (2019). *Identification of marine microplastics from the Eastern Harbor, Mediterranean coast of Egypt, using differential scanning calorimetry*. Marine Pollution Bulletin 142: 494–503.

Shabaka, S.H., Marey, R.S., Ghobashy, M. et al. (2020). *Thermal analysis and enhanced visual technique for assessment of microplastics in fish from an Urban Harbor, Mediterranean coast of Egypt*. Marine Pollution Bulletin 159: 111465.

Smith, M., Love, D.C., Rochman, C.M., and Neff, R.A. (2018). *Microplastics in seafood and the implications for human health*. Current Environmental Health Reports 5: 375–386. https://doi.org/10.1007/s40572-018-0206-z.

Tagg, A.S., Sapp, M., Harrison, J.P., and Ojeda, J.J. (2015). *Identification and quantification of microplastics in wastewater using focal plane array-based reflectance micro-FT-IR imaging*. Analytical Chemistry 87 (12): 6032–6040. https://doi.org/10.1021/acs.analchem.5b00495.

Tanaka, K. and Takada, H. (2016). *Microplastic fragments and microbeads in digestive tracts of planktivorous fish from urban coastal waters. Scientific Reports* 6: 1–8. https://doi.org/10.1038/srep34351.

Tata, T., Belabed, B.E., Bououdina, M., and Bellucci, S. (2020). *Occurrence and characterization of surface sediments microplastics and litter from North African coasts of Mediterranean Sea: preliminary research and first evidence. Science of the Total Environment* 713: 136664. https://doi.org/10.1016/j.scitotenv.2020.136664.

Themistocleous, K., Papoutsa, C., Michaelides, S. et al. (2020). *Investigating detection of floating plastic litter from space using sentinel-2 imagery. Remote Sensing* 12 (16): 1–18.

Topouzelis, K., Papageorgiou, D., Karagaitanakis, A. et al. (2020). *Remote sensing of sea surface artificial floating plastic targets with Sentinel-2 and unmanned aerial systems (plastic litter project 2019). Remote Sensing* 12 (12): 1–17.

Toumi, H., Abidli, S., and Bejaoui, M. (2019). *Microplastics in freshwater environment: the first evaluation in sediments from seven water streams surrounding the lagoon of Bizerte (northern Tunisia). Environmental Science and Pollution Research* 26 (14): 14673–14682. https://doi.org/10.1007/s11356-019-04695-0.

Tsangaris, C., Digka, N., Valente, T. et al. (2020). *Using* Boops boops (Osteichthyes) *to assess microplastics ingestion in the Mediterranean Sea. Marine Pollution Bulletin* 158: 111397. https://doi.org/10.1016/j.marpolbul.2020.111397.

Tunali, M., Uzoefuna, E.N., Tunali, M.M. et al. (2020). *Effect of microplastics and microplastic-metal combinations on growth and chlorophyll a concentration of* Chlorella vulgaris. *Science of the Total Environment* 743: 1–7.

Tziourrou, P., Megalovasilis, P., Tsounia, M., and Karapanagioti, H. (2019). *Characteristics of microplastics on two beaches affected by different land uses in Salamina Island in Saronikos Gulf, East Mediterranean. Marine Pollution Bulletin* 149: 110531. https://doi.org/10.1016/j.marpolbul.2019.110531.

Van Cauwenberghe, L., Devriese, L., Galgani, F. et al. (2015). *Microplastics in sediments: a review of techniques, occurrence and effects. Marine Environmental Research* 111: 5–17. https://doi.org/10.1016/j.marenvres.2015.06.007.

Van der Hal, N., Ariel, A., and Angel, D.L. (2017). *Exceptionally high abundances of microplastics in the oligotrophic Israeli Mediterranean coastal waters. Marine Pollution Bulletin* 116 (1–2): 151–155. https://doi.org/10.1016/j.marpolbul.2016.12.052.

Van der Hal, N., Yeruham, E., and Angel, D.L. (2018). *Dynamics in microplastic ingestion during the past six decades in herbivorous fish on the Mediterranean Israeli coast.* In: *Proceedings of the International Conference on Microplastic Pollution in the Mediterranean Sea* (eds. M. Cocca, E. Di Pace, M. Errico, et al.). Cham: Springer Water. Springer https://doi.org/10.1007/978-3-319-71279-6_21.

Veiga, J.M., Vlachogianni, T., Pahl, S. et al. (2016). *Enhancing public awareness and promoting co-responsibility for marine litter in Europe: the challenge of MARLISCO. Marine Pollution Bulletin* 102 (2): 309–315.

Vered, G., Kaplan, A., Avisar, D., and Shenkar, N. (2019). *Using solitary ascidians to assess microplastic and phthalate plasticizers pollution among marine biota: a case study of the Eastern Mediterranean and Red Sea. Marine Pollution Bulletin* 138: 618–625. https://doi.org/10.1016/j.marpolbul.2018.12.013.

Vlachogianni, T., Skocir, M., Constantin, P. et al. (2020). *Plastic pollution on the Mediterranean coastline: generating fit-for-purpose data to support decision-making via a participatory-science initiative. Science of the Total Environment* 711: 1–11.

Wakkaf, T., El Zrelli, R., Kedzierski, M. et al. (2020a). *Characterization of microplastics in the surface waters of an urban lagoon (Bizerte Lagoon, southern Mediterranean Sea): composition, density, distribution, and influence of environmental factors. Marine Pollution Bulletin* 160 (111): 625. https://doi.org/10.1016/j.marpolbul.2020.111625.

Wakkaf, T., El Zrelli, R., Kedzierski, M. et al. (2020b). *Microplastics in edible mussels from a southern Mediterranean lagoon: preliminary results on seawater-mussel transfer and implications for environmental protection and seafood safety. Marine Pollution Bulletin* 158: 111355.

WWF (World Wildlife Fund) (2019). No plastic in nature: Assessing plastic ingestion from nature to people. https://awsassets.panda.org/downloads/plastic_ingestion_press_singles. pdf (accessed 26 October 2021).

Yabanlı, M., Yozukmaz, A., Şener, İ., and Ölmez, Ö.T. (2019). *Microplastic pollution at the intersection of the Aegean and Mediterranean seas: a study of the Datça Peninsula (Turkey). Marine Pollution Bulletin* 145: 47–55. https://doi.org/10.1016/j.marpolbul.2019.05.003.

Yildiz, T. and Karakulak, F.S. (2016). *Types and extent of fishing gear losses and their causes in the artisanal fisheries of Istanbul, Turkey. Journal of Applied Ichthyology* 32 (3): 432–438.

Zayen, A., Sayadi, S., Chevalier, C. et al. (2020). *Microplastics in surface waters of the Gulf of Gabes, southern Mediterranean Sea: distribution, composition and influence of hydrodynamics. Estuarine, Coastal and Shelf Science* 242: 106832. https://doi.org/10.1016/j. ecss.2020.106832.

Zielinski, S., Botero, C.M., and Yanes, A. (2019). *To clean or not to clean? A critical review of beach cleaning methods and impacts. Marine Pollution Bulletin* 139: 390–401.

Zitouni, N., Bousserrhine, N., Belbekhouche, S. et al. (2020). *First report on the presence of small microplastics (≤ 3 Mm) in tissue of the commercial fish* Serranus scriba (Linnaeus. 1758) *from Tunisian coasts and associated cellular alterations. Environmental Pollution* 263: 114576. https://doi.org/10.1016/j.envpol.2020.114576.

8

Advanced Detection Techniques for Microplastics in Different Environmental Media

Arely Areanely Cruz-Salas[1], Sara Ojeda-Benítez[1], Juan Carlos Álvarez-Zeferino[2], Carolina Martínez-Salvador[3], Jocelyn Tapia-Fuentes[2], Beatriz Pérez-Aragón[2], and Alethia Vázquez-Morillas[2]

[1] *Universidad Autónoma de Baja California, Mexicali, Baja California, Mexico*
[2] *Universidad Autónoma Metropolitana, Mexico City, Mexico*
[3] *Technische Universität Desdren, Dresden, Gérmany*

8.1 Introduction

Since the 1950s, plastic waste has become one of the most important environmental problems, as well as one of the main sources of marine waste (Van Rensburg et al. 2020). By 2018, 360 million tons of different types of plastics were produced for packaging, agriculture, construction, electronic devices, and other applications (PlasticsEurope 2019). From 1950 to 2015, the total plastic production has been estimated to be around 8.3 billion metric tons, with an annual growth rate of 8.4% (Geyer et al. 2017). Up to 50% of these plastics are designed for single-use, and many of them are improperly managed at the end of their life cycle, thus hindering their recycling or recovery process (Kedzierski et al. 2020). Only 9% of the plastic waste produced in the last 70 years has been recycled, whereas 12% has been incinerated, and 79% has ended up in dumping sites, controlled final disposal sites, and in the environment. It has been estimated that 47% of the world's plastic waste is derived from packaging (UNEP 2018).

Because of their chemical structure, plastic wastes in the environment will degrade slowly (from decades to hundreds of years) through wind action and photodegradation, which lead to loss of mechanical properties and fragmentation into smaller pieces known as microplastics (MPs). For these fragments to be considered as such, they must be less than 5 mm in any of their dimensions (GESAMP 2015; UNEP 2018). MPs fall into two categories; primary and secondary. The first category consists of particles that are originally designed and produced with a size of <5 mm. This category includes microbeads used personal in care products. Secondary MPs group all fragments from the rupture of large plastics (macroplastics), and include synthetic textile fibers (GESAMP 2016; NOAA 2017).

In recent years, the presence of MPs has been detected in various environmental components, such as the soils, the atmosphere, and the marine environments. In soils, their main

Plastic and Microplastic in the Environment: Management and Health Risks, First Edition.
Edited by Arif Ahamad, Pardeep Singh, and Dhanesh Tiwary.
© 2022 John Wiley & Sons Ltd. Published 2022 by John Wiley & Sons Ltd.

sources are associated with agriculture, leaks from final disposal sites, and activities carried out in urban areas (Guo et al. 2020). The presence of MPs in the atmosphere has been scarcely researched, but the main sources might be synthetic textile fibers and emissions from plastic waste recycling plants (Enyoh et al. 2019; Gasperi et al. 2018). MPs have also been found in marine environments such as estuaries, mangroves, beaches, oceans, and intercontinental zones (Guo et al. 2020; Wang et al. 2020b). MPs reach marine ecosystems through different paths (terrestrial and marine), mostly driven by air currents, wastewater discharges, river flows, and sea currents. In the case of primary MPs, the main sources are port activities, wastewater discharges containing microbeads, and industrial activities of factories located near coastal areas. Secondary MPs, on the other hand, come mainly from plastic waste that was improperly disposed of in terrestrial sources, or from recreational activities related to tourism in coastal areas (GESAMP 2016; Lavender & Thompson 2014; NOAA 2017).

MPs cause multiple impacts to marine environments, such as water contamination from additives contained in plastics (flame-retardants, heavy metals, bisphenol, and phthalates, to name a few) (Carney Almroth & Eggert 2019; Tsang et al. 2017). Additionally, marine fauna might ingest MPs accidentally or by confusing them with food, causing intestinal tract obstructions and malnutrition, because of false feelings of satiety. The absorption of toxic components containing plastic parts causes endocrine dysfunction, reproductive problems, neurological damage, and premature death (Carney Almroth & Eggert 2019; GESAMP 2016; Tsang et al. 2017). Another problem associated with MPs intake is the movement through the trophic chain, where species at the bottom of the marine chain could increase the concentrations of MPs in the intestinal tract of their predators, thus causing a biomagnification effect on species found at the highest levels of the chain (GESAMP 2016). It is estimated that around 600 marine species are affected by marine waste, while at least 15% of species affected both by MP intake and plastic waste entanglement are endangered (UNEP 2018).

There is a wide variety of methodologies for the detection of MPs in different marine environments. Nevertheless, those methodologies are not standardized, so customized adaptations are often made, usually based on the equipment and infrastructure available for each research team. In this sense, this research aims to analyze sampling techniques of MPs in different marine compartments and the steps used in sampling processing in the laboratory. Afterward, highlighted practices and techniques are proposed when carrying MPs-related research.

8.2 Methodology

This section discusses the methodology applied for the search and selection of research articles on the presence of MPs in different marine ecosystems. The focus is to analyze the methodologies reported on selected material for both MPs sampling and laboratory sample processing.

8.2.1 Selection of Criteria and Search for Articles

The article search was conducted in the *ScienceDirect* database in January 2021. The first criterion defined for searching articles were: (i) to include only articles written in English; (ii) published in

Table 8.1 Marine environments targeted in this chapter.

Marine environmental compartments	
Surface seawater	Beaches
Seawater column	Marine fauna
Sea bottom sediment	Marine vegetation
Mangroves	Phytoplankton

Source: Authors' own elaboration based on search design.

2020; and (iii) exclusively research and short communications categories (review articles, book chapters, and congresses were excluded). The terms selected as keywords were "microplastics" followed by the name of any of the marine environments shown in Table 8.1, among these terms the boolean operator "+" was used. The magazines and areas included in this search were only those where marine environment research has been published, such as *Science of the Total Environment* or *Marine Pollution Bulletin*.

8.2.2 Item Selection

After filtering by applying the above criterion, the identified articles were revised to verify that they, in fact, met the previously established inclusion criteria. Only articles from indexed journals were selected. Articles focused on rivers, esters, lagoons, air, or drinking water were excluded, as well as those dealing with ingestion *in vivo* (for the case of fauna). Once duplicates were eliminated, the resulting total was 115 reviewed articles. Figure 8.1 presents the flowchart with the steps that were followed for finding and selecting articles.

8.3 Results

This section is divided into two different parts, the first one includes the description of the techniques used for sampling MPs in different marine environments (seawaters, sea sediments, beaches, mangroves, marine fauna, and marine vegetation), and the second is about the processing of samples in the laboratory to extract MPs.

Among the 115 analyzed articles summarized in this study, several findings are worth mentioning; the three most studied marine environments when discussing MPs (Figure 8.2) are *sea waters* (30%), *beaches* (24%), and *marine wildlife* (20%), while the least common research environment is *marine vegetation* (2%). This might be related to the already existing abundance of methodologies (not standardized) for those first compartments, while MPs in *marine vegetation* is something relatively new.

A total of 43 countries (Figure 8.3) have contributed to the 115 articles analyzed in this chapter. The highest percentage of publications have been conducted in China, India, and United Kingdom, with 21, 10, and 5%, respectively. Of these three countries, China has the most kilometers of coastline (14 500 km), which might be reflected in the research production amount; however, United Kingdom (12 429 km) has more kilometers of coastline than

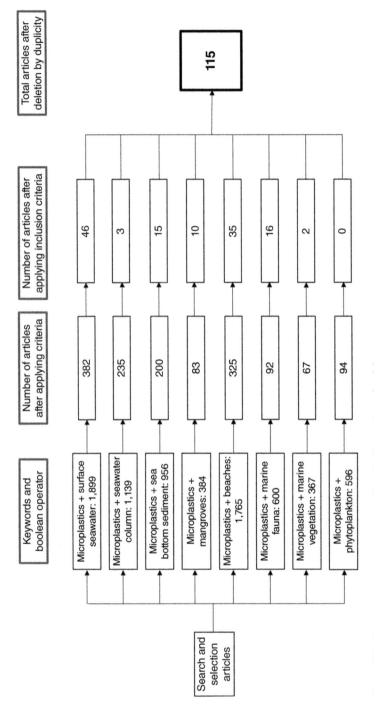

Figure 8.1 Step-by-step process to search and select reviewed articles.

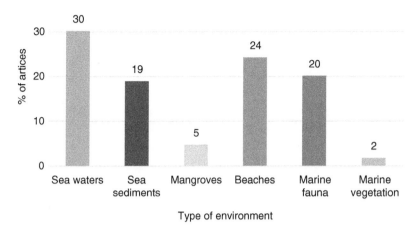

Figure 8.2 Type of studied marine environments for microplastics topic.

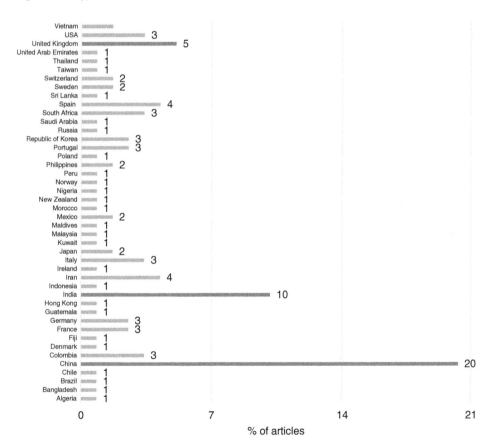

Figure 8.3 Countries to which microplastics studies belong.

India (7000 km) (CIA 2021), and its number of publications is smaller. This might indicate that other factors influence the research done by countries, such as the existing economic support available for this type of studies.

8.3.1 Sampling Techniques in Different Marine Environments

8.3.1.1 Seawater

Within this environment, the most researched compartment is *surface water* (90%), probably due to the technical issues related to the ease of sampling. For instance, to study the water column (7%), different equipment is used for sampling, more samplers are needed, and containers for sample storage must be gathered and stored, which also means more time required.

Number of Sites and Samples The number of sites that are usually selected to study seawater MPs ranges from 1–80 (Jones et al. 2020; Kwon et al. 2020), where 1–10 sites are the most common interval (51%). The location of such sampling sites or stations is causally related to the objective of the study; some of the criteria that have been considered are the presence of fishing or aquaculture activities (12%), proximity to wastewater treatment plants (12%), and tourist activities (7%). These represent potential sources of MPs pollution, so it is interesting to study the levels of MP contamination around them (Narmatha et al. 2020a; Nematollahi et al. 2020; Ory et al. 2020).

As for the number of samples taken per site and the depth at which they are taken, not all the analyzed studies reported them (25 and 56%, respectively), and for those that had reported them, it was observed that the variable depends on the type of water. In surface water, taking one sample per site is the most common practice (48%), and the depths range from 0 to 5 m (Rist et al. 2020; Wang et al. 2020a). However, there is currently no consensus on how deep surface waters must be considered, so deciding replicated samples and depths most likely would depend on the researcher's criteria. In the case of the water column, the number of samples is based on the depth of each sampling site, which in turn determines the depth at which the samples are taken (Jung et al. 2020; von Friesen et al. 2020).

Types of Samplers The samplers used for the study of MPs in seawater are diverse (Figure 8.4). The most common samplers are nets (67% trawl, plankton, neuston, and bongo nets), followed by pumps (13%), steel bucket (7.5%), and bottles (2%). Nets, pumps, and bottles are used in both surface water and water column. Network sampling consists of filtering water through the net as the vessel travels a certain distance, an action which is known as drag; the final part of the network is attached to a collector, collecting bag, or an end, which stores retained solids (Rist et al. 2020; Schönlau et al. 2020). When using this type of sampler, certain parameters are considered, such as sampling time, sample volume filtered by drag, and speed at which the vessel makes the journey. Values range in intervals of 5–60 minutes (Oliveira-Castro et al. 2020; Ramírez-Álvarez et al. 2020), 42.2–790 m^3 (Jones-Williams et al. 2020; Schönlau et al. 2020), and 0.5–5 knots (Lindeque et al. 2020; Mak et al. 2020), respectively.

Regarding the pumps, these can go from a simple pump that sucks water that is then filtered into a steel or net mesh (Jiang et al. 2020b) to integrated systems with a series of

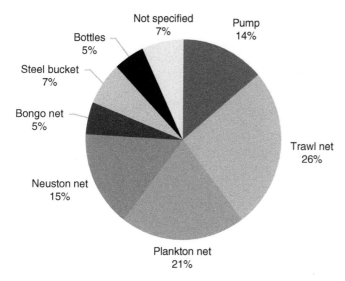

Figure 8.4 Samplers used in the study of microplastics in seawater.

stainless-steel filters that retain the different MP sizes present in the water (Rist et al. 2020; Schönlau et al. 2020). Researchers using this type of sampler reported filtered volumes of water ranging between 0.1 and 28 m³ (Jiang et al. 2020a; Schönlau et al. 2020) and sampling times from 1.5 to 138 minutes (Jung et al. 2020; Schönlau et al. 2020).

Bottles are also used when sampling. Glass containers are used to sample surface waters (Narmatha et al. 2020a) while in the water column, and plastic Niskin bottles are placed in a rosette sampler (Jung et al. 2020; von Friesen et al. 2020) through which the jam and closure of the bottle are activated to take the water sample at a certain depth. A logger measuring conductivity, temperature, and depth (CTD) (VLIZ 2021) can be connected to these systems.

Finally, the use of steel buckets allows researchers to take *surface water* samples manually and not set a speed or sampling duration time for the route (Jung et al. 2020; Md Amin et al. 2020; Ryan et al. 2020). Depending on the technical issues, the water sampled with Niskin or glass bottles and steel buckets, can be subjected to volume reduction through network filtering, or by transferring all the contents into a different container for further laboratory treatment.

Samples obtained with one of the instruments discussed previously are stored for further treatment, and for this purpose different types of containers (plastics, aluminum foil, glass) can be used depending on whether volume reduction is done in the field or not. It is best to make use of non-plastic utensils to avoid possible cross-contamination. So far, the glass containers have been the most used (Figure 8.5).

Sample Management To maintain samples, some researchers add 2.5 or 4% formalin (James et al. 2020; Liu et al. 2020b), 5% formaldehyde (Narmatha et al. 2020b), or 70% ethanol (Littman et al. 2020). These chemical compounds are used to fix zooplankton tissues that may be present in the samples and thus keep them intact for further laboratory analysis

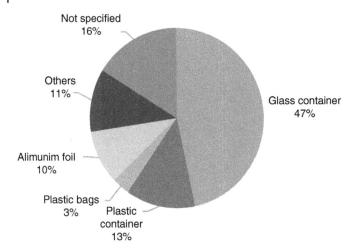

Figure 8.5 Types of used containers for storage of samples.

(Liu et al. 2020a). Additionally, ethanol is used to discolor the organisms present in such a way that MPs are more easily distinguished (Kovač et al. 2016). Among all the chemicals used, ethanol is the most recommended, because the others are highly toxic, their smell is unpleasant, and they are suspected to be carcinogenic (UTAS 2013).

8.3.1.2 Sea Sediments

The relevance of sediment sampling and monitoring relies on the idea that, ultimately, most pollutants might end up sinking into ocean sediments, and given the ubiquity of MPs, it is fair to assume that even deep-sea sediments might be a sinking reservoir for them (Courtene-Jones et al. 2020; Lechthaler et al. 2020; Peng et al. 2020). In this section, we describe sampled sediments in many categories, such as sediments gathered from water columns, sea bed (or floor) sediments, and deep-sea sediments. This discretization is made given that, at depths below 200 m, all samples could be considered as *deep-sea* sources (Lechthaler et al. 2020), and, from the technical point of view, sampling procedures might become more challenging.

There are some common features within all the studies. For instance, when it comes to *periodicity*, most sampling campaigns have been conducted once a year (31%; see Figure 8.6), while others have done only one sampling (27%). Around one-third of the studies (30%) do not state if they have any kind of periodicity, and just one study is part of a recognized major monitoring campaign, conducted in the Azores and the Canary Islands transect between 1999 and 2010 (longitudinal studies) (Reineccius et al. 2020).

Sampling Seasonality, Time, and Site Selection The *sampling time* and *seasonality* also vary between the studies, although both dry and wet seasons are proportionally represented (see Figure 8.7) and most studies are conducted in one season. Few studies (Ferreira et al. 2020; James et al. 2021; Khoironi et al. 2020; Oliveira-Castro et al. 2020; Suresh et al. 2020; Tsang et al. 2020) collected data for both seasons, and just two draw conclusions regarding seasonality and its correlation with MPs presence. In dry seasons there may be less vertical mobility due to slower currents, or an enhanced concentration due to less water coming

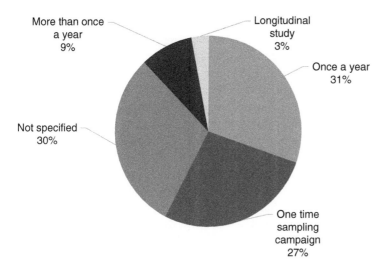

Figure 8.6 Sampling periodicity among reviewed studies. *Source:* Authors' collaboration based on data analysis.

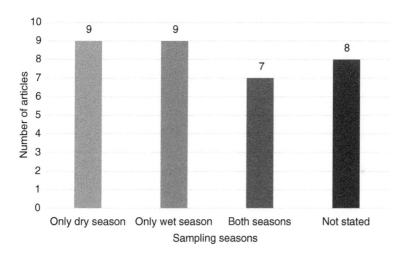

Figure 8.7 Sampling seasonality. *Source:* Authors' elaboration based on data analysis.

into a system (Tsang et al. 2020), while wet seasons may be related with higher MPs concentrations due to a flush out effect present in estuaries and rivers. On the other hand, seasonality may have different impacts in different compartments such as beach sediments, which suggest that wet seasons may just increase concentration due to sediment movement that stores MPs during dry seasons, thus liberating particles just by water flow action over rivers and estuaries (Lima et al. 2015; Mbedzi et al. 2020). Only eight did not specify sample collection date (nor year or season).

Another important factor, especially related to statistical significance, geographical representation and replicability are both *sampled sites* among study areas, as well as *replicates* within each sampled location. In that sense, most studies sampled no more than 10 sites within the study areas and most studies recorded less than five sites (40%), with an average of three replicates per site, although several studies do not state how many replicates they performed. This may not be due to the lack of data, but in some cases, additional information provided either in supplementary materials or specific methodological related articles were referenced, but they fell outside the parameters of this systematic research and thus were not considered.

The selection of *sampling sites* often has multiple reasons, but for most cases the need to analyze the impacts of human activities and pollution stressors is often cited (Cutroneo et al. 2020; Ferreira et al. 2020; Jang et al. 2020; Kazour & Amara 2020; Narmatha et al. 2020a; Nematollahi et al. 2020; Oliveira-Castro et al. 2020; Patterson et al. 2020; Ramírez-Álvarez et al. 2020; Renzi & Blašković 2020; Teng et al. 2020; Tsang et al. 2020; Wang et al. 2020a; Zheng et al. 2020), particularly in bay areas where rivers discharge, or where economic activities take place. Other times, the motivation relies on making pollution diagnoses over specific sites, be they protected or highly dynamic due to human activities (Alomar et al. 2020; Bakir et al. 2020; Lechthaler et al. 2020; Oliveira-Castro et al. 2020; Peng et al. 2020; Renzi et al. 2020). Sometimes the goal is to establish first glances into pollution where no studies have been done.

Types of Samplers *Samplers* were, for the most part, grabs and corers, which account for the majority of the sampling equipment (Figure 8.8). The most preferred tools were grabs, particularly Van Veen or Peterson types, and different types of corers, such as OSIL or box corers. The sampler capacity is directly related to available sample material for replicates, which constitutes a great advantage for large grabs and advance corer samplers, but they also need specific installations mounted onto ships to be operated, as they can weight up to 800 kg (OSIL 2021), which might be a disadvantage (Correia-Prata et al. 2019). Another major complication is that the resulting MPs concentration in sediments is largely dependent on meteorological features, a small disturbances on the sample might skew the results and diminish its representativity (Correia-Prata et al. 2019), although in some cases, the usage of different sampling tool deems no significant differences (Lechthaler et al. 2020).

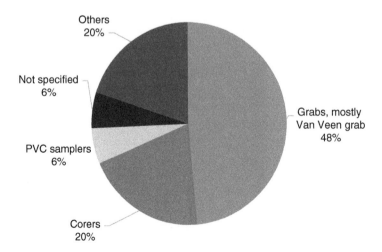

Figure 8.8   Sediment samplers used in reviewed article.

Even though just two studies used PVC corers (Laptenok et al. 2020; Zheng et al. 2020), they are worth mentioning because of the cross-contamination risk and the measures taken to prevent it. This is a commonly addressed issue and thus most studies decided to use either glass or stainless steel material when gathering samples. While Laptenok et al. (2020) did not state specific measures to avoid cross-contamination while sampling, Zheng et al. (2020) discarded the outer layer of the sediment that was in direct contact with the PVC corer. Related to the chosen samplers is the *volume of the sample* itself. Because in most cases samplers vary in size, model, and capacity, most studies did not report the volume collected in each sampling site (82% of studies). Around half of the studies reported *mass*, although in most cases it was the mass used to perform the experiments, and not the collected mass in the sampled sites, usually below 500 g. Just 20% of sampling procedures required more than 500 g to be performed.

Depth of Sampling   The natural features of the study site are important for the result; not only to assess to what extent pollution has reached sediments at the seafloor, but how vertical dynamics play a role in MPs fates and how sediments cause these pollutants to sink. In this sense, although most studies did not report *geographical samplings depths*, around seven studies can be classified as *deep-sea sediments* or *pelagic sediments*, if at least one of the sampled sites described in the articles were collected below the 200 m threshold (Lechthaler et al. 2020; Lyle 2016; Metaxas & Snelgrove 2019). These deep sample studies are shown in Figure 8.9. Those samples that are not deep-sea, may be located between 0 to 40 m depth, as can be seen in Figure 8.10.

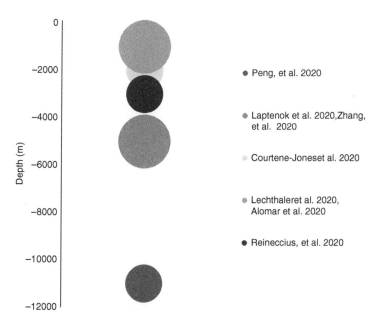

Figure 8.9   Depths of sampling in deep-sea sediments, which are those sampled below 200 m depth. For deep-sea sediments, most studies concentrate above the 5000 m mark, and just one was done at trench-depths of around 11 000 m. The sizes of the circles represent how many articles focused on each sampling depth.

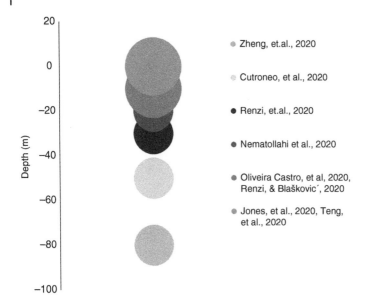

Figure 8.10 Depths of sampling in the seafloor or seabed sediments, which are those sampled above 200 m depth. Most sampled areas are located above the 40 m depth threshold.

Sediment sampling also considers *depth within the sampled cores*. Most studies do not describe the depth within the sediment sample (51%) and when stated, sampling has only been done within a few centimeters from the sea bottom downwards (42%), which stresses the presence of MPs in the seafloor without considering the vertical movement of MPs once they have sunk. Only two studies have done core discretization in soil horizons, and only one found MPs in the samples in amounts that allowed the researchers to correlate depth and concentration. While Lechthaler et al. (2020) only found fibers in two of the 11 sampled sites, and no reported correlation between horizons and concentration, Zheng et al. (2020) found that the concentration and size of MPs linearly decreased with depth. While concentration can be explained by sedimentation rate, fragmentation of plastic particles might help with sea sediments colloids, thus increasing their settling "likelihood."

Sample Management After sampling, storage treatment is also an important part of the procedure. The most common preservation and sampling treatment measurements are temperature control, fixing sediments with saline, dark storage, and material usage to avoid cross-contamination (rubber, aluminum foil, glass, etc.). Around 27% of studies did not report any specific measures to treat samples in the text, while the others reported different steps to preserve samples. For the remaining articles that did specify after-sampling treatment, 67% of the studies described samples either being frozen or stored at 4 °C (most were freezing samples), 13% received special treatment regarding light (dark storage), and just two studies fixed sediments using saline solutions (James et al. 2020; Reineccius et al. 2020). In the instances where storage material was described (38%), most used containers made of PE or HDPE, although in most cases steps to avoid cross-contamination were taken.

8.3.1.3 Beaches

Sand beaches have been the most studied marine ecosystem because they are easy to access, which means sampling economic costs are affordable, and no specialized equipment is required. This section discusses a review of the parameters that are registered in beach sampling on the reviewed articles.

Number and Selection of Beaches The number of beaches in each study is relevant in terms of spatial coverage. Sometimes it is necessary to analyze the presence of MPs in specific areas, such as natural protected areas, so a smaller number of beaches are chosen, while in other cases, it is necessary to determine the presence of MPs on a much larger scale, thus the number of selected sampled beaches increases considerably. In general, the average number of sampled beaches is 12.6, the median and mode are 8; however, the range of sampled sites may vary significantly, from 1 to 48 beaches, as shown in Figure 8.11.

Research related to MPs on beaches has different objectives. Only 24 of the 42 reviewed articles specified what the objectives to study the selected beaches were (see Figure 8.12).

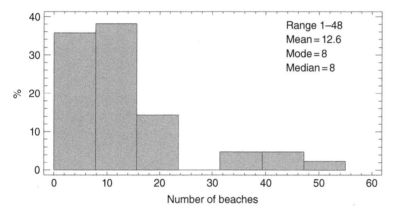

Figure 8.11 Histogram of the number of beaches sampled per study (n = 42).

Figure 8.12 Objectives of the study of MPs on beaches (n = 24).

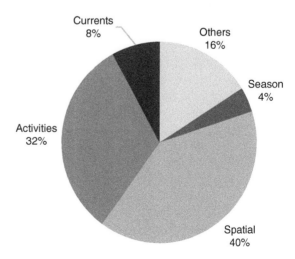

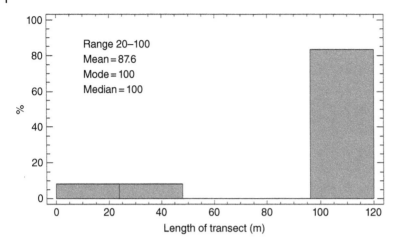

Figure 8.13 Histogram of the length of transect per study (n = 12).

So, the spatial objective was found to be the most common with 40% of mentions, on which the most important was to establish baselines of MPs pollution on the beaches studied. Subsequently, 32% presented the objective associated with the known relationships between MPs and anthropogenic activities, such as tourism or fishing, among others, to identify the origin of the MPs present on the beaches.

The selected transects length (Figure 8.13) on the beaches was another identified parameter in the reviewed articles; however, only 28.6% mention this criterion. Among those that specified lengths, transects range from a minimum value of 20 m to a maximum of 100 m, the latter being the most common.

Periodicity of Sampling The periodicity of sampling varies and is highly dependent on the purpose of the research. Regularly, when more than one sampling per year is made, it is aimed at assessing the presence of MPs in different seasons, which can help to establish climatic factors that would influence both the MPs' behavior and presence in the environment. The sampling interval ranges from one to five per year, with once a year being the most common value (see Figure 8.14).

Number and Selection of Sampling Points The beach zones selected for sampling are important because the tide dynamics may play an important role within MPs occurrence; thus areas where MPs are collected might be divided as "low tide," "intertidal zone," "high tide," and "backshore." Each focused zone shows information on different MPs concentration drivers; while choosing "backshore" might help to analyze the accumulation of MPs derived from high-energy events (tropical storms, hurricanes, among others) or poor waste management on the beach (Hosseini et al. 2020; Kor et al. 2020), "high tide" provides information related to the residues deposited by tides in a shorter time interval. In this, the latter area is the most studied (60%) (Bissen & Chawchai 2020; Patchaiyappan et al. 2020), as can be seen in Figure 8.15.

Regarding the number of samples taken on the beaches, the average value is 8.3, with a mode value of 3 and a median of 5. Some studies only take two samples per beach, and in

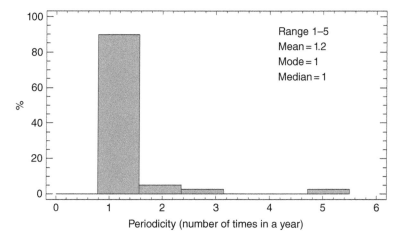

Figure 8.14 Histogram of the periodicity of studies (n = 42).

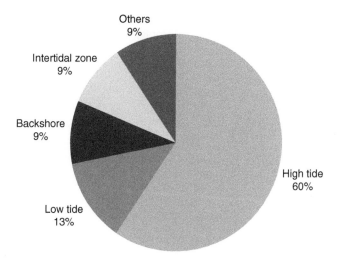

Figure 8.15 Distribution of sampling areas on beaches (n = 33).

one instance, a study reported 44 samples (Godoy et al. 2020). Number of sample's occurrence can be seen in Figure 8.16.

Samplers and Tools The sampler configuration (material, form, and dimension) is one of the most important parameters to consider (see Figure 8.17). Of the 42 articles consulted, only 33% of the studies described the sampler material, with metal being the most common, present in 57% of the studies, followed by wood, with 29%, and finally plastic, with 14%. It is important to mention that when using plastic samplers, it should be ensured that the sampler is in good condition to avoid cross-contamination, as well as to take a reference sample that may account for sampler material presence. The most common sampler is a square frame (78%), followed by a circle (14%). The form is relevant because when using a

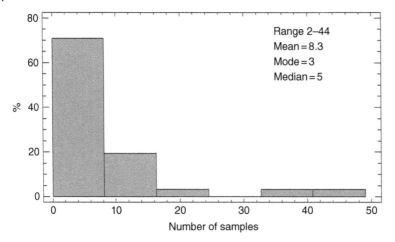

Figure 8.16 Histogram of the number of samples (n = 31).

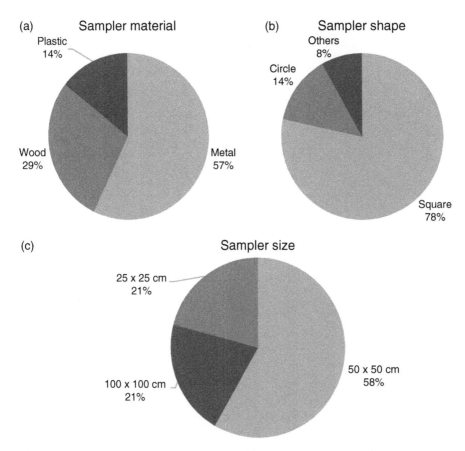

Figure 8.17 Distribution of (a) sampler material (n = 14), (b), sampler shape (n = 37), and (c) sampler size, only for quadrants (n = 36).

square frame, it is regularly mentioned that the samples are collected superficially. However, this is related to the subjective appreciation of the person collecting the sample, so the mass and depth of sand is not always the same, which causes a bias when the results are reported in terms of mass or volume.

On the other hand, the depth at which the samples were taken also is an important parameter to assess, given it might shed light on MPs dynamics and vertical transport within beach sediments (see Figure 8.18). The average value is 4.9 cm with a mode and median of 5 cm. Sometimes, the research aims to establish a temporary trend of MPs deposition, so samples are taken at different depths.

The sampling in all cases was performed with a metal spatula to avoid cross-contamination of the samples. Finally, the most used materials for sample storage are aluminum foil and glass, as it can be seen in Figure 8.19.

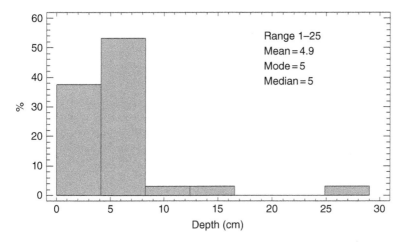

Figure 8.18 Histogram of the depth at which the samples are taken (n = 37).

Figure 8.19 Distribution of storage materials (n = 26).

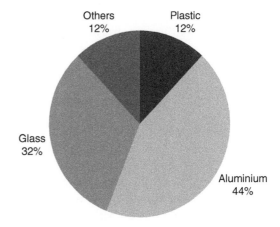

8.3.1.4 Mangroves

According to the results, only six articles related to the presence of MPs in mangrove sediments. However, interest in them grew after their biomass was proven to act as a barrier and a buffer area for marine waste retention, including macroplastics and MPs (Duan et al. 2020; Li et al. 2020a).

Selection of Areas and Sites The main selection criterion for researched areas was to diagnose the occurrence and disposal of plastic waste in mangroves. The selection criteria are related to the number of sampled sites. For example, in mangroves with 20 to 27 sampling sites (50%), the objectives were focused on obtaining information to allow the creation of conservation policies, analyze the effect of reforestation on MP presence, and identify sources of MPs consumption in marine fauna, such as snails (*Ellobium chinense*) (Deng et al. 2020; Li et al. 2020b; Zhou et al. 2020).

In the remaining 50% of researched areas, samples were obtained from three different sites within the mangrove area (Duan et al. 2020; Li et al. 2020a; Zuo et al. 2020). For most cases, replicates per site (83.3%) vary between 1–7, with an average of triplicate replicas. In one of the investigations alone, replicas were higher (13, 14, and 21) (Deng et al. 2020; Duan et al. 2020; Li et al. 2020a; Li et al. 2020b; Zhou et al. 2020; Zuo et al. 2020).

Sampling Depth All samples collected corresponded to surface sediments. The vast majority (83.3%) of mangrove sediments were collected at depths of between 0 and 5 cm, except for a set of samples whose collection depth was between 0 and 10 cm (Deng et al. 2020; Duan et al. 2020; Li et al. 2020a, b; Zhou et al. 2020; Zuo et al. 2020). Only one-sixth of the studies used composite samples (Duan et al. 2020).

Sampler Types and Sample Management Five of the six studies did not specify the type of sampler used, and just one described the dimensions of the quadrant (30 × 30 cm). Most of the articles described the materials used during the study; stainless steel shovels were mainly used to collect samples (83.3%, 5 out of 6), while in one study, the collection utensil used was not specified (Deng et al. 2020; Duan et al. 2020; Li et al. 2020a,b; Zhou et al. 2020; Zuo et al. 2020).

In approximately one-third of the investigations, storage and sample handling from field to laboratory facilities were not described. In the remaining 66%, aluminum containers and sampling bags were used (no bag material specified) (Deng et al. 2020; Li et al. 2020b; Zhou et al. 2020). Finally, the occurrence of MPs was also determined for adjacent environmental compartments, besides mangrove areas, such as beaches (Zhou et al. 2020), tidal and pore water, and marine fauna (Li et al. 2020b). Additionally, in one of the researches, physico-chemical parameters were also determined in mangrove vegetation (Li et al. 2020b).

8.3.1.5 Marine Fauna

In recent years, it has been demonstrated that marine organisms ingest MPs (Cole et al. 2013; Garcés-Ordóñez et al. 2020c; Rodriguez-Seijo et al. 2017; Rotjan et al. 2019), and in some cases their accumulation and transport through the food chain has been observed (Elizalde-Velázquez et al. 2020). Monitoring plastic pollution through analyzing MPs' presence in organisms may be a viable option. In this section, relevant aspects during the

Figure 8.20 Groups of sampled organisms in the articles reviewed.

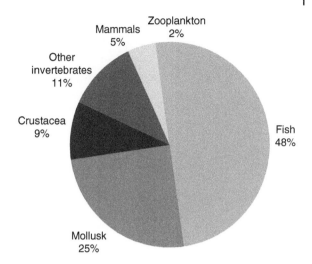

Zooplankton 2%

Mammals 5%

Other invertebrates 11%

Crustacea 9%

Fish 48%

Mollusk 25%

sampling of aquatic organisms for the search of MP will be addressed, as well as the most frequent steps performed for the extraction and processing of the targeted particles.

Selection of Species MPs sampling has been performed in animals of different trophic levels, from zooplankton (James et al. 2020) to marine mammals (Novillo et al. 2020). Of the reviewed articles, almost half (48%) sampled fishes, 25% mollusks (bivalves and snails), 9% crustaceans, and 11% other invertebrates (annelids, polychaetes, holothurians, and echinoderms; Figure 8.20). In general, fishes and mollusks are among the most studied groups, while mammals represent only 5% of the studies (Figure 8.20).

The variety of the focused species is diverse, despite the fact of being in the same group of organisms. For example, in a study carried out with fish, the number of recorded species was 32 (Huang et al. 2020a), while Bosshart et al. (2020) only worked with one species. The fact of including more species in the study implies a greater effort at the processing time of the samples. Nevertheless, having more species included in a study, even though more time-consuming, may shed light and serve as a source of useful comparisons and data that can describe the current state of a group of organisms in the food chain.

Number of Organisms The number of sampled organisms is also variable, and mainly dependent on the type of organism. For example, when sampling fish, the range varied from 37 to 653 specimens, while for mollusks it was 8–90, and for crustaceans 14–600 specimens. When sampling MPs in organisms, increasing the number of specimens sampled is useful for statistical analysis on the concentrations of MPs, since it increases the confidence interval of the results.

Sample Management When organisms are sampled for MPs, the advantage is that the organism itself serves as storage for gastric and intestinal contents, avoiding possible cross-contamination when dealing with sampling; unlike when doing it for water or sediments. Nevertheless, it is important to consider the type of material in which the sampled organisms are stored. As can be seen in Figure 8.21, in the reviewed articles the most used

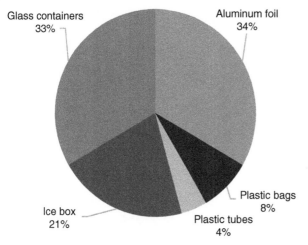

Figure 8.21 Storing containers used during organism sampling.

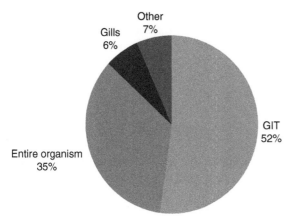

Figure 8.22 Organism section analyzed in the articles reviewed. (Note: GIT = Gastrointestinal tract).

materials for storage were aluminum and glass. The choice of non-plastic materials avoids possible cross-contamination. Despite this, some studies still reported plastic bags or tubes as the choice for storage materials of organisms collected in the field (Maaghloud et al. 2020; Naidoo et al. 2020).

Although plastic materials were used in some of the studies reviewed, other measures to prevent environmental contamination were taken, such as the use of blanks during analyses, the use of cotton clothing and nitrile gloves, or strict cleanliness (Alomar et al. 2020; Bosshart et al. 2020; Naidoo et al. 2020; Novillo et al. 2020).

Selection of Tissues, Organs, and Systems Concerning the tissue to be analyzed, the feeding strategy and the anatomy of each organism to be studied must be considered. Most of the collected and studied organisms were predators, so the gastrointestinal tract (GIT) was analyzed in 52% of the cases, which is a common practice when fish are studied (Bosshart et al. 2020; James et al. 2020, 2021).

The entire organism was analyzed in 35% of cases (Figure 8.22), discarding hard body parts such as shells or bones. This analysis occurred mostly in mollusks, as they possess soft tissue, and only in some fish (Naidoo et al. 2020). The study of different sections and organs may be relevant, depending on the objective of the study. For example, if it is desired to observe translocation of MPs, it would be convenient to analyze different tissues separately. However, if what is desired is to analyze the MPs presence within an organism, it is desirable to analyze the entire body.

8.3.1.6 Marine Vegetation

As with mangroves, the density of marine vegetation favors the accumulation of plastic waste of different sizes, even with a greater proportion than areas without vegetation. The presence of plastic waste in these ecosystems represents a risk to the food chain and the habitat of the marine species (Cozzolino et al. 2020; Huang et al. 2020b).

When it comes to sampling and analyzing MP in vegetation, it requires techniques focused on the biological characteristics of plant organisms. Studies in search of MP in vegetation are very scarce as we found from the literature search conducted; four articles were found, however, three of them sampled sediments in seagrass (Cozzolino et al. 2020; Huang et al. 2020b; Plee & Pomory 2020) and only one of them analyzed vegetation for MP (Jones et al. 2020).

The study of MP present in the vegetation analyzed blades of seagrass beds of the *Zostera marina*. The research was carried out only once a year. Ten sampling sites were chosen with six samples per site and the collected samples were stored in zipper-close bags (Jones et al. 2020).

8.3.2 Sample Processing in Laboratory

After field campaigns and sample collection, the samples undergo processing through various stages, where the main goal is to separate the existing MPs and then subject those to visual inspection and chemical characterization. The following steps and their sequence may vary depending on the type of sample and its characteristics, the main objectives of the study, the methodology followed or customized by the researchers, as well as the materials, equipment, and available reagents. In general, the steps of these tests are as follows (see Figure 8.23):

In the next section, we describe the most common steps for each type of sample (water, sediments, marine fauna, and flora).

8.3.2.1 Water

This section describes the processing of all water samples collected in the open sea, mangroves, and estuaries, as they share similar processing methods; even with sediments, vegetation, and marine fauna, processing steps are like those followed when analyzing water samples. The extraction and identification of MPs in the collected samples consist of

Figure 8.23 General procedure for sample processing in Laboratory to extract MPs.

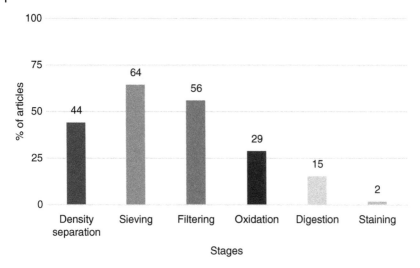

Figure 8.24 Percent of articles that include each step in the treatment of the water samples.

several stages with sieving and filtering as the most common steps (Figure 8.24); the order, application, and materials used are based on the main objectives of the study, the methodology followed or modified by the researchers, and the resources available to them in their respective areas of study.

Sieving   *Sieving* is probably the first step for most cases, all extraction of MPs in water samples is usually done in a laboratory; however, in some research (25%), and to reduce the volume sample to be transported, sieving is done during fieldwork. For water samples, most sieving meshes have diameters of <0.5 mm (43%). While in most cases only one mesh size is used, in about one-third of the articles (31%), a set of sieves with different diameter sizes is used to retain MPs of different sizes. Within these, the most used diameter ranges from 0.5–5 mm (Jiang et al. 2020b; Kor & Mehdinia 2020; Kwon et al. 2020; Sui et al. 2020).

Density Separation   Following sieving, *density separation* or *density test*, is applied to differentiate MPs from organic matter or other present particles. Some solutions (Figure 8.25) such as sodium chloride (NaCl), zinc chloride (ZnCl$_2$), and sodium iodide (NaI) at different densities (between 1.2–1.6 g/cm^3) are mainly used for this test. Some others have been used in a smaller proportion, such as Lithium metatungstate, salted water, canola oil, and sodium Polytungstate (Jiang et al. 2020b; Kor & Mehdinia 2020; Kwon et al. 2020; Sui et al. 2020). This test usually lasts between 12 and 24 hours (Goswami et al. 2020; Jiang et al. 2020b; Kazour & Amara 2020; Kwon et al. 2020; Narmatha et al. 2020a; Patterson et al. 2020).

Filtration   After density testing, *vacuum filtration* allows the removal of the entire liquid interface; be that the solutions used in the density test or some other liquid that had been used during the extraction of the MPs. Fiberglass is the most used type of filter paper (30%), followed by nitrocellulose (8%), polycarbonate paper (8%), or a nylon net, coupled with

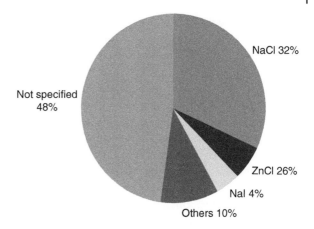

Figure 8.25 Distribution of the solutions used in density tests.

stainless-steel filters (8%) (Goswami et al. 2020; Jiang et al. 2020b; Kazour & Amara 2020; Kwon et al. 2020; Narmatha et al. 2020a; Patterson et al. 2020). Approximately 46% of articles do not specify the type of filter paper used. In addition to the type of filter, an important factor for this procedure is the pore size; the most used range being from 0.45–0.7 µm (20%). Some other sizes used are 10, 20, and 200 µm (Jang et al. 2020; Jones-Williams et al. 2020; Rist et al. 2020).

To complete this step, the filters are dried to remove any residual moisture, usually, by storing them in containers like Petri glass boxes (15%) and then leave them to dry either at room temperature (8%) or in the oven at 50 to 60 °C (7%), usually for about 24 hours (7%) (Jang et al. 2020; Jones-Williams et al. 2020; Rist et al. 2020).

Remotion of Organic Matter After filtration, some water samples may be subjected to an *oxidation step* which aims to remove any organic matter residues still present, and that may compromise MPs identification. Different chemical reagents are usually applied, such as hydrogen peroxide (H_2O_2) in various concentrations (<30%, 30%, and >30%), or H_2SO_2 at 30%, which is the most used solution (13%); the least common reagent is an iron solution at 0.05 M (9%) (Athapaththu et al. 2020; Ferreira et al. 2020).

Other parameters that are considered during oxidation are the time and temperature at which the process is carried out. The general range of these parameters are a temperature between 70 and 90 °C (Athapaththu et al. 2020; Ferreira et al. 2020; Kwon et al. 2020; Vilakati et al. 2020), and a time of 24 hours (Jiang et al. 2020b; Lechthaler et al. 2020; Liu et al. 2020b; Sui et al. 2020; Teng et al. 2020). Lower in proportion, temperatures between 50 and 70 °C (Jung et al. 2020) and times longer than 24 hours are usually reported (Narmatha et al. 2020a; Patterson et al. 2020).

In addition to oxidation, *digestion* might be part of the sample processing. In water samples (although unconventional), a digestion step may also be carried out aimed at disintegrating any remaining traces of other marine species such as zooplankton. However, this type of test has greater relevance and application when measuring and characterizing MPs in marine fauna (Frias et al. 2020; Wang et al. 2020a).

The reagents generally used are protease and lipase, potassium hydroxide (KOH) with a concentration of <30%, and hydrogen peroxide (H_2O_2) with a concentration of 30%. Unlike oxidation, the most used temperatures in this test range between 30–50 °C (8%); less time is also set (less than 24 hours, 5%) (Frias et al. 2020; Jiang et al. 2020a; Jones-Williams et al. 2020; Lindeque et al. 2020; Rist et al. 2020; Wang et al. 2020a; Zhang et al. 2020).

More than 80% of the research analyzed in this section does not specify the temperature and time at which oxidizing and digestion were carried out, and in approximately 66% of the studies, the time used in the separation by density is not specified either. This information in tests involving chemical reactions indicates the rate at which these reactions are carried out and the type of products that can be obtained from them. In physical tests such as density separation, the agitation and sedimentation time are important factors for the flotation of the MPs on the surface (Jung et al. 2020; Narmatha et al. 2020a; Patterson et al. 2020).

To ease identification, in some research dye *staining* is used to identify microfibers present in water samples. According to the results obtained, in only one of 59 researched found mentions of the use of Nile Red (10 mg/ml) as a dye in the staining process (Tošić et al. 2020).

Polymer Identification   Once all previous steps have been completed, *polymer identification* follows. The extracted particles and potential MPs are preliminarily identified to determine whether they are polymer particles or other material. Some of the most used types of equipment are stereo microscope (59%) and microscope (33%) (optical, dissection, binocular, trinocular); both are used with different magnifications to assess MPs presence in the dry samples, ranging from 40X to 120X.

After pre-identification, the chemical composition of MPs subsamples is analyzed. To perform this test, more sophisticated equipment is used, which can provide detection accuracies up to 98% (Jiang et al. 2020b; Jung et al. 2020). Fourier transform infrared (FTIR) analysis is the most widely used (53%). Other analyses used are μ-FTIR (16%), Raman (6%), and μ-Raman (8%).

8.3.2.2   Sediments

The types of sediments covered in this section include seabed, deep-sea, mangroves, beaches, and seagrass areas. All sediments were grouped given they share similar laboratory procedures. The stages used to process the samples are the same for most types of samples (see Figure 8.23, Section 8.3.2). From the following steps (density separation, filtering, drying, sieving, digestion, and staining), the first three are the most used (Figure 8.26). Processing is usually carried out in the laboratory (95% of the studies), however, there are cases where part is done in the field (sifted to have a volume reduction) and the rest in the laboratory (Garcés-Ordóñez et al. 2020a,b; González-Hernández et al. 2020).

The selection of some of the stages and the order in which they are performed varies between the different studies. For example, Patchaiyappan et al. (2020) contemplate all stages to treat beach sediments in the following order; drying → sieving → digestion → density separation → filtering → staining. On the other hand, Lechthaler et al. (2020) used five steps for seabed sediments (drying, sieving, density separation, filtering, digestion, filtering), but each step was performed twice. There are also some articles where only two steps were carried out, such as density separation and filtering (Rapp et al. 2020; Tsang et al. 2020), density separation and sieving (Robin et al. 2020), or sieving and drying

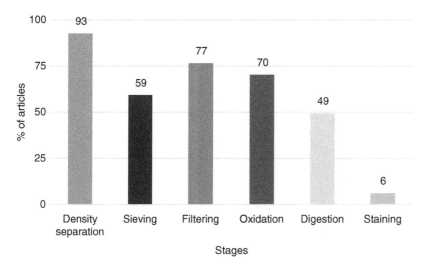

Figure 8.26 Percent of articles that include each step in the treatment of the sediment samples.

(Ramírez-Álvarez et al. 2020). Finally, sometimes only one step is conducted for MP isolation and analysis, be they sieving (Katsumi et al. 2020) or drying (Laptenok et al. 2020).

Sieving Regarding *sieving*, this step is usually performed before drying, either in the laboratory or in the field (González-Hernández et al. 2020; Kor et al. 2020), after drying (Mazariegos-Ortíz et al. 2020), or after density separation (Portz et al. 2020). The size of mesh used is between 0.0063 and 5 mm (Deng et al. 2020; Narmatha et al. 2020a), and the three most used mesh sizes are <0.5 mm (17%), 5 mm (16%), and 1–1.99 mm (15%). The mesh size determines the size of the studied MPs. Some articles use up to five meshes of different sizes, thus MPs of different sizes are obtained and analyzed (Deng et al. 2020; Saeed et al. 2020).

Drying *Drying* is usually the first step, except when a reduction in volume is done in the field by sieving (González-Hernández et al. 2020). The drying of the sample allows having dry-dough data and expressing concentrations on comparable units of MPs. In the analyzed studies, the drying temperature varies from room temperature (James et al. 2021) up to 105 °C (Zhou et al. 2020), with the most common ranging from 41–60 °C (37%). Drying time ranged from 12 hours (Zhou et al. 2020) to four weeks (Lo et al. 2020), of which the interval 24–48 hours (23%) is the most common. Both parameters can depend on how wet this sample is. Drying time also has a close relationship with the selected temperature; for example, drying samples at room temperature takes at least four weeks (Lo et al. 2020), while drying at 105 °C requires 12 hours (Zhou et al. 2020).

Density Separation For the *density separation* step, there are different solutions used to isolate MPs from sediments by differences in density (Table 8.2), but the most used is NaCl (50% of studies) because it is economical, easy to obtain, and environmentally friendly (Cutroneo et al. 2021). However, its disadvantage is that because of its low density (1.2 g/cm^3), the extraction of MPs from polymers denser than the solution cannot be

Table 8.2 Used solutions for density separation of microplastics from sediments samples.

Solution	Saturated density (g/cm^3)	References
Zinc chloride (ZnCl$_2$)	1.8	Kazour & Amara (2020)
Sodium iodide (NaI)	1.8	Peng et al. (2020)
Calcium chloride (CaCl$_2$)	1.6	Alvarez-Zeferino et al. (2020)
Potassium iodide (KI)	1.62	Aslam et al. (2020)
Sodium chloride (NaCl)	1.2	Jang et al. (2020)
Lithium metatungstate (Li$_2$O$_{13}$W$_4^{-24}$)	1.6	Jang et al. (2020)
Potassium formate (CHKO$_2$)	1.5	Li et al. (2020b)
Cesium chloride (CsCl)	1.8	Dodson et al. (2020)
Canola oil	Not specific	Lechthaler et al. (2020)

Source: Authors' collaboration.

guaranteed, for instance, when studying polyethylene terephthalate (PET, 1.37–1.45 g/cm^3), polyvinylchloride (PVC, 1.16–1.58 g/cm^3), polyester (1.24–2.3 g/cm^3), to name a few (Hidalgo-Ruz et al. 2012). Thus, some other solutions (Table 8.2), whose density is ≥ 1.5 g/cm^3 are also used. The main disadvantage is that they are more expensive than NaCl and some (ZnCl2, NaI, KI) are considered dangerous to human health and the aquatic environment (Cutroneo et al. 2021).

The volume of the solutions used for density separation is hardly reported, as it depends on the number of sediments to be analyzed, but it is recommended to be a reported parameter (Jeyasanta et al. 2020). A good suggestion regarding the amount of solution might be to use twice the volume of sediments, to allow MPs to float (Prata et al. 2020). Two other parameters that are considered when making density separation are agitation and rest times. For both cases, the reported values are diverse, ranging from 20 seconds (Reineccius et al. 2020) to 60 minutes (Prata et al. 2020), and from one minute (Alvarez-Zeferino et al. 2020) to 48 hours (Zheng et al. 2020). Besley et al. (2017) suggested that solution resting times should be at least five hours, so that the sand particles in the over-natant sink, thus ensuring that the MPs are completely separated from the sand.

Digestion As stated before, *digestion* allows removal of organic matter present in the sediments. Some researchers digest the samples either before or after *density separation* (Vetrimurugan et al. 2020; Wang et al. 2020a). However, 47% of the analyzed articles do not use this step. This may be because the number of organic residues in sediments is almost zero (Correia-Prata et al. 2019). From the studies that do digest samples, 42% used hydrogen peroxide (H$_2$O$_2$) as oxidizer at 10–35%, with H$_2$O$_2$ to 30% being more common (32% of studies), which is sometimes combined with Fe (II) as recommended by NOAA (National Oceanic and Atmospheric Administration) (Masura et al. 2015).

Digestion temperature when using H$_2$O$_2$ varies from room temperature (Rahman et al. 2020) to 100 °C (Duan et al. 2020), and the time ranges widely, between 30 minutes

(Patchaiyappan et al. 2020) and 10 days (Nematollahi et al. 2020). However, there are studies that handle this stage with two different temperatures for each repetition (Bridson et al. 2020; Duan et al. 2020).

Filtration After separation density or after digestion, a filtration step is performed to separate the solution from MP. The procedure may or may not include vacuum (Lechthaler et al. 2020; Vetrimurugan et al. 2020; Wang et al. 2020a). The pore size of the membrane used determines the smallest size of the MPs that can be studied.

In sediments, 75% of studies perform this step, the most used membranes are glass fibers and nitrocellulose (22 and 21%, respectively), and pore size ranges from 0.2–20 m (Bakir et al. 2020; Jang et al. 2020). Finally, membranes for drying after filtration are placed in Petri dishes (25% glass, 1% plastic, and 9% unknown material) or aluminum foil (1%) at a temperature ranging from room temperature (Narmatha et al. 2020a) to 60 °C (Duan et al. 2020), and a time between 30 minutes (Li et al. 2020b) to 48 hours (Deng et al. 2020).

The drying of the membrane or filter is usually the last stage if *staining* is not performed. When used, staining consists of placing a certain amount of dye in the filter and letting it react for a certain time, so that the present MPs are stained and easier to see during identification (Nel et al. 2020). If there is excess dye, the filter is rinsed with distilled water and then dried (Godoy et al. 2020). The most commonly used dye is Nile Red (5%) at concentrations of 5–1000 g/ml, and the reaction time varies from 5 minutes to 24 hours (Godoy et al. 2020; Nel et al. 2020). This technique is useful for analyzing small MPs because they are not visible to the naked eye. Therefore, staining is recommended when working with MPs sizes <1 mm (Erni-Cassola et al. 2017). However, staining is rarely carried out because both dye and inspection equipment are expensive.

Once the samples have gone through the last stage (*filtration* or *staining*) and the MPs have been removed, they are *inspected* to verify that they are plastic particles. The equipment used to do so is stereo microscope (79%), followed by optical microscope (11%), fluorescent microscope (2.7%), and UV spectrometer (1.3%). The latter two are used when particles have been dyed (Godoy et al. 2020; Patchaiyappan et al. 2020).

Identification of Polymers Subsequently, particles that may be MPs undergo a *chemical characterization* to ascertain the type of polymer. If they are not MPs, this technique helps to discard particles and offers a high degree of confidence when compared to the visual inspection technique. Most common identification tools to evaluate MP polymer are FTIR spectroscopy (coupled to a total reflectance attenuated equipment) (56%), Raman spectroscopy (11%), μ-FTIR (12%), and μ-Raman (6%); all of these are non-destructive techniques and require small amounts of samples (da Silva et al. 2018). The first two (FTIR and Raman) are used to analyze large MPs (>1 mm), while the latter are for smaller MPs (<1 mm) (Lo et al. 2020).

8.3.2.3 Marine Fauna
In this section, the techniques used to extract MPs from animals are described. Various steps were taken for the treatment of MPs. In general, digestion and filtration are the most applied procedures, whereas staining and sieving are less common (Figure 8.27).

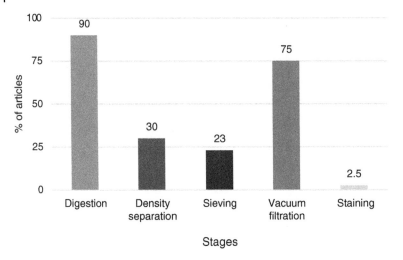

Figure 8.27 Percent of articles that include each step in the treatment of the marine fauna samples.

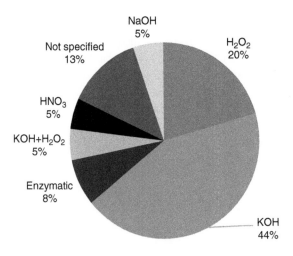

Figure 8.28 Digestion solutions applied in the studies reviewed.

Digestion To analyze MPs content, the first step is often to extract them, which may involve a digestion process, either chemical or enzymatic, where organic matter can be digested leaving recalcitrant materials with little damage. Tissue digestion can be performed with strong oxidizing acids such as HNO_3, or strong bases such as KOH. However, many of these can cause some damage to the polymers. When deciding which reagent to use, it is important to consider various aspects, such as the damage the substances may cause to the plastics, digestion time, temperature, budget, and safety of reagent handling (von Friesen et al. 2020).

In the reviewed studies, KOH was the most used substance (42%) to digest organic matter, followed by H_2O_2 at 20% (Figure 8.28). Although KOH was the most used reagent, it has been shown that it can damage some polymers, such as rayon, which is destroyed at 60 °C but not at 40 °C (Thiele et al. 2019). It has also been found that digestion with HNO_3, despite being quite effective with >98% loss of biological tissue (Claessens et al. 2013), may affect HDPE and PET (Thiele et al. 2019). On the other hand, enzymatic digestion represents a less aggressive option for MPs (von Friesen et al. 2020) but it is much more expensive (Thiele et al. 2019), laborious, and requires a longer digestion time (von Friesen et al. 2020), which might be the reason why this technique, despite its efficiency, was only used in 10% of the reviewed articles.

Management after Digestion   After performing a preliminary extraction of MPs, some researchers did separation by density in 28% of the cases. Within these, the most used reagent was NaCl (82%). Whether the authors applied the two steps or digestion alone, the next most-recurrent step was vacuum filtration; in fact, 75% of the studies performed this step. When vacuum filtration was employed, the most used filter was glass fiber with 28%, followed by cellulose and nitrocellulose (Figure 8.29). The pore sizes employed were <0.45 µm, 0.7–1.5 µm, ≥5 µm, in similar proportions (23, 28, and 23%, respectively).

After filtering, about half of the studies (45%) performed some filter drying process, either at room temperature or in an oven, between 30 and 70 °C. Following this step, a visual inspection was usually performed using a microscope or stereoscope to preliminarily assess the extracted particles; which was done in 98% of the cases. In this step, the use of a hot needle can be useful to corroborate the plastic nature of the extracted particles. This simple method, also called "hot needle test" or "hot point test," has been used in some studies with success (De Witte et al. 2014; Devriese et al. 2015; Vandermeersch et al. 2015) but in the studies reviewed, this method wasn't applied.

Polymer Identification   Once the plastic particles have been extracted, the next step is to know their chemical composition. For this purpose, different techniques and their variants are used, mainly Fourier Transformation Infrared Transmission spectroscopy (FTIR) and Raman spectroscopy. In the reviewed articles, 68% used the FTIR technique or one of its variants (µ-FTIR or ATR-FTIR), while 17% used Raman spectroscopy, and the rest did not mention having used any technique or they did not chemically identify the particles found in the study (see Figure 8.30).

Figure 8.29   Filter material chosen in the studies reviewed.

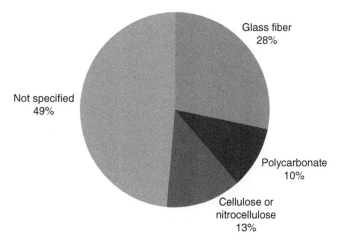

Glass fiber 28%

Not specified 49%

Polycarbonate 10%

Cellulose or nitrocellulose 13%

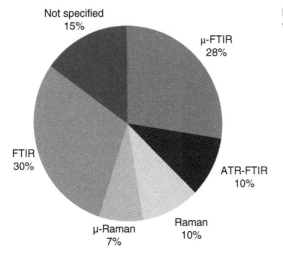

Figure 8.30 MP identification technique.

8.3.2.4 Marine Vegetation

For this section, only the procedures of the study that sampled vegetation (indicated in Section 8.3.1.6) will be discussed (Jones et al. 2020).

To extract the particles on the seagrass blade, a wash was performed with distilled water, followed by filtration on a 0.7 µm pore size filter. The filter was dried overnight in Petri dishes. Once the particles were extracted, an inspection was performed under a stereomicroscope. Subsequently, the ATR-FTIR technique was chosen for the analysis of possible MPs (Jones et al. 2020).

8.4 Conclusions and Future Perspectives

The results of this review show the existence of a wide variety of sampling and analysis methods for MP. The selection of techniques depends on the specific research's objectives, equipment availability, and the methods employed. Other relevant factors defining the characteristics of the selected method is the use of non-plastic equipment (stainless steel, glass), in order to avoid cross-contamination.

Water is the most studied component used in marine environments, mainly by sampling surface water, which requires less time, materials, and sophisticated sampling equipment. Although this type of sampling is by no doubt useful, other approaches, such as sampling along the water column may provide relevant data about vertical distribution, transport dynamics, and differences in MP distribution, colonization, and degradation. For any type of water sampling, detailed descriptions of the sampling procedure, navigation features, and volume sampled should be included in the articles, as this would improve comparability. A similar situation was found regarding sand beaches; while they have been extensively studied, discrepancies among sampling techniques hinder the possibility of comparison among different results, as well as replicability. Reported results would improve their usefulness by clearly describing all the parameters mentioned in this review, which would allow for making comparisons among results and developing a better understanding of MPs in environmental compartments.

In the sampling of marine sediments, the proportion of studies in deep waters is low, mainly due to the technical challenges involved. However, the relevance of these studies cannot be ignored, as they give direct evidence of the extent of marine pollution by plastics, as sediments are considered as one of their main sinks. As they are not usually affected in the short term by meteorological phenomena or anthropogenic activity, the systematic study of deep-water sediments can be a valuable tool to measure the extent of the presence of MPs in the natural environment.

Sampling in biotic compartments has specific challenges. In the case of fauna, representativity of the results will be highly influenced by the number or organisms analyzed. Other relevant factors may include the analysis of different species in the same area, which may provide information about the movement of MPs along the food chain, or the study of the same species in different seasons or life stages. In all cases, a basic need is the application of ethical codes in the collection and management of simples. The analysis of vegetation, such as mangroves and seagrass, are still scarce, and it is common for sampling of these areas to be focused on the extraction of sediments. The relationship between MPs and marine plants, as well as the systematic research about the presence of marine waste in marine vegetation, are under-studied fields that should receive more attention, in order to identify effects on this species, which usually act as the base of rich marine ecosystems.

On the other hand, processing of samples in the laboratory, while varying, usually include sieving and density separation, in order to reduce volume and extract MPs from the medium. Other steps, such as drying and digestion, must guarantee that the procedures do not damage or alter the MPs. In order to carry this out, preliminary studies should be performed before applying these techniques to the actual samples. When analyzing small MPs (<0.5 mm), staining can be a very useful technique, as it will allow avoiding overestimating the results. The use of specific analytical techniques for these smaller particles, such as micro FTIR or Raman, is highly recommended.

For both sampling and analysis, specific steps should be taken to avoid cross contamination, which may greatly alter the results, especially in low-concentration samples and when studying MPs smaller than 1 mm. These precautions, coupled with a procedure to eliminate organic matter, will prevent the misidentification of false positives, which may alter the results. While standardization of sampling and analytical techniques for MPs in marine environments is still a developing work, complete report of the methods and objective analysis of their advantages and limitations will allow us to understand, analyze, and compare the results of the many studies related to this topic that are published. This information must be also seen as a necessary tool to begin a global discussion that leads to consensus on how this relevant data should be gathered, analyzed, and reported.

References

Alomar, C., Deudero, S., Compa, M., and Guijarro, B. (2020). *Exploring the relation between plastic ingestion in species and its presence in seafloor bottoms. Marine Pollution Bulletin* 160: 111641. https://doi.org/10.1016/j.marpolbul.2020.111641.

Alvarez-Zeferino, J.C., Ojeda-Benítez, S., Cruz-Salas, A.A. et al. (2020). *Microplastics in Mexican beaches. Resources, Conservation and Recycling* 155 https://doi.org/10.1016/j.resconrec.2019.104633.

Aslam, H., Ali, T., Mortula, M.M., and Attaelmanan, A.G. (2020). *Evaluation of microplastics in beach sediments along the coast of Dubai, UAE. Marine Pollution Bulletin* 150: 110739. https://doi.org/10.1016/j.marpolbul.2019.110739.

Athapaththu, A.M.A.I.K., Thushari, G.G.N., Dias, P.C.B. et al. (2020). *Plastics in surface water of southern coastal belt of Sri Lanka (northern Indian Ocean): distribution and characterization by FTIR. Marine Pollution Bulletin* 161: 111750. https://doi.org/10.1016/j.marpolbul.2020.111750.

Bakir, A., Desender, M., Wilkinson, T. et al. (2020). *Occurrence and abundance of meso and microplastics in sediment, surface waters, and marine biota from the South Pacific region. Marine Pollution Bulletin* 160: 111572. https://doi.org/10.1016/j.marpolbul.2020.111572.

Besley, A., Vijver, M.G., Behrens, P., and Bosker, T. (2017). *A standardized method for sampling and extraction methods for quantifying microplastics in beach sand. Marine Pollution Bulletin* 114 (1): 77–83. https://doi.org/10.1016/J.MARPOLBUL.2016.08.055.

Bissen, R. and Chawchai, S. (2020). *Microplastics on beaches along the eastern Gulf of Thailand – a preliminary study. Marine Pollution Bulletin* 157 https://doi.org/10.1016/j.marpolbul.2020.111345.

Bosshart, S., Erni-Cassola, G., and Burkhardt-Holm, P. (2020). *Independence of microplastic ingestion from environmental load in the round goby* (Neogobius melanostomus) *from the Rhine river using high quality standards. Environmental Pollution* 267: 115664. https://doi.org/10.1016/j.envpol.2020.115664.

Bridson, J.H., Patel, M., Lewis, A. et al. (2020). *Microplastic contamination in Auckland (New Zealand) beach sediments. Marine Pollution Bulletin* 151 https://doi.org/10.1016/j.marpolbul.2019.110867.

Carney Almroth, B. and Eggert, H. (2019). *Marine plastic pollution: sources, impacts, and policy issues. Review of Environmental Economics and Policy* 13 (2): 317–326. https://doi.org/10.1093/reep/rez012.

CIA (U.S. Central Intelligency Agency). (2021). *The World Factbook. Field Listing – Coastline.* https://www.cia.gov/the-world-factbook/field/coastline.

Claessens, M., Van Cauwenberghe, L., Vandegehuchte, M.B., and Janssen, C.R. (2013). *New techniques for the detection of microplastics in sediments and field collected organisms. Marine Pollution Bulletin* 70 (1–2): 227–233. https://doi.org/10.1016/j.marpolbul.2013.03.009.

Cole, M., Lindeque, P., Fileman, E. et al. (2013). *Microplastic ingestion by zooplankton. Environmental Science and Technology* 47 (12): 6646–6655. https://doi.org/10.1021/es400663f.

Correia-Prata, J.P., da Costa, J., Duarte, A.C., and Rocha-Santos, T. (2019). *Methods for sampling and detection of microplastics in water and sediment: a critical review. TrAC Trends in Analytical Chemistry* 110: 150–159. https://doi.org/10.1016/J.TRAC.2018.10.029.

Courtene-Jones, W., Quinn, B., Ewins, C. et al. (2020). *Microplastic accumulation in deep-sea sediments from the Rockall Trough. Marine Pollution Bulletin* 154: 111092. https://doi.org/10.1016/j.marpolbul.2020.111092.

Cozzolino, L., Nicastro, K.R., Zardi, G.I., and de los Santos, C.B. (2020). *Species-specific plastic accumulation in the sediment and canopy of coastal vegetated habitats. Science of the Total Environment* 723: 138018. https://doi.org/10.1016/j.scitotenv.2020.138018.

Cutroneo, L., Cincinelli, A., Chelazzi, D. et al. (2020). *Baseline characterisation of microlitter in the sediment of torrents and the sea bottom in the Gulf of Tigullio (NW Italy). Regional Studies in Marine Science* 35: 101119. https://doi.org/10.1016/j.rsma.2020.101119.

Cutroneo, L., Reboa, A., Geneselli, I., and Capello, M. (2021). *Considerations on salts used for density separation in the extraction of microplastics from sediments. Marine Pollution Bulletin* 166: 112216. https://doi.org/10.1016/j.marpolbul.2021.112216.

De Witte, B., Devriese, L., Bekaert, K. et al. (2014). *Quality assessment of the blue mussel* (Mytilus edulis)*: comparison between commercial and wild types. Marine Pollution* 85: 146–155. https://doi.org/10.1016/j.marpolbul.2014.06.006.

Deng, J., Guo, P., Zhang, X. et al. (2020). *Microplastics and accumulated heavy metals in restored mangrove wetland surface sediments at Jinjiang Estuary (Fujian, China). Marine Pollution Bulletin* 159 https://doi.org/10.1016/j.marpolbul.2020.111482.

Devriese, L.I., Van Der Meulen, M.D., Maes, T. et al. (2015). *Microplastic contamination in brown shrimp* (Crangon crangon, Linnaeus 1758) *from coastal waters of the southern North Sea and channel area. Marine Pollution Bulletin* 98: 179–187. https://doi.org/10.1016/j. marpolbul.2015.06.051.

Dodson, G.Z., Shotorban, A.K., Hatcher, P.G. et al. (2020). *Microplastic fragment and fiber contamination of beach sediments from selected sites in Virginia and North Carolina, USA. Marine Pollution Bulletin* 151 https://doi.org/10.1016/j.marpolbul.2019.110869.

Duan, J., Han, J., Zhou, H. et al. (2020). *Development of a digestion method for determining microplastic pollution in vegetal-rich clayey mangrove sediments. Science of the Total Environment* 707: 136030. https://doi.org/10.1016/j.scitotenv.2019.136030.

Elizalde-Velázquez, A., Carcano, A.M., Crago, J. et al. (2020). *Translocation, trophic transfer, accumulation and depuration of polystyrene microplastics in* Daphnia magna *and* Pimephales promelas. *Environmental Pollution* 259 https://doi.org/10.1016/j.envpol.2020.113937.

Enyoh, C.E., Verla, A.W., Verla, E.N. et al. (2019). *Airborne microplastics: a review study on method for analysis, occurrence, movement and risks. Environmental Monitoring and Assessment* 191 (11): 1–17. https://doi.org/10.1007/s10661-019-7842-0.

Erni-Cassola, G., Gibson, M.I., Thompson, R.C., and Christie-Oleza, J.A. (2017). *Lost, but found with Nile Red: a novel method for detecting and quantifying small microplastics (1 mm to 20 μm) in environmental samples. Environmental Science & Technology* 51 (23): 13641–13648. https://doi.org/10.1021/acs.est.7b04512.

Ferreira, M., Thompson, J., Paris, A. et al. (2020). *Presence of microplastics in water, sediments and fish species in an urban coastal environment of Fiji, a Pacific small island developing state. Marine Pollution Bulletin* 153: 110991. https://doi.org/10.1016/j.marpolbul.2020.110991.

Frias, J.P.G.L., Lyashevska, O., Joyce, H. et al. (2020). *Floating microplastics in a coastal embayment: a multifaceted issue. Marine Pollution Bulletin* 158: 111361. https://doi. org/10.1016/j.marpolbul.2020.111361.

von Friesen, L.W., Granberg, M.E., Pavlova, O. et al. (2020). *Summer sea ice melt and wastewater are important local sources of microlitter to Svalbard waters. Environment International* 139: 105511. https://doi.org/10.1016/j.envint.2020.105511.

Garcés-Ordóñez, O., Espinosa Díaz, L.F., Pereira Cardoso, R., and Costa Muniz, M. (2020a). *The impact of tourism on marine litter pollution on Santa Marta beaches, Colombian Caribbean. Marine Pollution Bulletin* 160 (2) https://doi.org/10.1016/j. marpolbul.2020.111558.

Garcés-Ordóñez, O., Espinosa, L.F., Cardoso, R.P. et al. (2020b). *Plastic litter pollution along sandy beaches in the Caribbean and Pacific coast of Colombia. Environmental Pollution* 267 https://doi.org/10.1016/j.envpol.2020.115495.

Garcés-Ordóñez, O., Mejía-Esquivia, K.A., Sierra-Labastidas, T. et al. (2020c). *Prevalence of microplastic contamination in the digestive tract of fishes from mangrove ecosystem in Cispata, Colombian Caribbean. Marine Pollution Bulletin* 154 (2): 111085. https://doi.org/10.1016/j. marpolbul.2020.111085.

Gasperi, J., Wright, S.L., Dris, R. et al. (2018). *Microplastics in air: are we breathing it in? Current Opinion in Environmental Science and Health* 1: 1–5. https://doi.org/10.1016/j. coesh.2017.10.002.

GESAMP (Joint Group of Experts on the Scientific Aspects of Marine Environmental Protection) (2015). Sources, fate and effects of microplastics in the marine environment: a global assessment. In: *Reports and Studies GESAMP*, vol. 93 (eds. P.J. Kershaw and C.M. Rochman). IMO/FAO/UNESCO-IOC/UNIDO/WMO/IAEA/UN/UNEP/UNDP Joint Group of Experts on the Scientific Aspects of Marine Environmental Protection http:// ec.europa.eu/environment/marine/good-environmental-status/descriptor-10/pdf/ GESAMP_microplastics full study.pdf.

GESAMP (Joint Group of Experts on the Scientific Aspects of Marine Environmental Protection) (2016). Sources, fate and effects of microplastics in the marine environment: part two of a global assessment. In: *Reports and Studies GESAMP*, vol. 93 (eds. P.J. Kershaw and C.M. Rochman). IMO/FAO/UNESCO-IOC/UNIDO/WMO/IAEA/UN/UNEP/UNDP Joint Group of Experts on the Scientific Aspects of Marine Environmental Protection https://doi. org/10.13140/RG.2.1.3803.7925.

Geyer, R., Jambeck, J.R., and Law, K.L. (2017). *Production, use, and fate of all plastics ever made. Science Advances* 3 (7): e1700782. https://doi.org/10.1126/sciadv.1700782.

Godoy, V., Prata, J.C., Blázquez, G. et al. (2020). *Effects of distance to the sea and geomorphological characteristics on the quantity and distribution of microplastics in beach sediments of Granada (Spain). Science of the Total Environment* 746 https://doi.org/10.1016/j. scitotenv.2020.142023.

González-Hernández, M., Hernández-Sánchez, C., González-Sálamo, J. et al. (2020). *Monitoring of meso and microplastic debris in Playa Grande beach (Tenerife, Canary Islands, Spain) during a moon cycle. Marine Pollution Bulletin* 150 https://doi.org/10.1016/j. marpolbul.2019.110757.

Goswami, P., Vinithkumar, N.V., and Dharani, G. (2020). *First evidence of microplastics bioaccumulation by marine organisms in the Port Blair Bay, Andaman Islands. Marine Pollution Bulletin* 155: 111163. https://doi.org/10.1016/j.marpolbul.2020.111163.

Guo, J.J., Huang, X.P., Xiang, L. et al. (2020). *Source, migration and toxicology of microplastics in soil. Environment International* 137: 105263. https://doi.org/10.1016/j.envint.2019.105263.

Hidalgo-Ruz, V., Gutow, L., Thompson, R.C., and Thiel, M. (2012). *Microplastics in the marine environment: a review of the methods used for identification and quantification. Environmental Science & Technology* 46 (6): 3060–3075. https://doi.org/10.1021/es2031505.

Hosseini, R., Sayadi, M.H., Aazami, J., and Savabieasfehani, M. (2020). *Accumulation and distribution of microplastics in the sediment and coastal water samples of Chabahar Bay in the Oman Sea, Iran. Marine Pollution Bulletin* 160 https://doi.org/10.1016/j. marpolbul.2020.111682.

Huang, J.S., Koongolla, J.B., Li, H.X. et al. (2020a). *Microplastic accumulation in fish from Zhanjiang mangrove wetland, South China. Science of the Total Environment* 708: 134839. https://doi.org/10.1016/j.scitotenv.2019.134839.

Huang, Y., Xiao, X., Xu, C. et al. (2020b). *Seagrass beds acting as a trap of microplastics – emerging hotspot in the coastal region? Environmental Pollution* 257: 113450. https://doi.org/10.1016/j.envpol.2019.113450.

James, K., Vasant, K., Padua, S. et al. (2020). *An assessment of microplastics in the ecosystem and selected commercially important fishes off Kochi, south eastern Arabian Sea, India. Marine Pollution Bulletin* 154: 111027. https://doi.org/10.1016/j.marpolbul.2020.111027.

James, K., Vasant, K., Sikkander Batcha, S.M. et al. (2021). *Seasonal variability in the distribution of microplastics in the coastal ecosystems and in some commercially important fishes of the Gulf of Mannar and Palk Bay, Southeast coast of India. Regional Studies in Marine Science* 41: 101558. https://doi.org/10.1016/j.rsma.2020.101558.

Jang, M., Shim, W.J., Cho, Y. et al. (2020). *A close relationship between microplastic contamination and coastal area use pattern. Water Research* 171: 115400. https://doi.org/10.1016/j.watres.2019.115400.

Jeyasanta, K.I., Sathish, N., Patterson, J., and Edward, J.K.P. (2020). *Macro-, meso- and microplastic debris in the beaches of Tuticorin district, Southeast coast of India. Marine Pollution Bulletin* 154 https://doi.org/10.1016/j.marpolbul.2020.111055.

Jiang, Y., Yang, F., Zhao, Y., and Wang, J. (2020a). *Greenland Sea Gyre increases microplastic pollution in the surface waters of the Nordic Seas. Science of the Total Environment* 712: 136484. https://doi.org/10.1016/j.scitotenv.2019.136484.

Jiang, Y., Zhao, Y., Wang, X. et al. (2020b). *Characterization of microplastics in the surface seawater of the South Yellow Sea as affected by season. Science of the Total Environment* 724: 138375. https://doi.org/10.1016/j.scitotenv.2020.138375.

Jones, K.L., Hartl, M.G.J., Bell, M.C., and Capper, A. (2020). *Microplastic accumulation in a Zostera marina L. bed at Deerness Sound, Orkney, Scotland. Marine Pollution Bulletin* 152: 110883. https://doi.org/10.1016/j.marpolbul.2020.110883.

Jones-Williams, K., Galloway, T., Cole, M. et al. (2020). *Close encounters – microplastic availability to pelagic amphipods in sub-antarctic and antarctic surface waters. Environment International* 140: 105792. https://doi.org/10.1016/j.envint.2020.105792.

Jung, J.-W., Park, J.-W., Eo, S. et al. (2020). *Ecological risk assessment of microplastics in coastal, shelf, and deep sea waters with a consideration of environmentally relevant size and shape. Environmental Pollution* 270: 116217. https://doi.org/10.1016/j.envpol.2020.116217.

Katsumi, N., Kusube, T., Nagao, S., and Okochi, H. (2020). *The role of coated fertilizer used in paddy fields as a source of microplastics in the marine environment. Marine Pollution Bulletin* 161 https://doi.org/10.1016/j.marpolbul.2020.111727.

Kazour, M. and Amara, R. (2020). *Is blue mussel caging an efficient method for monitoring environmental microplastics pollution? Science of the Total Environment* 710: 135649. https://doi.org/10.1016/j.scitotenv.2019.135649.

Kedzierski, M., Frère, D., Le Maguer, G., and Bruzaud, S. (2020). *Why is there plastic packaging in the natural environment? Understanding the roots of our individual plastic waste management behaviours. Science of the Total Environment* 740: 139985. https://doi.org/10.1016/j.scitotenv.2020.139985.

Khoironi, A., Hadiyanto, H., Anggoro, S., and Sudarno, S. (2020). *Evaluation of polypropylene plastic degradation and microplastic identification in sediments at Tambak Lorok coastal area, Semarang, Indonesia. Marine Pollution Bulletin* 151: 110868. https://doi.org/10.1016/j.marpolbul.2019.110868.

Kor, K. and Mehdinia, A. (2020). *Neustonic microplastic pollution in the Persian Gulf. Marine Pollution Bulletin* 150: 110665. https://doi.org/10.1016/j.marpolbul.2019.110665.

Kor, K., Ghazilou, A., and Ershadifar, H. (2020). *Microplastic pollution in the littoral sediments of the northern part of the Oman Sea. Marine Pollution Bulletin* 155: 111166. https://doi.org/10.1016/j.marpolbul.2020.111166.

Kovač, M., Palatinus, A., Koren, Š. et al. (2016). *Protocol for microplastics sampling on the sea surface and sample analysis. Journal of Visualized Experiments: JoVE* 118: 55161. https://doi.org/10.3791/55161.

Kwon, O.Y., Kang, J.H., Hong, S.H., and Shim, W.J. (2020). *Spatial distribution of microplastic in the surface waters along the coast of Korea. Marine Pollution Bulletin* 155: 110729. https://doi.org/10.1016/j.marpolbul.2019.110729.

Laptenok, S.P., Martin, C., Genchi, L. et al. (2020). *Stimulated Raman microspectroscopy as a new method to classify microfibers from environmental samples. Environmental Pollution* 267: 115640. https://doi.org/10.1016/j.envpol.2020.115640.

Lavender, K. and Thompson, R. (2014). *Microplastics in the seas. Science* 345 (6193): 144–145. https://doi.org/10.1126/science.1254065.

Lechthaler, S., Schwarzbauer, J., Reicherter, K. et al. (2020). *Regional study of microplastics in surface waters and deep sea sediments south of the Algarve Coast. Regional Studies in Marine Science* 40: 101488. https://doi.org/10.1016/j.rsma.2020.101488.

Li, R., Yu, L., Chai, M. et al. (2020a). *The distribution, characteristics and ecological risks of microplastics in the mangroves of Southern China. Science of the Total Environment* 708: 135025. https://doi.org/10.1016/j.scitotenv.2019.135025.

Li, R., Zhang, S., Zhang, L. et al. (2020b). *Field study of the microplastic pollution in sea snails* (Ellobium chinense) *from mangrove forest and their relationships with microplastics in water/ sediment located on the north of Beibu Gulf. Environmental Pollution* 263: 114368. https://doi.org/10.1016/j.envpol.2020.114368.

Lima, A.R.A., Barletta, M., and Costa, M.F. (2015). *Seasonal distribution and interactions between plankton and microplastics in a tropical estuary. Estuarine, Coastal and Shelf Science* 165: 213–225. https://doi.org/10.1016/j.ecss.2015.05.018.

Lindeque, P.K., Cole, M., Coppock, R.L. et al. (2020). *Are we underestimating microplastic abundance in the marine environment? A comparison of microplastic capture with nets of different mesh-size. Environmental Pollution* 265: 114721. https://doi.org/10.1016/j.envpol.2020.114721.

Littman, R.A., Fiorenza, E.A., Wenger, A.S. et al. (2020). *Coastal urbanization influences human pathogens and microdebris contamination in seafood. Science of the Total Environment* 736: 139081. https://doi.org/10.1016/j.scitotenv.2020.139081.

Liu, T., Zhao, Y., Zhu, M. et al. (2020a). *Seasonal variation of micro- and meso-plastics in the seawater of Jiaozhou Bay, the Yellow Sea. Marine Pollution Bulletin* 152: 110922. https://doi.org/10.1016/j.marpolbul.2020.110922.

Liu, Y., Li, Z., Jalón-Rojas, I. et al. (2020b). *Assessing the potential risk and relationship between microplastics and phthalates in surface seawater of a heavily human-impacted metropolitan bay in northern China. Ecotoxicology and Environmental Safety* 204: 111067. https://doi.org/10.1016/j.ecoenv.2020.111067.

Lo, H.S., Lee, Y.K., Po, B.H.K. et al. (2020). *Impacts of Typhoon Mangkhut in 2018 on the deposition of marine debris and microplastics on beaches in Hong Kong. Science of the Total Environment* 716: 137172. https://doi.org/10.1016/j.scitotenv.2020.137172.

Lyle, M. (2016). Deep-sea sediments. In: *Encyclopedia of Marine Geosciences* (eds. J. Harff et al.), 143–208. Springer Science+ Business Media https://doi.org/10.1007/978-94-007-6238-1.

Maaghloud, H., Houssa, R., Ouansafi, S. et al. (2020). *Ingestion of microplastics by pelagic fish from the Moroccan Central Atlantic coast. Environmental Pollution* 261: 114194. https://doi.org/10.1016/j.envpol.2020.114194.

Mak, C.W., Tsang, Y.Y., Leung, M.M.L. et al. (2020). *Microplastics from effluents of sewage treatment works and stormwater discharging into the Victoria Harbor, Hong Kong. Marine Pollution Bulletin* 157: 111181. https://doi.org/10.1016/j.marpolbul.2020.111181.

Masura, J., Baker, J., Foster, G., and Arthur, C. (2015). *Laboratory Methods for the Analysis of Microplastics in the Marine Environment: Recommendations for quantifying synthetic particles in waters and sediments*. NOAA Marine Debris Program, National Oceanic and Atmospheric Administration, U.S. Department of Commerce; Technical Memorandum NOS-OR&R-48, July 2015. https://marinedebris.noaa.gov/sites/default/files/publications-files/noaa_microplastics_methods_manual.pdf (accessed 26 October 2021).

Mazariegos-Ortíz, C., de los Ángeles Rosales, M., Carrillo-Ovalle, L. et al. (2020). *First evidence of microplastic pollution in the El Quetzalito sand beach of the Guatemalan Caribbean. Marine Pollution Bulletin* 156: 1–6. https://doi.org/10.1016/j.marpolbul.2020.111220.

Mbedzi, R., Cuthbert, R.N., Wasserman, R.J. et al. (2020). *Spatiotemporal variation in microplastic contamination along a subtropical reservoir shoreline. Environmental Science and Pollution Research* 27 (19): 23880–23887. https://doi.org/10.1007/s11356-020-08640-4.

Md Amin, R., Sohaimi, E.S., Anuar, S.T., and Bachok, Z. (2020). *Microplastic ingestion by zooplankton in Terengganu coastal waters, southern South China Sea. Marine Pollution Bulletin* 150: 110616. https://doi.org/10.1016/j.marpolbul.2019.110616.

Metaxas, A. and Snelgrove, P. (2019). Caring for the Deep Sea. In: *The Future of Ocean Governance and Capacity Development* (eds. D. Werle, P.R. Boudreau, M.R. Brooks, et al.), 245–251. https://doi.org/10.1163/9789004380271_041.

Naidoo, T., Sershen, Thompson, R.C., and Rajkaran, A. (2020). *Quantification and characterisation of microplastics ingested by selected juvenile fish species associated with mangroves in KwaZulu-Natal, South Africa. Environmental Pollution* 257: 113635. https://doi.org/10.1016/j.envpol.2019.113635.

Narmatha, M., Immaculate, K., and Patterson, J. (2020a). *Monitoring of microplastics in the clam* Donax cuneatus *and its habitat in Tuticorin coast of Gulf of Mannar (GoM), India. Environmental Pollution* 266: 115219. https://doi.org/10.1016/j.envpol.2020.115219.

Narmatha, M., Immaculate, K., and Patterson, J. (2020b). *Occurrence of microplastics in epipelagic and mesopelagic fishes from Tuticorin, Southeast coast of India. Science of the Total Environment* 720: 137614. https://doi.org/10.1016/j.scitotenv.2020.137614.

Nel, H.A., Sambrook Smith, G.H., Harmer, R. et al. (2020). *Citizen science reveals microplastic hotspots within tidal estuaries and the remote Scilly Islands, United Kingdom. Marine Pollution Bulletin* 161 https://doi.org/10.1016/j.marpolbul.2020.111776.

Nematollahi, M.J., Moore, F., Keshavarzi, B. et al. (2020). *Microplastic particles in sediments and waters, south of Caspian Sea: frequency, distribution, characteristics, and chemical composition. Ecotoxicology and Environmental Safety* 206: 111137. https://doi.org/10.1016/j.ecoenv.2020.111137.

NOAA (National Oceanic and Atmospheric Administration). (2017). *What are microplastics?* https://oceanservice.noaa.gov/facts/microplastics.html.

Novillo, O., Raga, J.A., and Tomás, J. (2020). *Evaluating the presence of microplastics in striped dolphins* (Stenella coeruleoalba) *stranded in the Western Mediterranean Sea. Marine Pollution Bulletin* 160: 111557. https://doi.org/10.1016/j.marpolbul.2020.111557.

Oliveira-Castro, R., Lopez da Silva, M., Marques, M.R.C., and Vieira de Araújo, F. (2020). *Spatio-temporal evaluation of macro, meso and microplastics in surface waters, bottom and beach sediments of two embayments in Niterói, RJ, Brazil. Marine Pollution Bulletin* 160: 111537. https://doi.org/10.1016/j.marpolbul.2020.111537.

Ory, N.C., Lehmann, A., Javidpour, J. et al. (2020). *Factors influencing the spatial and temporal distribution of microplastics at the sea surface – a year-long monitoring case study from the urban Kiel Fjord, southwest Baltic Sea. Science of the Total Environment* 736: 139493. https://doi.org/10.1016/j.scitotenv.2020.139493.

OSIL (Ocean Scientific International). (2021). *Sampling Equipment.* Data Buoy Platforms | Water Column & Seabed Sampling. https://osil.com/product-category/sampling-equipment.

Patchaiyappan, A., Ahmed, S.Z., Dowarah, K. et al. (2020). *Occurrence, distribution and composition of microplastics in the sediments of South Andaman beaches. Marine Pollution Bulletin* 156 https://doi.org/10.1016/j.marpolbul.2020.111227.

Patterson, J., Jeyasanta, K.I., Sathish, N. et al. (2020). *Microplastic and heavy metal distributions in an Indian coral reef ecosystem. Science of the Total Environment* 744: 140706. https://doi.org/10.1016/j.scitotenv.2020.140706.

Peng, G., Bellerby, R., Zhang, F. et al. (2020). *The ocean's ultimate trashcan: Hadal trenches as major depositories for plastic pollution. Water Research* 168: 115121. https://doi.org/10.1016/j.watres.2019.115121.

PlasticsEurope. (2019). *Plastics – The Facts 2019. An analysis of European plastics production, demand and waste data.* https://www.plasticseurope.org/en/resources/publications/1804-plastics-facts-2019.

Plee, T.A. and Pomory, C.M. (2020). *Microplastics in sandy environments in the Florida Keys and the panhandle of Florida, and the ingestion by sea cucumbers* (Echinodermata: Holothuroidea) *and sand dollars* (Echinodermata: Echinoidea). *Marine Pollution Bulletin* 158 https://doi.org/10.1016/j.marpolbul.2020.111437.

Portz, L., Manzolli, R.P., Herrera, G.V. et al. (2020). *Marine litter arrived: distribution and potential sources on an unpopulated atoll in the Seaflower Biosphere Reserve, Caribbean Sea. Marine Pollution Bulletin* 157: 111323. https://doi.org/10.1016/j.marpolbul.2020.111323.

Prata, J.C., Reis, V., Paço, A. et al. (2020). *Effects of spatial and seasonal factors on the characteristics and carbonyl index of (micro)plastics in a sandy beach in Aveiro, Portugal. Science of the Total Environment* 709: 135892. https://doi.org/10.1016/j.scitotenv.2019.135892.

Rahman, S.M.A., Robin, G.S., Momotaj, M. et al. (2020). *Occurrence and spatial distribution of microplastics in beach sediments of Cox's Bazar, Bangladesh. Marine Pollution Bulletin* 160 https://doi.org/10.1016/j.marpolbul.2020.111587.

Ramírez-Álvarez, N., Rios Mendoza, L.M., Macías-Zamora, J.V. et al. (2020). *Microplastics: sources and distribution in surface waters and sediments of Todos Santos Bay, Mexico. Science of the Total Environment* 703: 134838. https://doi.org/10.1016/j.scitotenv.2019.134838.

Rapp, J., Herrera, A., Martinez, I. et al. (2020). *Study of plastic pollution and its potential sources on Gran Canaria Island beaches (Canary Islands, Spain). Marine Pollution Bulletin* 153 https://doi.org/10.1016/j.marpolbul.2020.110967.

Reineccius, J., Appelt, J.S., Hinrichs, T. et al. (2020). *Abundance and characteristics of microfibers detected in sediment trap material from the deep subtropical North Atlantic Ocean. Science of the Total Environment* 738: 140354. https://doi.org/10.1016/j. scitotenv.2020.140354.

Renzi, M. and Blašković, A. (2020). *Chemical fingerprint of plastic litter in sediments and holothurians from Croatia: assessment & relation to different environmental factors. Marine Pollution Bulletin* 153: 110994. https://doi.org/10.1016/j.marpolbul.2020.110994.

Renzi, M., Blašković, A., Broccoli, A. et al. (2020). *Chemical composition of microplastic in sediments and protected detritivores from different marine habitats (Salina Island). Marine Pollution Bulletin* 152: 110918. https://doi.org/10.1016/j.marpolbul.2020.110918.

Rist, S., Vianello, A., Winding, M.H.S. et al. (2020). *Quantification of plankton-sized microplastics in a productive coastal Arctic marine ecosystem. Environmental Pollution* 266: 115248. https://doi.org/10.1016/j.envpol.2020.115248.

Robin, R.S., Karthik, R., Purvaja, R. et al. (2020). *Holistic assessment of microplastics in various coastal environmental matrices, southwest coast of India. Science of the Total Environment* 703 https://doi.org/10.1016/j.scitotenv.2019.134947.

Rodriguez-Seijo, A., Lourenço, J., Rocha-Santos, T.A.P. et al. (2017). *Histopathological and molecular effects of microplastics in Eisenia andrei Bouché. Environmental Pollution* 220: 495–503. https://doi.org/10.1016/j.envpol.2016.09.092.

Rotjan, R.D., Sharp, K.H., Gauthier, A.E. et al. (2019). *Patterns, dynamics and consequences of microplastic ingestion by the temperate coral,* Astrangia poculata. *Proceedings of the Royal Society B: Biological Sciences* 286 (1905): 1–9. https://doi.org/10.1098/rspb.2019.0726.

Ryan, P.G., Suaria, G., Perold, V. et al. (2020). *Sampling microfibres at the sea surface: the effects of mesh size, sample volume and water depth. Environmental Pollution* 258: 113413. https:// doi.org/10.1016/j.envpol.2019.113413.

Saeed, T., Al-Jandal, N., Al-Mutairi, A., and Taqi, H. (2020). *Microplastics in Kuwait marine environment: results of first survey. Marine Pollution Bulletin* 152: 110880. https://doi. org/10.1016/j.marpolbul.2019.110880.

Schönlau, C., Karlsson, T.M., Rotander, A. et al. (2020). *Microplastics in sea-surface waters surrounding Sweden sampled by manta trawl and in-situ pump. Marine Pollution Bulletin* 153: 111019. https://doi.org/10.1016/j.marpolbul.2020.111019.

da Silva, M.L., Oliveira Castro, R., Souza Sales, A., and Vieira de Araújo, F. (2018). *Marine debris on beaches of Arraial do Cabo, RJ, Brazil: An important coastal tourist destination. Marine Pollution Bulletin* 130: 153–158. https://doi.org/10.1016/J.MARPOLBUL.2018.03.026.

Sui, M., Lu, Y., Wang, Q. et al. (2020). *Distribution patterns of microplastics in various tissues of the Zhikong scallop* (Chlamys farreri) *and in the surrounding culture seawater. Marine Pollution Bulletin* 160: 111595. https://doi.org/10.1016/j.marpolbul.2020.111595.

Suresh, A., Vijayaraghavan, G., Saranya, K.S. et al. (2020). *Microplastics distribution and contamination from the Cochin coastal zone, India. Regional Studies in Marine Science* 40: 101533. https://doi.org/10.1016/j.rsma.2020.101533.

Teng, J., Zhao, J., Zhang, C. et al. (2020). *A systems analysis of microplastic pollution in Laizhou Bay, China. Science of the Total Environment* 745: 140815. https://doi.org/10.1016/j. scitotenv.2020.140815.

Thiele, C.J., Hudson, M.D., and Russell, A.E. (2019). *Evaluation of existing methods to extract microplastics from bivalve tissue: adapted KOH digestion protocol improves filtration at* single-digit pore size. *Marine Pollution Bulletin* 142: 384–393. https://doi.org/10.1016/j.marpolbul.2019.03.003.

Tošić, T.N., Vruggink, M., and Vesman, A. (2020). *Microplastics quantification in surface waters of the Barents, Kara and White seas. Marine Pollution Bulletin* 161: 111745. https://doi.org/10.1016/j.marpolbul.2020.111745.

Tsang, Y.Y., Mak, C.W., Liebich, C. et al. (2017). *Microplastic pollution in the marine waters and sediments of Hong Kong. Marine Pollution Bulletin* 115 (1–2): 20–28. https://doi.org/10.1016/j.marpolbul.2016.11.003.

Tsang, Y.Y., Mak, C.W., Liebich, C. et al. (2020). *Spatial and temporal variations of coastal microplastic pollution in Hong Kong. Marine Pollution Bulletin* 161: 111765. https://doi.org/10.1016/j.marpolbul.2020.111765.

UNEP (United Nations Environment Programme). (2018). *Single-Use Plastics: A Roadmap for Sustainability.* https://www.unep.org/ietc/resources/publication/single-use-plastics-roadmap-sustainability

UTAS (University of Tasmania Australia). (2013). *Zooplankton Sampling.* https://www.imas.utas.edu.au/zooplankton/about/zooplankton-sampling

Van Rensburg, M.L., Nkomo, S.L., and Dube, T. (2020). *The 'plastic waste era'; social perceptions towards single-use plastic consumption and impacts on the marine environment in Durban, South Africa. Applied Geography* 114: 102132. https://doi.org/10.1016/j.apgeog.2019.102132.

Vandermeersch, G., Van Cauwenberghe, L., Janssen, C.R. et al. (2015). *A critical view on microplastic quantification in aquatic organisms. Environmental Research* 143: 46–55. https://doi.org/10.1016/j.envres.2015.07.016.

Vetrimurugan, E., Jonathan, M.P., Sarkar, S.K. et al. (2020). *Occurrence, distribution and provenance of micro plastics: a large scale quantitative analysis of beach sediments from southeastern coast of South Africa. Science of the Total Environment* 746: 141103. https://doi.org/10.1016/j.scitotenv.2020.141103.

Vilakati, B., Sivasankar, V., Mamba, B.B. et al. (2020). *Characterization of plastic micro particles in the Atlantic Ocean seashore of Cape Town, South Africa and mass spectrometry analysis of pyrolyzate products. Environmental Pollution* 265: 114859. https://doi.org/10.1016/j.envpol.2020.114859.

VLIZ (Vlaams Instituut voor de Zee) (2021). *Niskin bottle.* http://www.vliz.be/en/Niskinbottle.

Wang, T., Hu, M., Song, L. et al. (2020a). *Coastal zone use influences the spatial distribution of microplastics in Hangzhou Bay, China. Environmental Pollution* 266: 115137. https://doi.org/10.1016/j.envpol.2020.115137.

Wang, W., Ge, J., Yu, X., and Li, H. (2020b). *Environmental fate and impacts of microplastics in soil ecosystems: progress and perspective. Science of the Total Environment* 708: 134841. https://doi.org/10.1016/j.scitotenv.2019.134841.

Zhang, W., Zhang, S., Zhao, Q. et al. (2020). *Spatio-temporal distribution of plastic and microplastic debris in the surface water of the Bohai Sea, China. Marine Pollution Bulletin* 158: 111343. https://doi.org/10.1016/j.marpolbul.2020.111343.

Zheng, Y., Li, J., Cao, W. et al. (2020). *Vertical distribution of microplastics in bay sediment reflecting effects of sedimentation dynamics and anthropogenic activities. Marine Pollution Bulletin* 152: 110885. https://doi.org/10.1016/j.marpolbul.2020.110885.

Zhou, Q., Tu, C., Fu, C. et al. (2020). *Characteristics and distribution of microplastics in the coastal mangrove sediments of China*. Science of the Total Environment 703: 134807. https://doi.org/10.1016/j.scitotenv.2019.134807.

Zuo, L., Sun, Y., Li, H. et al. (2020). *Microplastics in mangrove sediments of the Pearl River Estuary, South China: correlation with halogenated flame retardants' levels*. Science of the Total Environment 725: 138344. https://doi.org/10.1016/j.scitotenv.2020.138344.

9

Bio-Based and Biodegradable Plastics as Alternatives to Conventional Plastics

Bhabesh Kumar Choudhury[1], Rupjyoti Haloi[2], Kaushik Kumar Bharadwaj[3], Sanchayita Rajkhowa[4], and Jyotirmoy Sarma[5]

[1] Department of Chemistry, University of Guwahati, Guwahati, Assam, India
[2] Department of Electrical Engineering, Assam Engineering College, Jalukbari, Guwahati, Assam, India
[3] Department of Bioengineering and Technology, GUIST, University of Guwahati, Guwahati, Assam, India
[4] Department of Chemistry, Jorhat Institute of Science and Technology, Jorhat, Assam, India
[5] Department of Chemistry, Assam Kaziranga University, Jorhat, Assam, India

9.1 Introduction

According to reported sources, around 83 000 million tons of plastics were generated until the end of 2017 (Lakhawat et al. 2020). Among all the municipal wastes, plastics contribute a large share; discarded packages made of plastics are among major waste management challenges. Although the recyclability of waste increased in the recent years, the recyclability of plastics are found to be low (Song et al. 2009). Plastic materials are more popular due to their wide range of applications, properties, and performance over metals or wooden materials (Alvarez-Chavez et al. 2011). The use of plastics in large excess and generation of its waste leads to both environmental and economic crises around the globe.

Among environmental problems, the land filling by plastic waste is a major concern (Philp et al. 2013). With increase in plastic waste generation the land filling capacity is gradually shrinking. To manage this waste, strong legislation is important. Other critical problems that arise from plastics are due to its non-biodegradability, accumulation in oceans, greenhouse gas generation during incineration, and burning (Bezirhan & Ozsoy 2015). Environmental and economic burdens on society due plastic waste have generated immense concern.

The benefits of conventional or non-biodegradable plastic are well recognized, but the disadvantages mentioned have made us think about alternatives such as biodegradable plastics. Around 8% of crude oil and gas produced globally are used in polymer synthesis in commodity polymers such as polyethylene (PE) and polystyrene (PS) (Lambert & Wagner 2017). The short period of use of plastic materials is the reason for large plastic waste production. Out of the total 83 000 tons of plastics produced in the period 1905–2015, 25 000 tons are still in use, while rest have been discarded as waste (Geyer et al. 2017).

Plastic and Microplastic in the Environment: Management and Health Risks, First Edition.
Edited by Arif Ahamad, Pardeep Singh, and Dhanesh Tiwary.
© 2022 John Wiley & Sons Ltd. Published 2022 by John Wiley & Sons Ltd.

Among this only 9% has been recycled. Through a directive of the parliament of EU [Directive (EU) 2019/904] released on 5 June 2019, they emphasized reduction the effect of some plastic products on the environment, however 50% of their plastic waste is still disposed of this way.

Only few countries, namely Germany, Netherlands, Denmark, and Austria have been able to recycle 80–100% of plastics (Emadian et al. 2017). Compared to developed countries, the developing countries such as China, Indonesia, Sri lanka, Vietnam, and Philippines are using more plastics due to increasing urbanization, and at more than 50% are leading in plastic waste contribution to the marine environment (Emadian et al. 2017). By the year 2050, the global population will be around 9 billion, which will cause the demand in plastics to rise. At same time, the waste generation will also increase. To mitigate such environmental problems, the production of bioplastics has risen in popularity due to its biodegradability.

Conventional plastics are synthesized from petroleum sources. The crisis in oil and petroleum since 1970s has alarmed active communities around the globe. Although only 4–5% of global oil production is consumed in plastic resin production (Queiroz & Queiroz 2008), the growing demand will shift the percentage of requirements toward higher consumption, which will create more oil crises in the future.

One of the important examples of major oil consumption is found in plastic bottle production. Due to rapid growth of beverage industries, beverage bottles are in high demand. To fulfill this demand a large share of global oil production is in use to synthesize bottles (Wang et al. 2019). As oil is a non-renewable resource and beverage bottles are mostly disposable, it creates both economic and environmental burdens. To save oil, recycling plastic bottles has become a concern in bottle industries. Recycling involves identification and sorting of plastics based on polythene terephthalate or other impurities, and color (Yaradoddi et al. 2017). These processes are time consuming and non-economic.

The plastic pollution over the globe has reached such a height that United Nations Environment Programme (UNEP) decided to make the theme of world environment day 2018 "Beat the Plastic Pollution." UNEP published the report on global plastic pollution in 2018; the theme requested the state parties of UN India to take action on individuals, governments, public, and private sectors to find solutions to reduce the heavy burdens of plastic pollution on natural places, wildlife, and our health (UN India 2018).

Hence, the demand of alternative, biodegradable plastic, which is less harmful for the environment, is continuously increasing. It is believed that biological originated or bio-based, biodegradable plastic can be an eco-friendly, environment friendly alternative to conventional petrochemical-based plastics. Biopolymers are effectively decomposed by enzymatic activities by microorganisms, so it is beneficial to use biopolymers instead of conventional plastics.

The biodegradable plastics or bioplastic industries are still in the developmental stages, which allows more scope to explore its resources as well as its market. The terms bioplastic or biodegradable plastics here will be considered as materials that contain biopolymers in variable content.

In this chapter, we have focused on various aspects of bio-based and biodegradable plastics (BBBPs) as alternatives to conventional plastics.

9.2 Definition and Classification of Plastics

Plastics are synthetic polymeric material derived from monomers obtained in petrochemicals. The monomers may be one molecule or more than one molecule that repeats to make the polymer. Polymer chemists use only eight elements including hydrogen, carbon, nitrogen, oxygen, silicon, fluorine, chlorine, and sulfur to produce more than 100 types of plastics (Alauddin et al. 1995). To become plastic a material has to achieve a convenient and useful range of plasticity, so plasticizing materials need to be added (Karrer 1930). Resin is the most important constituent of plastic, however other materials such as fillers, solvents, and colorants are also required. According to the American Society for Testing and Materials (ASTM), plastic is defined as a finished solid material that contains one or more organic polymeric substances of large molecular weight that can be given shape of various finished material (Lambert & Wagner 2017; ASTM International 2019).

Alternatively, bioplastic is a common term used to define polymers derived from renewable biological sources that is degradable under different conditions, but the International Union for Pure and Applied Chemistry (IUPAC) encourages the use of the term 'bio-based polymer' instead of bioplastics. The bio-based plastic inclusively defines a plastic that contains carbonaceous material derived from agricultural, plant, animal, fungi, microorganism, forestry, or marine material found in nature. Biodegradable plastics are plastics that degrade by the action of bacteria, fungi, algae, etc. (Meereboer et al. 2020).

Plastics have numerous grades and varieties; presently, approximately 20 different groups of plastics are available on the market. As plastics possess qualities like low cost, light weight, durability, resistance from corrosion, and efficient thermal and electrical insulation it became a versatile material that encourages large-scale production, with many economical and societal benefits. According to a report, Bakelite is the first synthetic plastic developed by Belgian scientist Leo Baekeland in 1907 (Thompson et al. 2009a). The mass production of various plastics prevailed after 1950. In the past, plastics were mainly in use to make lightweight components for airplanes, vehicles that could help in fuel consumption, and sterile medical components. Today however, the world's communities are using plastics mainly in disposable items like packaging that produces more and more plastic debris (Hopewell et al. 2009).

Another enemy of the environment is microplastics (MPs), which are particles with <5 mm diameter (Du et al. 2021). MPs have both primary and secondary sources. Primary sources are those where plastics are synthesized in micron size for the various applications such as plastic beads in cosmetics, toothpastes, etc.; secondary MPs are generated from fragmentation of MPs. Thus, plastics are spread over the world.

E. Kerrer in his article mentioned different categories of plastics. These are as follows:

1) Thermoplastic: The stability of such plastics depends on temperature. They are moldable using heat.
2) Chemo-plastics: Materials which achieve plasticity by some chemical reactions.
3) Mechano-plastics: Materials which get plasticity by mechanical agitation.
4) Cheo-plastics: Plastics which are obtained by mixing metals with materials in various ways. Dental amalgams used in dentistry are example of such plastics.
5) Solvoplastics: Plastics which are obtained by adding plasticizing materials by means of solvation.

According to the data from American Chemistry Council, plastics are available in six forms; polyethylene terephthalate (PET, PETE), high-density polyethylene (HDPE), low-density polyethylene (LDPE), polyvinyl chloride (PVC), polypropylene (PP), and polystyrene (PS). Those plastics involving resin in the formation are kept outside these six categories (American Chemistry Council 1988).

Biodegradable plastics are classified as (i) carbohydrate based bio-plastics; (ii) protein based bio-plastics; (iii) lipid- or poly-hydroxy alkanoate based (PAH) bio-plastics. The term bioplastic is commonly used to discuss both biodegradable and bio-based plastics.

Though bioplastics are derived from renewable sources such as polysaccharides, proteins, lipids, or products from microorganisms, it is also synthesized chemically from bio-derived products such as polylactic acid (PLA), polybutalene succicinate (PBS), etc. (Batori et al. 2018). In addition to all mentioned bio-based plastics, there is a category of bio-based, but not biodegradable plastics, known as "drop-ins" such as bio-PET, bio-PE, etc. which have similarities in their structures with plastics derived from petrochemicals.

More properly, bioplastic (bio-based or biodegradable) can be categorized as follows.

1) Renewable-resource based bioplastics: This category includes all plastics derived from renewable resources such as cellulose, proteins, chitosan, etc. Recently PE, PP, and nylon-like renewable biopolymers were derived successfully (Reddy et al. 2013).
2) Petroleum-based bio-plastics: These are derived from petroleum sources, but at the same time, they are biodegradable as some functionalities are introduced to them. For example, polycaprolactone is placed in this category (Reddy et al. 2013).
3) Bio-plastics from mixed sources: These are produced by using both bio-based and petroleum sources. For example poly(trimethylene tetraphthalate) (PTT) which is derived from terephthalic acid and biologically derived 1,3-propanediol is placed in this category (Reddy et al. 2013).

It is not necessary that all biologically derived plastics are biodegradable. The biodegradability is dependent on chemical nature but not on the source. For example, the 100% bio-derived PE is not biodegradable but polybutylene adipate terephthalate (PBAT), which is 100% petro-derived plastic, is biodegradable.

Although there are various types of plastic materials available, they can be simply categorized as shown Figure 9.1. The thermoplastic mentioned in the figure is of more importance as it can be melted, softened, reshaped, and recycled according to the requirements.

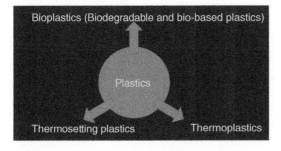

Figure 9.1 The general classification of plastics.

On the other hand, the thermosetting plastics are opposite the thermoplastics. The plastics mentioned in this section fall into these three categories mentioned in Figure 9.1.

9.3 Current Status of Conventional Plastics and Effect on Environment

Due to its immense scopes of utility, the plastic industry has grown at a rapid pace worldwide since 1950. Worldwide, there are currently more than 70 million people are engaged in these industries. Figure 9.2 shows the graphical representation of plastic production in the world from 1950 to 2015 (Ritchie & Roser 2018). This shows that the production of plastics has consistently increased from 1950, reaching a value of approximately 400 million tons in 2015.

Plastic has penetrated into the lives of all people through a wide range of products. Starting with industries dealing with basic household products and moving to various heavy industries, plastics are used as one of the most common raw materials. Conventional plastics are moldable, organic solids which are generally synthetic. These organic solids are polymers with higher molecular mass. Commonly these are derived from petrochemicals or fossil fuels (Nielsen et al. 2020).

Studies have revealed that the majority of plastics are manufactured for a one-time use. This results in accumulations of huge volumes of plastic waste all over the world. The current means of disposal of this waste is to either release them into the sea or directly into the soil. As the conventional plastics are PE, PP, and PS, etc. types of substances, these do not easily decompose in soil or in water, and they may sustain for several hundreds of years without being decomposed. This is becoming the most critical challenge of using conventional plastics today, and it has created potential threats for the environment, wildlife, and humans.

Pollution caused by the accumulation of plastics garbage are commonly visible in developing countries. This is because the waste management systems in these nations are not properly managed. The developed countries also experience similar effects caused by

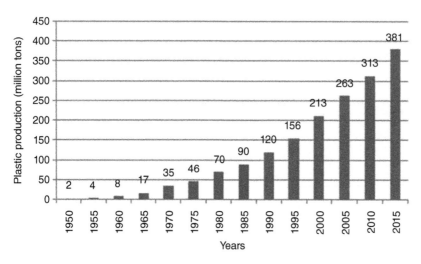

Figure 9.2 Global production of plastic from 1950 to 2015.

plastics pollution, although they concentrate on recycling the plastic waste in a proper way. Before 1980, the awareness of recycling plastic waste did not exist. During those years, almost 100% of the waste generated from plastic was discarded into the environment. After 1980 the process of incineration and recycling of plastic waste were started. Waste management techniques play a vital role in resisting its odd impacts as not all plastic wastes are recyclable or compostable. In the process of crushing these wastes, plastics are broken into smaller or tiny pieces, these particles are referred to as MPs. This is how the plastics enter into the food chains of many living creatures starting from microscopic animals to large animals. It is estimated by the Natural Environment website that approximately 1 lak (0.1 million) sea turtles and other marine animals succumb to death every year as they get strangled in plastic waste bags or consume them as food. Most of the time plastic in the ocean looks like seaweed, or jellyfish in the ocean. Thus, marine species like turtles consume them as food and face death (WWF 2021). Entanglement, ingestion, and interaction are the ways by which the plastic debris affects the wildlife. Entanglement affects various species of turtles, sea species, sea whales, seabirds, fishes, and other vertebrates. This is caused mainly due to abandoned plastic materials used for fishing in the sea. Ingestion may occur intentionally or unintentionally; any living species or organism in the sea may consume plastic indirectly while eating prey species. Millions of land-based animals such elephants, zebras, camels, leopards, and cattle are also affected by plastic waste.

The flow chain of plastic waste accumulated in the seawater is started from the garbage stored on land and from those wastes discarded in the smaller rivers. Although the majority of these accumulated wastes remain in the coastal areas, these may travel all over the world with the flow of seawater. Studies show that the coastal countries release plastic waste of approximately 8 million tons into the oceans every year. If the rate of accumulation of plastic waste remains at the current pace, it has been predicted that the mass of accumulated plastics in the oceans will overcome the total mass of fishes living in the ocean water of the planet by 2050.

A majority of plastics are manufactured from fossil fuels. Each stage of processing of the fossil fuels result in serious concerns for the human health. The process of extraction of fuels releases various toxic materials into the water and the surrounding air. These toxic substances have major impacts on the normal functions of various organs of humans. The process of refining the fossil fuels for manufacturing the plastics also produces carcinogenic and similar substances having higher degrees of toxicity. Exposure to these toxins causes various disorders, weakening the immune system of the associated persons. Plastics are being transformed into various products based on the needs of the consumers. After use, disposal of these products generates the plastic debris or waste.

All modern waste management systems adopted by various countries release several hazardous products into the water, air, and soils. These products include micro- or nanoplastics, which have proved to be a major concern for human health. These substances are accumulated in the tissues of animals and other species living nearby. There are various ways by which these particles are ingested by humans such as inhalation of air and oral consumption of water and other marine food items. Thus, they enter into the human bodies living in the nearby communities through their food chain. The toxicity accumulated in the human body through the said processes has affected the nervous systems, reproductive systems, and the functions of various sensory organs of humans.

As per reports, 79% of plastic waste reaches landfills, while 12% is incinerated and 9% is recycled, globally (Ortiz et al. 2020). If plastic waste generation continues at the current rate, by 2050 there will be 12 billion tons of plastic litter in landfills, which accounts for 15% of our carbon budget and 20% of oil consumption, globally (World Economic Forum 2016).

According to the report from the World Economic Forum titled "The New Plastics Economy: Rethinking the Future of Plastics," plastic and plastic packaging are an unavoidable part of world economy. At present, 26% of total plastic production is engaged in plastic packaging. Plastic packaging can contribute to the economy by reducing food waste and reducing fuel consumption, as plastic packages are of lower weight, but these packaging materials become solid waste after use. Thus, 95% of plastic materials that cost 80–120 billion dollars are lost annually. Every year 8 million tons of plastics are added to the oceans, which is equivalent to one truck of garbage per minute. It is expected that if no action is taken at this time, by 2030 it will be two trucks, and by 2050 three trucks per minute of garbage in the ocean. The burning of plastics generates greenhouse gases. As per the "Handbook on Measurement, Reporting and Verification," the United Nations Framework Convention (COP 21) has decided to limit the carbon budget of the world so that global warming is restricted to a maximum increase of 2 °C by 2100 (UNFCCC 2021).

Studies show that single-use plastic amounts to one-third of the total plastic produced globally. After use, they become hazards to environment. However, as of now, there is no binding agreement among the nations to limit the plastic pollution (Ortiz et al. 2020); so a definite and strict regulation is important to restrict the use overuse of plastic.

Figure 9.3 represents the global production of plastics in different regions, where we can see that the North American Free Trade Agreement (NAFTA) zone alone produces 23% of

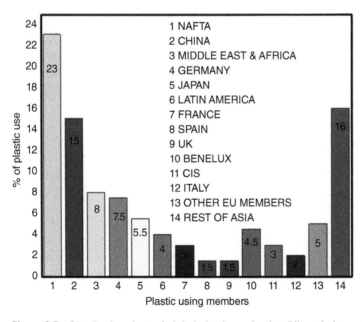

Figure 9.3 Distribution chart of global plastic production (Nkwachukwu et al. 2013).

global plastics. In Asia, China leads in plastic production with 15% of the global plastic production share (World Economic Forum 2016). The production is proportional to consumption; if consumption of plastic increases there will be an increased demand of plastic production. With increase in consumption, the generation of plastic will increase to become even larger amounts of plastic wastes. The term BENELUX indicates three countries namely Belgium, Netherlands, and Luxemburg, while the term CIS indicates Commonwealth of Independent states.

We can take India as an example for a case study to understand the pattern of plastic consumption. According to a report, there is 13 times the increase in the consumption of plastics in India from 1990 to 2006, which is clear from Figure 9.4. In the case of consumption, PP and high-density polythene (HDPE) represent the largest share. PE, PP, and polyvinylchloride (PVC) comprise 80% of the commodity produced in India. In 1990–1991, the per capita consumption of plastic in India was 0.8 kg, which reached the mark of 3.5 kg in 2000, and the projected estimate of per capita consumption of plastic by Indian in 2021 is 10.8 kg (Banerjee & Srivastava 2014).

From the total plastics consumed, packaging materials alone are the largest with 42% of plastic consumed in India. On the other hand, industrial goods, building and construction materials, and consumer products comprise 13, 14, and 24% of total plastics consumption, respectively. The other 7% of the total plastics consumed are shared by other sectors. With this pattern of plastic consumption, India is generating around 1.3 MT of plastics, which is equal to 36% of the annual plastic consumption of India. However, the recycling of plastic from plastic waste is very low in India, which is a major concern that inspires us to investigate biodegradable or bio-based plastic as an alternative to conventional plastics.

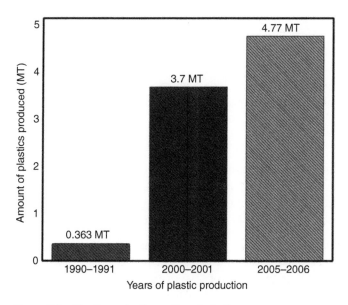

Figure 9.4 Plastic production pattern in India.

9.4   Advantages and Disadvantages of Conventional Plastics

Although we are finding alternatives to conventional plastics, there are many advantages of using conventional plastics over bioplastics. In this section, we are dedicated to discussing the pros and cons of conventional plastics. The use of plastics brought a new era our lives. At the very beginning, use of plastic in daily life was like a blessing. The advantages and disadvantages of using plastics are outlined next.

9.4.1   Advantages

1) Plastic has extraordinary stability, and strength-to-weight ratio that makes it useful for application in electronics, transport, etc. The light weight of plastics allows the production of lightweight transport including aircraft (Klemes et al. 2021).
2) Plastics are regarded as suitable materials for use in packaging as it is durable, flexible, and hygienic. Plastic containers, bottles, cups, drums, boxes, trays, baby package, protective packaging, are also made.
3) In health care plastic has remarkable use. Plastics are used for producing disposable medical tools such as syringes, saline pipes, packaging, and medical surgical implants for joints (American Chemistry Council 1988).
    a) Since the outbreak of COVID-19 pandemic in December 2019 in Wuhan, China the use of Personal Protective Equipment (PPE) has emerged as effective safety gear to prevent transmission from person to person. The PPE kits contain more than 50% of plastics such as PVC and PP (Kumar et al. 2021). Thus, plastic is irreplaceable in health sector at present.
4) Conventional plastics are low cost in comparison to bioplastics.
5) There is a misunderstanding that bio-plastic means biodegradable, but this is not the case. Biodegradability depends on humidity, temperature, and action of microorganisms (Bezirhan & Ozsoy 2015). In comparison to bio-plastics, the conventional plastics are more recyclable as they are not easily degraded.
6) Other advantages of conventional plastics include high resistance to corrosion, high flexibility, etc. (Awasthi et al. 2017), which made the conventional plastics so popular.

9.4.2   Disadvantages

Despite the revolution brought by the use of plastics, plastics have created huge problems in waste management. Reports show that 5–15% of total municipal solid waste, by weight, is occupied by plastic alone, which is 20–30% of the total volumetric proportion (WWF 2021). Thus, we must focus to see the disadvantages offered by plastics. The disadvantages are:

1) The report of Swatch Bharat Mission (Urban) has estimated that approximately 70% of plastics used in packaging becomes waste in a very short period time (Ministry of Housing and Urban Affairs 2019). In India, 60% of plastics are recycled but the remaining 40% (around 9400 tons) is either land filled or added to water sources somehow. This is not a positive management at all.

2) It is seen that plastic wastes never degrade automatically. Although plastics are recycled, the recycled products are found to be more harmful as they contain additives or colors. A plastic can be recycled about 2–3 times, after which they lose their functionality due to thermal pressure, so recycling is not a permanent solution for plastic waste management (Ortiz et al. 2020).

3) As plastics are not degraded easily, they immediately start polluting land or water once they are disposed of. After many years, the plastics break into small pieces to form MPs. If plastics are landfilled, due to some chemical processes, toxic chemicals are released into soil and finally reach the water (Ortiz et al. 2020; Plastic Pollution Coalition 2018).

4) If plastics are deposited in the oceans, they become a threat to aquatic lives. The plastics floating in the oceans decreases the dissolved oxygen level in the water. It is seen that most of the oceanic creatures eat plastics as food, which become the reason for decline in their populations. Due to ingestion and suffocation, larger aquatic birds and fishes have faced death, and the presence of MPs have been found inside phytoplankton (Ortiz et al. 2020).

5) The burning of plastic is harmful as it releases retardants called halogens, which may cause cancer, endometriosis, neurological damage, endocrine disruption, birth defects, child developmental disorder, asthma, multiple organ damage, reproductive damage, etc. (Ortiz et al. 2020).

9.5 Current Status of Biodegradable Plastics and Effect on Environment

As a sustainable alternative to conventional plastic, bioplastic has attracted researchers and the market. As the fossil fuels are gradually decreasing and global-warming is increasing, so the global market has shifted toward the bio-economy (Storz & Vorlop 2013). Currently, around 0.5% of the total global plastic production uses bio-based or bio-plastics. Germany is the largest plastic producer in the EU with a large plastic market. On one side, the fossil or non-renewable resources are diminishing, while the price of plastic production is increasing, with the renewable energy debate for alternative source escalating. Although plastic has become irreplaceable in our lives, it is our responsibility to consider the negative impacts of plastics on the environment, and this situation leads the scientific community to depend on biomass.

Poly(hydroxyalkanoate) (PAH) was the first bio-based plastic product produced by Great Britain in the 1980s, but it could not replace the market of conventional non-biodegradable plastic due to many drawbacks for useful application. Polylactic acid is the first purely bio-based plastic synthesized and produced in large quantities by the American company Nature LLC in 2001.

Some of the popular bio-plastics are discussed in following segment.

a) Polylactic acid (PLA): Most widely-used bioplastic, which has qualities such as it is renewable, biodegradable, and biocompatible. It is prepared by ring opening or direct polymerization of lactide or lactic acid. PLA or its advanced composite polymer are found in markets in packaging, biomedical application, textile industries. Although

Mitsubishi, Toyota from Japan, Biomer of Germany, etc. produce PLA, the US is the largest producer of PLA from corn.

b) Polyhydroxyalkanoates (PHA): These are derived from various carbon sources by the action of microorganisms. Recent reports suggest that cost-effective PHA can be produced by accretion of PHA in transgenic plants such as Arabidopsis. Poly(hydroxyl butarate-cohydroxyvalerate) (PHBV) and polyhydroxybutarate (PHB) fall in this category. Besides application of PHA in packaging, fiber, and biomedical uses, it can be used for production of biofuels and drug synthesis. More than 20 companies are found globally which are devoted to commercialize PHA and its other derivatives.

c) Bio-based polymides: These are also known as nylons, having amide groups as and integral part of polymer chain. These polymers find applications in flexible electronics, automotives, packaging, etc. In conventional polymerization, nylons are produced from diamines and dibasic acids, but castor oil is used to synthesize bio-based nylons. Nylon 10, 10 is a bio-based nylon.

d) Bio-based poly(butyl succinate) (Bio-PBS): This is prepared by the polymerization of succinic acids and 1,4-butandiol. Currently, bio-PBSs are prepared using starch, corn steep liquor, cereals, etc. by the action of bacteria such as *Actinobacillus succinogenes*, *Mannheimia succiniciproducens*, and *Anaerobiospirillum*.

e) Bio-based polyethylene (Bio-PE) and Bio-based polypropylene (Bio-PP): Polythene is a widely-used plastic produced from polymerization of ethane, which consumes a large share of petroleum stock. Now bio-PE is replacing PE and PP. Braskem is the first company to produce bio-based PE commercially.

f) Bio-based polyethylene terephthalate (PET): PETs are polymers produced from the polymerization of terephthalic acids and ethylene glycol. Currently, bio-based ethylene glycol is used to produce bio-based bio-PET. Coca-Cola is presently using bio-PET bottles to pack their cold drinks, and other companies such as Toyota Tsusho Corporation, Japan, and Futura Polysters in India are the largest producers of bio-based PET.

9.6 How Plastic Degrades

The process by which the functionality of the polymers are altered by bond breaking or other chemical transformation is called polymer degradation. The factors involved may be environmental including light, heat, or moisture, or chemical or biological factors. The first case of plastic (lignin and paraffin) degradation was reported by Fush and Jen-hou in 1961, and was due to bacterial action (Shahnawaz et al. 2019). According to them, plastics up to 4800 molecular weight are degradable from biological action. The plastics are degraded in three ways, which are bioremediation, photodegradation, and oxy-photodegradation. While the photodegradation requires light sources and oxy-photodegradation require both heat and oxygen supply, bioremediation requires the action of microorganisms in the decay process. Thus, bioremediation is the most accepted process of plastic waste management.

The biodegradation of plastic is reported as a four-step mechanism. Initially the microbes become attached to the surface of the polymer; next the microbes assume it is food and start consuming as carbon source. Then in the primary degradation the polymer converts into pieces of low molecular weight, which is followed by final degradation of polymer. During these actions, the microbes secret enzymes, which catalyze the entire process.

In the final process of degradation the plastics converts to CO_2, H_2O, and biomass under aerobic conditions, but under anaerobic conditions the plastic may produce CH_4 as an additional product. On the other hand, under sulphidogenic environments it has the possibility to produce H_2S as an additional product.

It is also reported that fungi can grow faster than bacteria and can survive in low nutrient and moisture conditions as well as acidic pH, which facilitates better degradation of polymers. Although we are advocating for biodegradation of plastic, it is not 100% efficient. The actions of microorganisms can reduce the mechanical properties of plastics up to 98% (Muthukumar & Veerappillai 2015). It is believed that a small part of the polymers are absorbed into the biomass and natural products (Shah et al. 2008).

The process of biodegradation of plastics by microorganisms in soil varies based on soil conditions, as different microorganisms have different optimal growth conditions in soil. The biodegradation of polymers is also controlled by factors such nature of polymer, nature of microorganisms, nature of pre-treatment, etc. Other factors that govern the polymer degradation are functional groups, molecular weight, tacticity, mobility, structure of polymer, plasticizers, or additives present in the plastics.

9.7 Advantages and Disadvantages of Bio-Based Plastics

As unlimited exploitation of non-renewable fossil resources for plastic production has become a cause of concern as they will soon run out if we continue in the same trend; and bio-plastics have started to become the alternative to conventional plastics. To popularize the use of bio-plastics, researchers have creatively designed a tool called "Plastic spectrum" as shown in Figure 9.5.

The plastic spectrum places the bioplastics on the right side of the spectrum, which infers that they are produced from renewable resources and less harmful for environment in comparison to conventional plastics. This will create more awareness among people to procure bioplastics. However, this has both positive and negative impacts, and in this section we will discuss a few advantages and disadvantages.

Advantages are as follows.

1) Low carbon footprint: Once a plastic is produced, the CO_2 is sequestered forever, as it is not biodegradable. Plastics produced from biological sources sequester the carbon

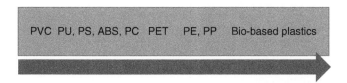

Figure 9.5 Plastics spectrum (PVC: polyvinyl chloride; PU: polyurethane; PS: polystyrene; ABS: acrylonitrile butadiene styrene; PC: polycarbonate; PET: polyethylene terephthalate; PE: polyethylene; PP: polypropylene) (Bezirhan & Ozsoy 2015).

dioxide captured by plants by photosynthesis, which upon degradation returns to the environment as CO_2 and H_2O, which continue to balance the carbon.

2) Easily available resources: Bio-plastics are made from renewable resources, so there are plenty of resources for the production of bioplastics.
3) Energy efficient: Conventional plastics consume more energy in production than bio-plastics. As bioplastic is independent of petroleum energy consumption, there is less chance of price fluctuation in bioplastics.
4) Ecologically safe: Bioplastics produce less environmentally hazardous pollutants.

Most probable disadvantages of bioplastics are as follows.

1) Bioplastics are costly in comparison to conventional plastics, as large-scale production has not yet been accomplished.
2) There are recyclability issues with bioplastics; one cannot recover bioplastics properly if they are mixed with conventional plastics.
3) Raw material competition: If bioplastic production depends on agricultural bi-products, then there is a possibility of competition between food and raw material for bioplastics.
4) Lack of legislation for use of plastics: It is reported that there was a target for the production of 6.7 million tons of bioplastics until 2018, however many countries are still lacking the proper legislation for production, usage, and waste management.

9.8 National and International Agreements and Conventions to Control Use of Plastics

Awareness regarding the negative impacts of plastic waste is not sufficient to control the exploited use of plastics; international or national laws and agreements are necessary. In 1992 the United Nations brought forth a legislation called the United Nations Conventions on the Law of Sea (UNCLOS) that entered into action in 1994 (Costa et al. 2020), which tries to regulate the use of oceans and other aspects of oceanic resources. It regulates issues such as navigational rights, rights on resources present on the seabed, territorial sea limits, conservation of marine environment, etc. The law has no specific provision to control plastic pollution but it has considered plastic as hazardous as other pollutants in marine environment (article 46). Additionally, article 210 requests signatory states to develop rules and regulations to prevent, reduce, and control marine pollution by dumping; but there are some drawbacks, as not all the states are signatories. For example, the US has not signed the UNCLOS; thus, the control of plastic pollution cannot be effective. UNEP and National Oceanic and Atmospheric Administration (NOAA) of the US have developed an agenda to control plastic pollution; but its action is non-binding in nature to meet the challenges of pollution. In Nairobi, the gathering of UNEP in 2017 passed a draft resolution to limit the littering of oceans by MPs of <5 mm. The Group of 7 (G7) and Group of 20 (G20) had also addressed the issues of plastic pollution; they stressed making strategies to promote awareness to promote sustainable waste management.

In 2019, the EU passed regulations to control and reduce plastic pollution. In India, there is only one law that prohibits use of plastic bags below 50 μm; but there is no strict law that totally bans the use of such bags. However, many large companies and vendors impose charges on plastic bags to discourage people from using them. National Green Tribunal of India has put a ban on disposable plastics such as cutlery, but we have not observed strict action on the implementation of such a ban in reality. China implemented a law in 2008 to charge for plastic bags; thus, in two years, 50% of plastic bag users have reduced the use of disposable plastic bags. Sweden is known for having the best waste recycling system; their recycling system is so strict that less than 1% of wastes go into landfills. The "No Plastic Ban, Instead more Plastic Recycling" law was imposed. (They encourage their people to follow the policy of "No plastic ban, instead more plastic recycling".)

Now they are running out of garbage and urge other countries to donate garbage. Rwanda, an African country has imposed a total ban on disposable plastic use. France is the first country to ban all daily use plastics and aims to cut the use of plastic bags 50% by 2025; they passed the "Plastic Ban" law in 2016 (UNEP 2018).

9.9   The Future of Plastics

When we are advocating for the use of biodegradable plastic or bio-based plastic it does not mean that the use conventional plastic will cease immediately. Humans have availed the benefits of using polymers since 1600 BCE when the Mesopotamians prepared balls, figurines, and bands using natural rubber (Andrady & Neal 2009). Since the discovery of PP in 1954 by Giulio Natta, and its commercial production in 1957, to the present, it is the most-used plastic globally. We will not suddenly be able to replace it totally with bio-plastic. As the plastic pollution is increasing, there is immense scope of growing markets for alternative plastics. As mentioned by Andrady & Neal (2009) in a report, the futurist Ray Hammond stated in a publication titled "The World in 2030" that there will be an exponential growth in technological development by 2030, where plastic will play a remarkable role (Hammond 2012). To minimize the spending of fossil resources we are going to depend mostly on biomass. A life without plastic becomes unavoidable, and to get rid of it will take time.

In the near future, plastic will play a significant role in the advancement in medical equipment, including organ and tissue implantation (Thompson et al. 2009b). The Boeing 787 Dreamliner aircraft already uses the lightweight component, which will reduce the fuel consumption and reduce carbon emissions, and there will be more developments in the future with the use of BDBPs.

9.10   Conclusions

From various studies it has been revealed that plastic has become a global hazard; and only biodegradable plastic or bio-based plastics can be the real alternative. If consumption of conventional plastics is reduced, it will save fossil resources for the future use. The conventional plastic production consumes nearly 4% of total global oil production; if it continues there will be shortage of petroleum in the future. The time to find efficient alternatives to petroleum-based plastics is upon us. We have to create awareness among the masses about economic, social, and individual health risks from any form of plastic. Currently, there is no law that binds all global community together to restrict the use of disposable plastics. The pollution is a transboundary issue. Everyone has to work together to mitigate such issues.

Sweden is an ideal example of proper and efficient waste management. We have to take their ideas to reduce any form of waste generation and recycling; merely creating laws is not sufficient. We have to spread more and more awareness to bring the world citizens into this issue. To propagate the use of bio-based and biodegradable plastics we have to create an environment among the masses where people can choose them logically, thinking of a safe future and environment. Any developments which progress without disturbing the environment are called sustainable development; and sustainable development must be promoted.

It is observed that both bio-based and biodegradable plastic has a bright future. They have the possibility to bring growth in economy and development in environmental conditions. Current technological breakthroughs will definitely help in the expansion of the market of bioplastics. The policymakers of every nation must keep in mind that one positive action to control the plastic pollution can save the planet for next million year.

References

Alauddin, M., Choudhury, I.A., Baradie, M.A.E., and Hashmi, M.S.J. (1995). *Plastics and their machining: a review. Journal of Material Processing Technology* 54: 40–46.

Alvarez-Chavez, C.R., Edwards, S., Moure-Eraso, R.L., and Geiser, K. (2011). *Sustainability of bio-based plastics: general comparative analysis and recommendations for improvement. Journal of Cleaner Production* 23 (1): 46–47.

American Chemistry Council (1988). Plastic packaging resins. https://plastics. americanchemistry.com/Plastic-Resin-Codes-PDF (accessed 26 October 2021).

Andrady, A.L. and Neal, M.A. (2009). *Applications and societal benefits of plastics. Philosophical Transactions of the Royal Society B* 364: 1977–1984. https://doi.org/10.1098/rstb.2008.0304.

ASTM International (2019). Standard specification for labelling of plastics designed to be aerobically composted in municipal or industrial facilities, Designation: D6400–19.

Awasthi, A.K., Shivasankar, M., and Majumdar, S. (2017). *Plastic solid waste utilization technologies: a review. IOP Conference Series: Materials Science and Engineering* 263: 022024. https://doi.org/10.1088/1757-899X/263/2/022024.

Banerjee, T. and Srivastava, R.K. (2014). Plastics waste management in India: an integrated solid waste management approach. In: *Handbook of Environmental and Waste Management*, vol. 2 (eds. Y.-T. Hung, L.K. Wang and N. Shammas), 11–24. World Scientific Publishing Co., Singapore.

Batori, V., Akesson, D., Zamani, A. et al. (2018). *Anaerobic degradation of bioplastics: a review. Waste Management* 80: 406–413. https://doi.org/10.1016/j.wasman.2018.09.040.

Bezirhan, E. and Ozsoy, H. (2015). *A review: investigation of bioplastics. Journal of Civil Engineering and Architecture* 9: 188–192. https://doi.org/10.17265/1934-7359/2015.02.007.

Costa, J.P., Mouneyra, C., Costa, M. et al. (2020). *The role of legislation, regulatory initiatives and guidelines on the control of plastic pollution. Frontiers in Environmental Science* 8: 104. https://doi.org/10.3389/fenvs.2020.00104.

Du, S., Zhu, R., Cai, Y. et al. (2021). *Environmental fate and impacts of microplastics in aquatic ecosystems: a review. RSC Advances* 11: 15762. https://doi.org/10.1039/d1ra00880c.

Emadian, S.M., Onay, T.T., and Demirel, B. (2017). Biodegradation of bioplastics in natural environments. *Waste Management* 59: 526–536. https://doi.org/10.1016/j.wasman.2016.10.006.

Geyer, R., Jambeck, J.R., and Law, K.L. (2017). *Production, use, and fate of all plastics ever made. Science Advances* 3: e1700782.

Hammond, R. (ed.) (2012). The World in 2030. Editions Yago, ISBN-13: 978-2916209180.

Hopewell, J., Dvorak, R., and Kosior, E. (2009). *Plastics recycling: challenges and opportunities. Philosophical Transactions of the Royal Society B* 364: 2115–2126. https://doi.org/10.1098/rstb.2008.0311.

Karrer, E. (1930). *Classification of plastics and definition of certain properties. Journal of Rheology* 1: 290–296. https://doi.org/10.1122/1.2116319.

Klemes, J.J., Fan, Y.V., and Jiang, P. (2021). *Plastics: friends or foes? The circularity and plastic waste footprint. Energy Sources, Part A: Recovery, Utilization, and Environmental Effects* 43: 1549–1565. https://doi.org/10.1080/15567036.2020.1801906.

Kumar, H., Azad, A., Gupta, A. et al. (2021). *COVID-19 creating another problem? Sustainable solution for PPE disposal through LCA approach, environment. Development and Sustainability* 23: 9418–9432. https://doi.org/10.1007/s10668-020-01033-0.

Lakhawat, S.S., Singh, D.D., Kumar, S., and Kumar, V. (2020). *Bioplastic feasibility: plastic waste disaster management. Journal of Critical Reviews* 7 (5): 260–264. ISSN:2394-5125. doi:https://doi.org/10.31838/jcr.07.05.46.

Lambert, S. and Wagner, M. (2017). *Environmental performance of bio-based andbiodegradable plastics: the road ahead. Chemical Society Reviews* 46: 6855. https://doi.org/10.1039/c7cs00149e.

Meereboer, K.W., Misra, M., and Mohanty, A.K. (2020). *Review of recent advances in the biodegradability of polyhydroxyalkanoate (PHA) bioplastics and their composites. Green Chem.* 22: 5519–5558. https://doi.org/10.1039/d0gc01647k.

Ministry of Housing and Urban Affairs. (2019). Swachh Bharat Mission (Urban) Plastic Waste Management; Issues, Solutions and Case Studies, March.

Muthukumar, A. and Veerappillai, S. (2015). *Biodegradation of plastics – a brief review. International Journal of Pharmaceutical Sciences Review and Research* 32: 204–209. www.globalresearchonline.net.

Nielsen, T.D., Hasselbalch, J., Holmberg, K., and Stripple, J. (2020). *Politics and the plastic crisis: a review throughout the plasticlife cycle. WIREs Energy and Environment* 9: 360. https://doi.org/10.1002/wene.360.

Nkwachukwu, O.I., Chima, C.H., Ikenna, A.O., and Albert, L. (2013). *Focus on potential environmental issues on plastic world towards a sustainable plastic recycling in developing countries. International Journal of Industrial Chemistry* 4: 34. http://www.industchem.com/content/4/1/34.

Ortiz, A.A., Sucozhanay, D., Vanegas, P., and Moscoso, A.M. (2020). *A regional response to a global problem: single use plastics regulation in the countries of the Pacific Alliance. Sustainability* 12: 8093. https://doi.org/10.3390/su12198093.

Philp, J.C., Ritchie, R.J., and Guy, K. (2013). *Biobased plastics in a bioeconomy. Trends in Biotechnology* 31 (2): 65–67.

Plastic Pollution Coalition. (2018). https://plasticpollutioncoalition.zendesk.com/hc/en-us/articles/222813127-Why-is-plastic-harmful. (Accessed on 29-05-21).

Queiroz, A.U.B. and Queiroz, F.P.C. (2008). *Innovation and industrial trends in bioplastics. Polymer Reviews* 49 (2): 65–78. https://doi.org/10.1080/15583720902834759.

Reddy, M.M., Vivekanandhan, S., Misra, M. et al. (2013). *Biobased plastics and bionanocomposites: current status and future opportunities. Progress in Polymer Science* 38 (10–11): 1653–1689. https://doi.org/10.1016/j.progpolymsci.2013.05.006.

Ritchie, H. and Roser, M. (2018). Plastic Pollution. Published online at our world in data. http://Data.org. Retrieved from: https://ourworldindata.org/plastic-pollution.

Shah, A.A., Hasan, F., Hameed, A., and Ahmed, S. (2008). *Biological degradation of plastics: a comprehensive review. Biotechnology Advances* 26: 246–265. https://doi.org/10.1016/j.biotechadv.2007.12.005.

Shahnawaz, M., Sangale, M., and Avinash, A. (eds.). (2019). *Case studies and recent update of plastic waste degradation*Bioremediation Technology for Plastic Waste. Springer Singapore https://doi.org/10.1007/978-981-13-7492-0_4.

Song, J.H., Murphy, R.J., Narayan, R., and Davies, G.B.H. (2009). *Biodegradable and compostable alternatives to conventional plastics. Philosophical Transactions of the Royal Society B* 364: 2127–2139. https://doi.org/10.1098/rstb.2008.0289.

Storz, H. and Vorlop, K.D. (2013). *Bio-based plastics: status, challenges and trends. Appl. Agric. Forestry Res.* 63: 321–332. https://doi.org/10.3220/LBF_2013_321-332.

Thompson, R.C., Swan, S.H., Moore, C.J., and Saal, F.S.V. (2009a). *Our plastic age. Philosophical Transactions of the Royal Society B* 364: 1973–1976. https://doi.org/10.1098/rstb.2009.0054.

Thompson, R.C., Moore, C.J., Saal, F.S.V., and Swan, S.H. (2009b). *Plastics, the environment and human health: current consensus and future trends. Philosophical Transactions of the Royal Society B* 364: 2153–2166. https://doi.org/10.1098/rstb.2009.0053.

UN India (2018). Our planet is drowning in plastic pollution–it's time for a change! Source: *Banning single-use plastic: lessons and experiences from countries*, UN Environment report (2018). https://www.unep.org/interactive/beat-plastic-pollution (accessed 31 May 2021).

UNEP (2018). New report offers global outlook on efforts to beat plastic pollution. Press release, June 5, New Delhi. https://www.unep.org/news-and-stories/press-release/new-report-offers-global-outlook-efforts-beat-plastic-pollution. (accessed 31 May 2021).

UNFCCC (2021). Handbook on measurement, reporting and verification http://unfccc.int/key steps/cancun_agreements/items/6132.php (accessed 24 May 2021).

Wang, Z., Peng, B., Huang, Y., and Sun, G. (2019). *Classification for plastic bottles recycling based on image recognition. Waste Management* 88: 170–181. https://doi.org/10.1016/j.wasman.2019.03.032.

World Economic Forum (2016). The new plastics economy – rethinking the future of plastics. http://www3.weforum.org/docs/WEF_The_New_Plastics_Economy.pdf (accessed 24 May 2021).

WWF (World Wildlife Federation). (2021). Plastic pollution is killing sea turtles: Here's how. www.wwf.org.au/news/blogs/plastic-pollution-is-killing-sea-turtles-heres-how. (Accessed on 31–05-2021).

Yaradoddi, J. S., Hugar, S., Banapurmath, N. R., et al. (2017). Alternative and renewable bio-based and biodegradable plastics. In: https://doi.org/10.1007/978-3-319-48281-1_150-1

10

Biodegradable Plastics: New Challenges and Possibilities toward Green Sustainable Development

Rabindra Kumar and Sumeer Razdan

Central Instrumentation Laboratory, Central University of Punjab, Bathinda, India

10.1 Introduction

Over the past five to six decades, conventional plastics have been extensively used as a replacement for wrapping materials, wood, utensils for various applications in the kitchen, daily use in households, automobile parts, motorbikes, and other manufacturing sectors worldwide. The conventional plastics are more popular due to their resilience, light weight, steadiness, and cost-effectiveness. In the present scenario, most of the world population is dependent on conventional plastic materials that have no effective arrangements for either discarding or degrading safely into the environment. Due to the extensive use of conventional plastics, collateral environmental damage has become a global phenomenon. The simple lifecycle of plastic and plastic products is shown in Figure 10.1. According to the global plastics and composites industry report 2021, approximately 370 million tons of plastic were consumed in 2019, and plastic waste collection was estimated to be 45–50% of the total consumption considered as disposable due to single-use plastic products and packaging resources (Global Plastics and Composites Industry Report 2021). A comparison of features is shown in Table 10.1 between bioplastics and conventional petroleum-based plastics.

Presently, due to the environmental and safety concerns the scenario has changed with the focus now on using bio-based alternatives instead of conventional plastics. The recent developments in the past few years for plastics based on biodegradable materials known as bio-based and biodegradable plastics with their environment-friendly advantages have come into force. Recently the use of bioplastics in our day-to-day life has increased in different ways. Still, the latest reports indicated that only four million tons of bioplastics are produced annually (Van den Oever et al. 2017) as compared to total plastic production. Therefore, the research needs the development of biodegradable products in different areas, as shown in Figure 10.2.

Bioplastics are used extensively in a wide range of applications such as packaging materials, electronic applications, medical products, fibers, agricultural and other fields, as shown

Plastic and Microplastic in the Environment: Management and Health Risks, First Edition.
Edited by Arif Ahamad, Pardeep Singh, and Dhanesh Tiwary.
© 2022 John Wiley & Sons Ltd. Published 2022 by John Wiley & Sons Ltd.

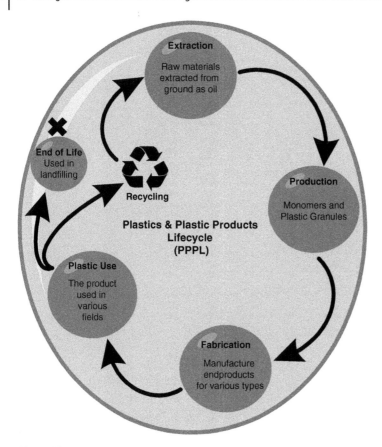

Figure 10.1 A simplified schematic representation of Plastics & Plastic Products Lifecycle (PPPL).

Table 10.1 A comparison of features of bioplastics and conventional petroleum-based plastics.

Features	Bioplastics	Conventional petroleum-based plastics
Renewable	Yes, or partially	No
Sustainable	Yes	No
Breaks down in the environment	Biodegradable and/or compostable	Some degradable by polymer oxidation
Polymer range	Limited but growing	Extensive
Greenhouse gas emissions	Usually, low	Relatively high
Fossil fuel usage	Usually, low	Relatively high
Arable land use	Currently low	None

Source: Harding et al. (2017).

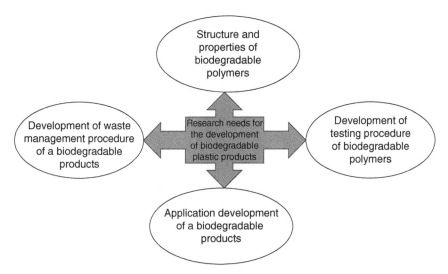

Figure 10.2 An approach toward research and development of biodegradable plastics.

in Table 10.2. Cellulose-based biopolymers are used to manufacture packaging materials, sponge cloths, and electronic applications (Vaverková & Adamcová 2015; Bilo et al. 2018). The rigid materials (especially medical products, carrier bags etc.) are manufactured from starch-based biopolymers (Soroudi & Jakubowicz 2013; Bilo et al. 2018; Calabro et al. 2020). PLA-based biopolymers have a broad spectrum of applicability in various fields such as packaging, building and constructions, textiles, and electronic applications (Vink et al. 2004; Soroudi & Jakubowicz 2013; Karamanlioglu et al. 2017). However, PHA-based polymers are extensively used in the manufacturing sector of agriculture and medical sectors (Hottle et al. 2013; Patel et al. 2019). Thus, extensive research is needed to explore the development method of cost-effective bioplastics under natural and eco-friendly conditions (Folino et al. 2020).

10.1.1 The Environmental Impact of Conventional Plastics

Conventional plastics are a more serious threat as it contains plasticizer phthalate and is known as carcinogenic. The source of phthalate is hot plastic containers such as freshly cooked hot food packed in plastic containers, which can reach our blood through the consumed food (Kumar & Pastore 2007; Ragossnig & Schneider 2017).

Polymers are the main constituents of any conventional plastics, as well as bio-based and biodegradable plastics (Figure 10.3). In conventional, nondegradable plastics, the polymers generally used are polyethelene (PE) and polyvinyl chloride (PVC).

10.1.2 Classification and Types of Biopolymers

Biodegradable polymers are classified (Hassan et al. 2019) based on their natural and synthetic origins (Figure 10.4). Natural origin biopolymers include polymers produced by

Table 10.2 Bioplastic sources and its use in various fields.

Serial No.	Bioplastic source	Use of bioplastics	References
1	Cellulose-based bioplastics	Packaging material	Bilo et al. (2018)
		Sponge cloths	Vaverková & Adamcová (2015)
		Electronic applications	Bilo et al. (2018)
2	Starch-based bioplastics	Films	Hottle et al. (2013)
		Rigid materials (plates, cutlery, and foams)	Hottle et al. (2013) Rincones et al. (2009)
		Packaging	Rincones et al. (2009); Soroudi & Jakubowicz (2013)
		Medical products	Rincones et al. (2009); Soroudi & Jakubowicz (2013)
		Agriculture field	Adhikari et al. (2016); Soroudi & Jakubowicz (2013); Accinelli et al. (2019)
		Carrier bags	Calabrò and Grosso (2018); Calabro et al. (2020)
3	PLA-based bioplastics	Packaging	Soroudi & Jakubowicz (2013); Hottle et al. (2013), Vink et al. (2004)
		Disposal or durable goods	Hottle et al. (2013)
		Electronic applications	Karamanlioglu et al. (2017)
		Fibers	Karamanlioglu et al. (2017)
		Medical products	Soroudi & Jakubowicz (2013)
		Buildings and constructions	Vink et al. (2004)
		Agricultural field	Soroudi & Jakubowicz (2013); Karamanlioglu et al. (2017)
		Textiles and films	Hottle et al. (2017)
4	PHA-based bioplastics	Medical products	Soroudi & Jakubowicz (2013); Hottle et al. (2013); Patel et al. (2019); Bhatia et al. (2019); Vu et al. (2020)
		Agriculture	Soroudi & Jakubowicz (2013); Hottle et al. (2013)
		Packaging	Soroudi & Jakubowicz (2013); Vu et al. (2020)
5	Petroleum-based bioplastics	Medical products	Soroudi & Jakubowicz (2013)
		Packaging	Soroudi & Jakubowicz (2013)
		Agriculture	Adhikari et al. (2016)

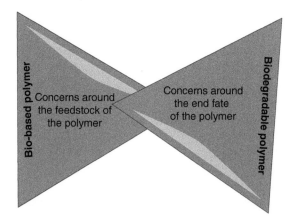

Figure 10.3 Differences between the bio-based and biodegradable polymers.

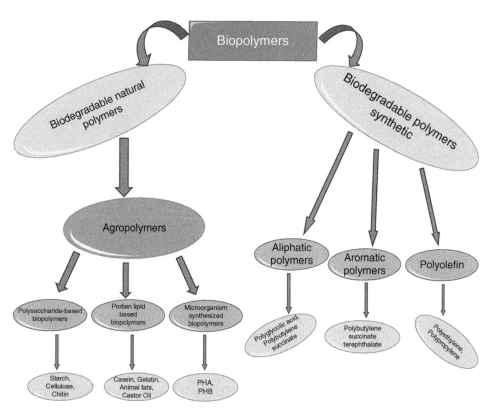

Figure 10.4 Classification of biodegradable polymers based on their natural and synthetic origins.

microbial systems such as PHA, PHB, recombinant protein polymers, and microbial polysaccharides. Biopolymers are synthesized by plants and higher organisms such as starch, plant cellulose, lignin, chitin, hyaluronic acid, etc. are also polymers of natural origin. Polymers produced by chemical polymerization of biological starting materials such as polymers synthesized from lactic acid and polyamino acids are also considered as natural biopolymers. The synthetic origin biopolymers are derived from petroleum sources and consist of synthetic monomers such as polyvinyl alcohols, aromatic and aliphatic polystyrenes, and engineered polyolefins. While natural biopolymers have desirable properties such as low weight, biodegradation ability, and enhanced barrier performance, they lack properties such as less chemical resistance, low fatigue, low mechanical properties, less durability, and restricted processing prospects (Younes 2017).

Bio-based and biodegradable plastics from renewable resources are made from biomass materials and are biodegradable in nature. These types of plastics include starch blend and modified thermoplastic starch, other bio-plastic polymers, and polylactic acid (PLA) or polyhydroxyalkonate (PHA). Bio-plastics and biodegradable plastics are also produced from fossil resources such as polycaprolactone (PCL), polybutylene succinate (PBS), and polybuthylene adipate terephthalate (PBAT). Biopolyethylene is produced from ethylene, which is made up by bioethanol (Rujnić-Sokele & Pilipović 2017).

10.2 Biopolymers of Microbial Systems

10.2.1 Microbial Polyesters: Polyhydroxyalkanoates

Polyhydroxyalkanoates (PHAs) are present in the cytoplasm of a microbial cell as granules and act as energy reserve material in the cell. PHA's have properties similar to petroleum-based polymers, and their mechanical characteristics can be tailored to rubber and crystalline plastic properties (Snell et al. 2015). PHA is originally brittle in nature, but polyhydroxy butyrate forms Poly-3-hydroxybutyrate-co-3-hydroxyvalerate (PHBV), which has equivalent properties of polyethylene and polypropylene with enhanced elasticity and strength. PHBV is produced when the bacterium *Alcaligenes eutrophus* is grown in the presence of calibrated amounts of glucose and propionic acid (Yu 2007). The raw material undergoes fermentation, and with the growth of bacteria, there is an accumulation of PHA inside the cell. The cell disruption releases the polymer, and after centrifugation and drying, the biopolymer is collected.

10.2.2 Recombinant Protein Polymers

Recombinant DNA technology is a desirable method for producing protein biopolymers or polypeptides. The protein polymers synthesized via this method are highly pure, have specific molecular weight, and have a high degree of uniformity. Genetically tailored polymers produced via rDNA include elastomers, electro optical materials, modified silk, adhesives, etc. Recombinant polymer proteins in microorganisms are produced by introducing genes encoding proteins/polymers of interest in the suitable expression vector, which is transformed into the microorganism (Casal et al. 2013). As the microorganisms grow under suitable conditions, the translation of the introduced gene is coded for the recombinant polymer protein.

10.2.3 The Microbial Polysaccharides

10.2.3.1 Bacterial Cellulose

Cellulose is a natural product traditionally extracted from various plant species. It consists of repeating glucose units joined by $\beta(1,4)$ glycosidic bonds. It can be synthesized in various bacterial species such as *Acetobacter* in the form of a polymer product via fermentation, producing highly pure and clean products. Cellulose is produced in bacterial cells by fermentation in an agitated state. The starting constituents include glucose, corn solution, and iron as chelators, along with various salts and growth enhancers. High polymer production rates occur when the growth medium contains glucose, salts, corn steep liquor, iron chelators, and various productivity enhancers. At the end of the process, bacterial cells are lysed via hot treatment and water insoluble cellulose, into the form of fibers having plastic-like thickening properties with efficient mechanical strength (Wang et al. 2019).

10.2.3.2 Xanthan

The xanthan polysaccharide consists of glucose, mannose, and gluouronic groups joined together to form a natural polymer. It also contains acetyl (Ac) and pyruvate units in the mannose structure. Xanthan polymer is chiefly produced by *Xanthomonas campestris*, utilizing corn and molasses syrup. It is recovered during the polymerization process via alcohol precipitation and removal of the bacterial cells (Ahmad et al. 2015).

10.2.3.3 Dextrans: Phullan and Glucans

Dextrans are a class of branched glucan polysaccaharides having glycosidic bonds between C–1 → C–6" atoms, which are formed by the condensation of glucose molecules. It is accumulated in the extracellular space by dextran sucrases. The dextrans are produced by the conversion of sucrose from sugarcane and sugar beet during the fermentation process by the microorganisms *Leuconstoc mesenteroides* (Esmaeilnejad-Moghadam et al. 2019). Dextrans are usually synthesized via fermentation or enzymatic filtration procedures, and the latter method produces intensified, uniform product which can be further purified.

Phullan is a class of microbial polysaccharide synthesized extracellularly by various yeast species. The polysaccharide consists of three glucose sugars linked with each other, with the repeat units linked with each other in a branched manner. The polysaccharide is a maltotriose trimer consisting of α-$(1 \rightarrow 6)$-linked $(1 \rightarrow 4)$-α-d-triglucosides.

The Phullan is produced by cultivating cultures of *Aureobasidium pullulans* via fermentation. The fermentation broth consists of starch hydrolysate, peptone, phosphate, and other basal salts. The aeration and stirring is maintained between 25 and 30 °C. The fermentation is followed by downstream processing which includes filtration of *A. pullulans* cells, which are later separated and purified using organic solvents and precipitation methods.

10.3 Biopolymers of Plants and Higher Organisms

10.3.1 Starch

Starch is a useful plant-based polymer with the potential to develop as an alternative to traditional plastics. The polymer consists of amylose; an amylopectin which vary in their branching and size. Amylose is comprised of α-(1–4)-linked glucose residues, while amylopectin consists of α-(1–4)-linked d-glucose backbone and few α-(1–6)-linked branches. Starch polymer is obtained from rice, corn, barley, potato, and wheat, which are used to manufacture adhesives, paper, diapers, stabilizers, thickening agents, and soil conditioners (Shrestha & Halley 2014).

Numerous starch-based blends are used to manufacture biodegradable garbage and polymer bags. Due to the linear structure, amylose offers superior properties for the production of biodegradable starch polymers. Prototypes of bio-plastics with enhanced mechanical strength and storage properties have been made from transgenically produced starch with predominant amylose content. Starch has no structural function, but the amylose and amylopectin content can be altered to produce starch polymer with desirable properties from mutants of rice, barley, and maize. Maize mutants, for higher amylose content, potato mutants with high amylopectin (via antisense), and amylose production of starch from barley have been obtained.

10.3.2 Cellulose

Cellulose is one of the chief constituents of plant cell walls (40–50% by weight). It is a homopolymer with repeat units of D-glucose units connected by $\beta(1 \rightarrow 4)$ glycosidic bonds. Due to the immense regularity and hydrogen bonding among the adjacent chains, a highly packed crystalline structure is formed creating hindrance in the exhibition of plastic properties (Shaghaleh et al. 2018).

This bottleneck may be alleviated by substitution of hydroxyl groups with macro groups such as acetate or nitrate. This results in the reduction of intermolecular forces, leading to enhanced solubility and increased plasticity. Similarly, cellophane can be synthesized by addition of glycerol and ethylene glycerols to regenerate cellulose. Cellulose-based polymers are primarily synthesized in two ways; first via breakdown or simplification of cellulose into polymerizable monomer entites, and second via incorporation of different forms of cellulose such as nanocellulose, natural fibers, and cellulose derivates as fillers in biopolymers. Recently some of the prominent cellulose-based polymer includes polylactic acid (PLA) based polymer matrices consisting of nanocellulose. The polymers are nontoxic, biodegradable, tough, and have great mechanical strength. Polymer matrices of cellulose, cellulose derivatives, aero gels, and nitrocellulose paper are inexpensive, bio-sustainable products with efficient electrical conductivity and are useful for energy storage and electrochemical devices.

10.3.3 Lignin

Lignin is one of the largest renewable bio-resources, with properties such as high carbon content and biodegradability, thus making it an important candidate for incorporating into the biocomposites. Due to these properties, acetylated lignins have been reported to improve the strength of thermoplastics. Lignin is incorporated into the polymeric matrix, such as plasticizers, UV light stabilizers, flame retardants, and flow enhancers to lessen the production cost, reduce plastic, and improve desirable properties. Lignin has been actively used in fiber biocomposites, and is used as a coupling agent to improve the mechanical and tensile strength of biocomposites and enhance the affinity between thermoset matrix and fibers. Lignin-based biocomposites have been used as alternatives for applications in 3D printing (Pradhan et al. 2021). The kraft softwood lignin and organosolve lignin have been extensively used to synthesize filaments for fused deposition modeling.

10.3.4 Chitin and Chitosan

Chitin is the second most abundant material as a renewable polysaccharide on Earth after cellulose (Kim 2011). Chitin is a semi-crystalline polysaccharide, hydrophilic in nature, with antimicrobial properties (Figure 10.5). It is used as fibers in composite materials to support natural polymers. However, chitosan has a tendency solubulize in acidic media and is easy to use in the manufacturing of sponges, fiber, film, and beads with different conformations. The advantage of chitosan derivatives may be due to the presence of $-NH_2$ group on the C–2 position of the D-glucosamine unit (Rinaudo 2014). Recent reports

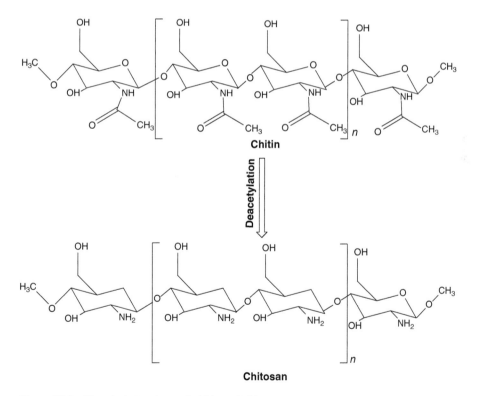

Figure 10.5 Chemical structures of chitin and chitosan.

of chitosan and its derivatives have been categorized into various applications in pharmaceutical sector, medical sector, textile, food, paper industries, agriculture, and environmental applications (Bakshi et al. 2020).

10.3.5   Polylactic Acid

Polylactic acid (PLA) is derived from biodegradable resources such as sorghum wheat, straw, and corn (Zhao et al. 2020), and has good thermoplastic properties making it a good material for making fibers and films. It is decomposed into carbon dioxide and water by various microorganisims. Due to the complexity and lack of economic feasibility of synthetic methods, the method of choice for the production of PLA remains via fermentation of sugars and subsequent conversion into monomers by hydrolysis. The three main methods for the synthesis of PLA include direct condensation polymerization, azeotrophic dehydration condensation, and lactide ring opening polymerization (Lasprilla et al. 2012). The condensation polymerization synthesis uses esterification accelerators and coupling agents to amplify the chain. The process involves condensation coupled with dehydration of carboxyl and hydroxyl groups to produce polylactic acid of low molecular weight. The purification is done via triphosgene for removing by-products along with adjuvants. The azeotrophic method does not use adjuvants but does use dibasic acids and glycols as solvents. To remove condensed water, distillation of lactic acid is done under the reduced pressure, and subsequently, diphenyl ether and catalyst are added to the reaction. After passing through molecular sieve, polymers can be precipitated, dissolved, or separated for further purification. Due to the elimination of water, the boiling point of the solvent increases, which causes extensive polymerization.

The ring opening polymerization involves solvent-free dehydration. In this process, lactic acid condensation causes the removal of condensed water, mesolactic acid, and low molecular polymers via recyclization, to obtain pure lactide with high molecular weight. Synthesis of lactide of high purity is followed by treatment with cation and anion initiators. Alkoxides and metal carboxylates are used to produce low toxicity polylactides.

10.3.5.1   Properties of PLA

PLA has excellent physical properties such as good ductility, tensile strength, and plasticity, which enables its use in the manufacture of products such as foam moldings, injections, and metal extrusion molds (Balakrishnan et al. 2012). It has been used to synthesize sutures and drug release packets as it has excellent properties of heat resistance. PLA has excellent virtues with respect to heat resistance, shine, transparency, and texture. The crystallinity decides the quantum of PLA, which dissolves at high temperatures in solvents such as benzene or dichloromethane.

The crystallization and mechanical strength depends upon the molecular weight and composition of the main chain in terms of stereochemical properties. The biodegradability of PLA is also affected by its crystalline structure; more crystallinity will cause increase in degradation time, while low crystallinity PLA will take a few weeks to degrade.

Glass temperature has a direct effect on processing and usability of PLA based material, while crystallinity does not affect the glass temperature, impact resistance is severely affected by amplifying molecular weight and crystallinity.

10.3.5.2 Improvements in PLA

PLA possesses desirable properties such as stretchability and biocompatibility, but full commercial usability is impaired due to properties such as hydrophobicity and low impact toughness. To circumvent this modified materials have recently been added and blended during the synthesis of these polymers (Simmons et al. 2019). Materials such as polyglycolic acid with high melting temperature and glass temperature are added to PLA to improve malleability, hydrophilicity, and flexibility. The combination of lactide and ethylene glycol is used in making surgical sutures of biocompatible nature. Similarly, polymers formed in combination with ε-caprolactone and pure I-lactide leads to elasticity, high flexibility, and crystallization with high melting point.

The blending improves PLA characteristics, enhances mechanical and physical attributes, and imparts desirable thermodynamic characteristics (Saini et al. 2016). Plasticizers such as citric acid, succinic acid, tartaric acid, and oxalates added with PLA increase its thermodynamic and mechanical properties. Other materials like montmorillonite (MMT) and PEG have been blended with PLA to produce tough plastic and agglomerated structures (Mohapatra et al. 2014). Similarly polymers made of PEO and PLA have efficient mechanical properties as compared to ε-caprolactone and PLA polymers. PEO imparts high hydrophilicity to this polymer, leading to increased molecular weight degradation and hydrolyzation. However, by using polypropylene oxide as an alternative to PEO, high molecular weight copolymers have been obtained (Sanabria-DeLong et al. 2008).

10.3.5.3 Application

Biocompatibility enables the use of PLA in the medical field. The intermediate product, lactic acid, is easily metabolized in the body and is non-toxic and non-inflammatory in nature.

PLA has applicability in the use of artificial blood comprising of RBC transporting oxygen (Sen Gupta 2017). Oxygen enters microcapsules via membranes, where it combines with hemoglobin and can be transported to various cells of the body. Copolymers of polylactic acid (PLA) and polyethylene glycol (PEG) enable an embedding rate of hemoglobin up to 90% (Jain et al. 2016). The PEG also has protective effects and helps to maintain the protein activity of hemoglobin during storage. PLA has been used in the drug delivery of naltrexone, pacitaxel, and doxorubin. The PLA nanoparticles enable it to easily disintegrate and load into the tumor microenvironment. Various types of hydrogels (Diblock, triblock, graft, and multiblock copolymers) have been developed as sustained drug delivery systems (Domenek & Ducruet 2016).

10.4 Factors Affecting the Rate of Degradation of Bio-plastics and Biodegradable Plastics

Biodegradable plastics have established use in many short service life applications where biodegradability is a useful feature (European Bioplastics 2008) such as compostable waste bags, biodegradable mulch film, catering products for large events, or service for packing snack foods, film packing, and bottles made from PLA for non-sparking beverages and dairy products. However, the recycling and reprocessing of biodegradable plastics waste may bring new methods of treatment and quality issues to recycling processes. Bio-plastics and biodegradable plastics as a municipal waste may result in complications for existing plastic recycling systems (La Mantia et al. 2013). Bio-plastics and biodegradable plastics are eco-friendly, but there are also some limitations such as high producing/manufacturing costs and low or poor mechanical tendency (Jain & Tiwari 2015; Thakur et al. 2017).

In the early 1960s, Merrick and Doudoroff (1961) first reported the microbial degradation of polymers. The rate of degradation of bioplastics is dependent on environmental factors such as temperature, moisture, acidic nature, etc. and the chemical nature of the polymers affects (Kale et al. 2007). Other factors are crystalline in nature, including molecular weight, composition of co-polymer, and size of the biopolymer can affect the rate of degradation of biodegradable plastics (Ho et al. 1999; Li et al. 2016).

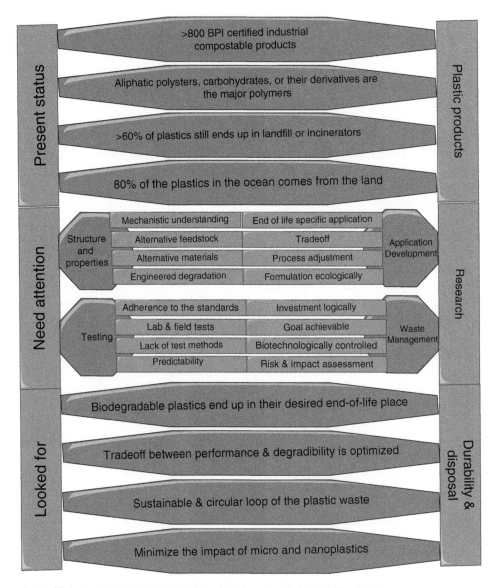

Figure 10.6 A schematic representation of roadmap to biodegradable plastics.

The classification of biodegradable processes is shown in Figure 10.6. This can be seen in only a few biodegradable polymers which are biodegradable in nature in all conditions, such as PHA and chemical pulp. Basically, the bacterial degradation process is categorized as anaerobic and aerobic biodegradation. Anaerobic conditions are mainly subject for the degradation and decomposition of polyhydroxyalkanoets, phosphinocarboxylic acids, and polylactide. The rate of biodegradation of polylactic acid (PLA) and polyhydroxybutyrate were much higher in anaerobic conditions than in aerobic conditions (Delidovich et al. 2016). However, in aerobic conditions, bacterial degradation has the ability to work with oxygen and mineralization; CO_2 and CH_4 are produced as the end products. Aerobic conditions for biodegradation of the product are primarily influenced by the nature through bacteria. Sukkhum et al. (2012) reported that after isolation, the bacteria became more proficient in trial conditions having low temperature and high pressure. Thus, several standardized methods are used to establish the disintegration and degradation of bio-plastics and biodegradable plastics. However, the most effective norm is the ISO 14855 (International Standards Organization). This is an incredible method to measure biopolymer biodegradability rate and degree of disintegration under the controlled temperature, humidity, and aeration (Petit et al. 2015).

Another recycling option for bio-plastics and biodegradable plastics is chemical recycling. In this process, the biopolymer wastes will be remelted and regranulated for the manufacture of new products. Biopolymer wastes are also converted in the form of building blocks, and this type of conversion is known as feedstock recycling. The chemical recycling process is known as the Loopla Process, in which the PLA is hydrolysed into lactic acid (Looplife 2016).

10.5 Future Aspects and Challenges for Development of Bio-based and Biodegradable Plastics

In the near future, bio-based and biodegradable polymers have tremendous potential due to their biodegradable properties, making it safe for both environmental and human use. However, the cost of production of bio-based and biodegradable plastics is high as compared to conventional plastics. The classification of biodegradable plastics is shown in Figure 10.7, which clearly indicates research outcomes in the present scenario and futuristic approaches toward the development of biopolymers.

A behavioral change of consumers buying of bio-plastics is an effective method to control conventional plastics worldwide (Muthusamy & Pramasivam 2019). Bio-plastics and biodegradable plastics are repeatedly apparent as the only possible solution for a safer environment. The wastes generated by these types of plastics are biodegradable in nature and utilized for various purposes, such as landfilling. In the search of sustainable development for the reduction of environmental impacts, bio-plastics and biodegradable plastics from renewable resources reasonably represent the best future prospects. However, sustainability issues need to be communicated to society and markets about the future of plastic awareness and use. Additionally, the prospect of more utilization of bio-plastics and biodegradable plastics can be further researched to enhance the livelihood of humankind. Developments toward cost cutting methods for recycling may create new opportunities of

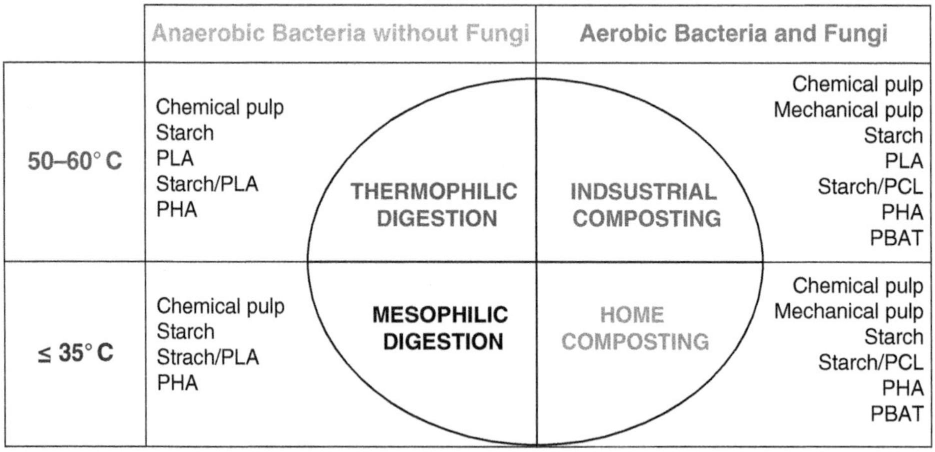

	Anaerobic Bacteria without Fungi		Aerobic Bacteria and Fungi	
50–60° C	Chemical pulp Starch PLA Starch/PLA PHA	THERMOPHILIC DIGESTION	INDSUSTRIAL COMPOSTING	Chemical pulp Mechanical pulp Starch PLA Starch/PCL PHA PBAT
≤ 35° C	Chemical pulp Starch Strach/PLA PHA	MESOPHILIC DIGESTION	HOME COMPOSTING	Chemical pulp Mechanical pulp Starch Starch/PCL PHA PBAT

Figure 10.7 Classification of biodegradable processes (DeWilde, 2013) (PBAT: Polybutylene Adipate Terephthalate; PCL: Polycaprolactone; PHA: Polyhydroxyalkanoate; PLA: Polylactic Acid).

applications in different fields such as agriculture, and pharmaceutical industries, and many more polymeric materials based on biodegradables are the strongest competitiveness to beat conventional plastic in the future (Raza et al. 2018).

As we know, bioplastics have high calorific value and can be used to produce energy in conventional plastic waste incineration facilities because they are very similar to wood; therefore recovery of energy is feasible from bioplastic wastes through incineration (Davis & Song 2006; Eriksson & Finnveden 2009; Tyskeng & Finnveden 2010; European Bioplastics 2008).

10.6   Conclusions

The use of bio-based and biodegradable plastics over conventional plastics is needed today to protect the environment and for human safety concerns. Based on recent trends in the development of biodegradable polymers and their production challenges, polymer chemistry will play an important role in developing biodegradable and bio-plastics, with the help of genetic engineering, chemical engineering, and biotechnology. The increased use of bio-plastics can be attributed to the development of emerging technology for using natural biopolymers such as lignin, chitin, polysaccharides, and proteins of microbial origin as biopolymers, and utilizing them in various ways for the development of sustainable biodegradable bio-plastics. In summary, the replacement of conventional plastics in whole or part by bio-based and biodegradable plastics strongly advocates toward a healthier environment and the safety of human health.

Acknowledgement

We acknowledge the Central University of Punjab, Bathinda, India for providing the opportunity to write this chapter. Further, due to the limited focus of the article, some references could not be included, for which we apologize.

References

Accinelli, C., Abbas, H.K., Shier, W.T. et al. (2019). *Degradation of microplastic seed film-coating fragments in soil. Chemosphere* 226: 645–650.

Adhikari, D., Mukai, M., Kubota, K. et al. (2016). *Degradation of bioplastics in soil and their degradation effects on environmental microorganisms. Journal of Agricultural Chemistry and Environment* 5 (1): 23.

Ahmad, N.H., Mustafa, S., and Che Man, Y.B. (2015). *Microbial polysaccharides and their modification approaches: a review. International Journal of Food Properties* 18 (2): 332–347.

Bakshi, P.S., Selvakumar, D., Kadirvelu, K., and Kumar, N.S. (2020). *Chitosan as an environment-friendly biomaterial–a review on recent modifications and applications. International journal of biological macromolecules* 150: 1072–1083.

Balakrishnan, H., Hassan, A., Imran, M., and Wahit, M.U. (2012). *Toughening of polylactic acid nanocomposites: a short review. Polymer-Plastics Technology and Engineering* 51 (2): 175–192.

Bhatia, S.K., Wadhwa, P., Bhatia, R.K. et al. (2019). Strategy for biosynthesis of polyhydroxyalkanoates polymers/copolymers and their application in drug delivery. In: *Biotechnological Applications of Polyhydroxyalkanoates*, 13–34. Singapore: Springer.

Bilo, F., Pandini, S., Sartore, L. et al. (2018). *A sustainable bioplastic obtained from rice straw. Journal of Cleaner Production* 200: 357–368.

Calabrò, P.S. and Grosso, M. (2018). *Bioplastics and waste management. Waste Management* 78: 800–801.

Calabrò, P. S., Folino, A., Karageorgiou, A., & Komilis, D. (2020). Biodegradation of wasted bioplastics in natural and industrial environments: A review. Sustainability, 12(15), 6030.

Casal, M., Cunha, A.M., and Machado, R. (2013). *Future trends for recombinant protein-based polymers: the case study of development and application of silk elastin-like polymers.* In: *Bio Based Plastics: Materials and Applications* (ed. S. Kabaci), 311–329. USA: Wiley.

Davis, G. and Song, J.H. (2006). *Biodegradable packaging based on raw materials from crops and their impact on waste management. Industrial Crops and Products* 23 (2): 147–161.

Delidovich, I., Hausoul, P.J., Deng, L. et al. (2016). *Alternative monomers based on lignocellulose and their use for polymer production. Chemical Reviews* 116 (3): 1540–1599.

DeWilde, B. (2013). Experiences on 20 years of biopolymer testing and certification: Challenges and new developments. In: *3rd international PLASTiCE conference, COBRO-Packaging Research Institute, Warsaw*, 1–2.

Domenek, S. and Ducruet, V. (2016). Characteristics and applications of PLA. In: *Biodegradable and Biobased Polymers for Environmental and Biomedical Applications* (ed. S. Kalia and L.R. Avérous), 171–224. USA: Wiley.

Eriksson, O. and Finnveden, G. (2009). *Plastic waste as a fuel-CO2-neutral or not? Energy & Environmental Science* 2 (9): 907–914.

Esmaeilnejad-Moghadam, B., Mokarram, R.R., Hejazi, M.A. et al. (2019). *Low molecular weight dextran production by Leuconostoc mesenteroides strains: optimization of a new culture medium and the rheological assessments. Bioactive Carbohydrates and Dietary Fibre* 18: 100181.

European Bioplastics (2008). *Bioplastics FAQ.* European Bioplastics. https://www.europeanbioplastics.org/?s=European+Bioplastics%2C+June+2008 (accessed 07 July 2021).

Folino, A., Karageorgiou, A., Calabrò, P.S., and Komilis, D. (2020). *Biodegradation of wasted bioplastics in natural and industrial environments: a review. Sustainability* 12 (15): 6030.

Global Plastics and Composites Industry Report (2021). Convergence and Collaboration to Usher Circular Economy in the Plastics and Composites Industry. https://www.researchandmarkets.com/r/boemdl (accessed 10 November 2021).

Harding, K.G., Gounden, T., and Pretorius, S. (2017). *"Biodegradable" plastics: a myth of marketing? Procedia Manufacturing* 7: 106–110.

Hassan, M.E.S., Bai, J., and Dou, D.Q. (2019). *Biopolymers; definition, classification and applications. Egyptian Journal of Chemistry* 62 (9): 1725–1737.

Ho, K.L.G., Pometto, A.L., and Hinz, P.N. (1999). *Effects of temperature and relative humidity on polylactic acid plastic degradation. Journal of Environmental Polymer Degradation* 7 (2): 83–92.

Hottle, T.A., Bilec, M.M., and Landis, A.E. (2013). *Sustainability assessments of bio-based polymers. Polymer Degradation and Stability* 98 (9): 1898–1907.

Hottle, T.A., Bilec, M.M., and Landis, A.E. (2017). *Biopolymer production and end of life comparisons using life cycle assessment. Resources, Conservation and Recycling* 122: 295–306.

Jain, R. and Tiwari, A. (2015). *Biosynthesis of planet friendly bioplastics using renewable carbon source. Journal of Environmental Health Science and Engineering* 13 (1): 1–5.

Jain, A., Kunduru, K.R., Basu, A. et al. (2016). *Injectable formulations of poly(lactic acid) and its copolymers in clinical use. Advanced Drug Delivery Reviews* 107: 213–227.

Kale, G., Kijchavengkul, T., Auras, R. et al. (2007). *Compostability of bioplastic packaging materials: an overview. Macromolecular Bioscience* 7 (3): 255–277.

Karamanlioglu, M., Preziosi, R., and Robson, G.D. (2017). *Abiotic and biotic environmental degradation of the bioplastic polymer poly(lactic acid): a review. Polymer Degradation and Stability* 137: 122–130.

Kim, S.K. (ed.) (2011). Cosmeceutical applications of chitosan and its derivatives. In: *Chitin, Chitosan, Oligosaccharides and their Derivatives*, 1e, 241. Boca Raton, FL: CRC Press.

Kumar, A. and Pastore, P. (2007). *Lead and cadmium in soft plastic toys. Current Science* 93 (25): 818–822.

La Mantia, F.P., Botta, L., and Scaffaro, R. (2013). *The effects of PLA in PET recycling systems. Macplas International at K.*

Lasprilla, A.J., Martinez, G.A., Lunelli, B.H. et al. (2012). *Poly-lactic acid synthesis for application in biomedical devices—a review. Biotechnology Advances* 30 (1): 321–328.

Li, W.C., Tse, H.F., and Fok, L. (2016). *Plastic waste in the marine environment: a review of sources, occurrence and effects. Science of the Total Environment* 566: 333–349.

Li, G., Zhao, M., Xu, F. et al. (2020). *Synthesis and biological application of polylactic acid. Molecules* 25 (21): 5023.

Looplife. (2016). Title-Bio-polymer http://www.looplife-polymers.eu/drupal/bio; (accessed 8 July, 2021).

Merrick, J.M. and Doudoroff, M. (1961). *Enzymatic synthesis of poly-β-hydroxybutyric acid in bacteria. Nature* 189 (4768): 890–892.

Mohapatra, A.K., Mohanty, S., and Nayak, S.K. (2014). *Effect of PEG on PLA/PEG blend and its nanocomposites: a study of thermo-mechanical and morphological characterization. Polymer Composites* 35 (2): 283–293.

Muthusamy, M.S. and Pramasivam, S. (2019). *Bioplastics–an eco-friendly alternative to petrochemical plastics. Current World Environment* 14 (1): 49.

Patel, S.K., Sandeep, K., Singh, M. et al. (2019). Biotechnological application of polyhydroxyalkanoates and their composites as anti-microbials agents. In: *Biotechnological Applications of Polyhydroxyalkanoates* (ed. V.C. Kalia), 207–225. Singapore: Springer.

Petit, M.G., Correa, Z., and Sabino, M.A. (2015). *Degradation of a Polycaprolactone/eggshell biocomposite in a bioreactor. Journal of Polymers and the Environment* 23 (1): 11–20.

Pradhan, R.A., Rahman, S.S., Qureshi, A., and Ullah, A. (2021). *Biopolymers: opportunities and challenges for 3D printing.* In: *Biopolymers and their Industrial Applications* (ed. S. Thomas, S. Gopi and A. Amlraj), 281–303. The Netherlands: Elsevier.

Ragossnig, A.M. and Schneider, D.R. (2017). *What is the right level of recycling of plastic waste? Waste Management & Research: The Journal for a Sustainable circular Economy* 35 (2): 129–131.

Raza, Z.A., Abid, S., and Banat, I.M. (2018). *Polyhydroxyalkanoates: characteristics, production, recent developments and applications. International Biodeterioration & Biodegradation* 126: 45–56.

Rinaudo, M. (2014). *Materials based on chitin and chitosan.* In: *Bio-Based Plastics: Materials and Applications* (ed. C.V. Stevens), 63–80. John.

Rincones, J., Zeidler, A.F., Grassi, M.C.B. et al. (2009). *The golden bridge for nature: the new biology applied to bioplastics. Journal of Macromolecular Science®, Part C: Polymer Reviews* 49 (2): 85–106.

Rujnić-Sokele, M. and Pilipović, A. (2017). *Challenges and opportunities of biodegradable plastics: a mini review. Waste Management & Research* 35 (2): 132–140.

Saini, P., Arora, M., and Kumar, M.R. (2016). *Poly(lactic acid) blends in biomedical applications. Advanced Drug Delivery Reviews* 107: 47–59.

Sanabria-DeLong, N., Crosby, A.J., and Tew, G.N. (2008). *Photo-cross-linked PLA-PEO-PLA hydrogels from self-assembled physical networks: mechanical properties and influence of assumed constitutive relationships. Biomacromolecules* 9 (10): 2784–2791.

Sen Gupta, A. (2017). *Bio-inspired nanomedicine strategies for artificial blood components. Wiley Interdisciplinary Reviews: Nanomedicine and Nanobiotechnology* 9 (6): e1464.

Shaghaleh, H., Xu, X., and Wang, S. (2018). *Current progress in production of biopolymeric materials based on cellulose, cellulose nanofibers, and cellulose derivatives. RSC Advances* 8 (2): 825–842.

Shrestha, A.K. and Halley, P.J. (2014). *Starch modification to develop novel starch-biopolymer blends: state of art and perspectives.* In: *Starch Polymers* (ed. P.J. Halley and L.R. Avérous), 105–143. Burlington, MA: Elsevier.

Simmons, H., Tiwary, P., Colwell, J.E., and Kontopoulou, M. (2019). *Improvements in the crystallinity and mechanical properties of PLA by nucleation and annealing. Polymer Degradation and Stability* 166: 248–257.

Snell, K.D., Singh, V., and Brumbley, S.M. (2015). *Production of novel biopolymers in plants: recent technological advances and future prospects. Current Opinion in Biotechnology* 32: 68–75.

Soroudi, A. and Jakubowicz, I. (2013). *Recycling of bioplastics, their blends and biocomposites: a review. European Polymer Journal* 49 (10): 2839–2858.

Sukkhum, S., Tokuyama, S., and Kitpreechavanich, V. (2012). *Poly(L-lactide)-degrading enzyme production by Actinomadura keratinilytica T16-1 in 3 L airlift bioreactor and its degradation ability for biological recycle. Journal of Microbiology and Biotechnology* 22 (1): 92–99.

Thakur, S., Govender, P.P., Mamo, M.A. et al. (2017). *Recent progress in gelatin hydrogel nanocomposites for water purification and beyond. Vacuum* 146: 396–408.

Tyskeng, S. and Finnveden, G. (2010). *Comparing energy use and environmental impacts of recycling and waste incineration. Journal of Environmental Engineering* 136 (8): 744–748.

Van den Oever, M., Molenveld, K., van der Zee, M., and Bos, H. (2017). *Bio-based and biodegradable plastics: facts and figures: focus on food packaging in the Netherlands* Wageningen Food & Biobased Research. 1722.

Vaverková, M.D. and Adamcová, D. (2015). Biodegrability of bioplastic materials in a controlled composting environment. *Journal of Ecological Engineering* 16 (3): 155–160.

Vink, E.T., Rábago, K.R., Glassner, D.A. et al. (2004). *The sustainability of NatureWorks™ polylactide polymers and Ingeo™ polylactide fibers: an update of the future. Macromolecular Bioscience* 4 (6): 551–564.

Vu, D.H., Åkesson, D., Taherzadeh, M.J., and Ferreira, J.A. (2020). *Recycling strategies for polyhydroxyalkanoate-based waste materials: an overview. Bioresource Technology* 298: 122393.

Wang, J., Tavakoli, J., and Tang, Y. (2019). *Bacterial cellulose production, properties and applications with different culture methods–a review. Carbohydrate Polymers* 219: 63–76. http://www.european-bioplastics.org.

Younes, B. (2017). *Classification, characterization, and the production processes of biopolymers used in the textiles industry. The Journal of the Textile Institute* 108 (5): 674–682.

Yu, J. (2007). Microbial production of bioplastics from renewable resources. In: *Bioprocessing for Value-Added Products from Renewable Resources*, 585–610. Elsevier.

Zhao, X., Hu, H., Wang, X. et al. (2020). *Super tough poly(lactic acid) blends: a comprehensive review. RSC Advances* 10 (22): 13316–13368.

11

Current Trends, Challenges, and Opportunities for Plastic Recycling

Yong Chen and Steplinpaulselvin Selvinsimpson

School of Environmental Science and Engineering, Huazhong University of Science and Technology, Wuhan, China

11.1 Introduction: The Pollution Problem Involving Plastic

Plastic is strewn everywhere; it's in the food we eat, the water we drink, the fisheries we depend on, and even our own bodies. Plastic garbage is piling up at alarming rates, putting the environment and human health in jeopardy. Microplastics (MPs), for example, have been detected in places as widespread as the Mariana Trench and human placentas. Plastic is becoming a significant topic of concern, despite the fact that the specific consequences on human health are still unknown (Browning et al. 2021; Bucci et al. 2020; Chiba et al. 2018; Ragusa et al. 2021). Plastics are also lightweight, long-lasting, and low-cost materials. Plastics have steadily supplanted other materials such as wood, metal, and glass as the ubiquitous materials of the world economy due to their chemical qualities and inexpensive cost. Plastics demand has been gradually rising, with global output hitting 359 million tons (Mt) in 2018 and a 355 billion euro industry turnover in 2017 (PlasticsEurope 2019). Plastic trash, on the other hand, has serious environmental consequences (Jiang 2018), including the destruction of natural systems (Ryberg et al. 2019), major greenhouse gas emissions, carbon feedstock depletion, and the circulation of toxic chemicals (Cottafava et al. 2021). Despite the fact that plastics have proven to be effective in areas such as food security and healthcare, the environmental and human health repercussions of their use are concerning. The environmental consequences of unmanaged or poorly managed plastic garbage entering the world's oceans have been well documented. As a result, immediate action is required to stem the flow of plastic trash into the surroundings (Browning et al. 2021).

In this sense, existing plastic waste management should be reevaluated in light of promising recycling technologies. Excessive land use, soil and groundwater depletion, air pollution, habitat loss, and the production of damaging greenhouse gases, to name a few, are all risks associated with landfilling, among the most common solid plastic waste (SPW) disposal methods (Dogu et al. 2021; Singh et al. 2017; Zhou et al. 2016). The energy recovery path, on the other hand, which is another popular choice, is less environmentally and economically friendly. Unfortunately, mechanical recycling lacks a truly sustainable option for

Plastic and Microplastic in the Environment: Management and Health Risks, First Edition.
Edited by Arif Ahamad, Pardeep Singh, and Dhanesh Tiwary.
© 2022 John Wiley & Sons Ltd. Published 2022 by John Wiley & Sons Ltd.

closing the loop, and in the vast majority of circumstances, re-use is only achievable to a small extent. As a result, a novel recycling approach is required for the process to be successful (Dogu et al. 2021).

11.2 Sources, Types, and Transportation of Plastics in the Environment

11.2.1 Plastics Sources and Types

Plastic has started to take the place of glass, wood, and metal in a variety of applications, and has become an integral part of human life. Their worldwide production has increased dramatically as a result of their valuable properties. (Thompson et al. 2009). In 2017, global production came close to 350 Mts (PlasticsEurope 2018). As global production and consumption of plastics expands, MP pollution in the atmosphere is expected to increase, posing a major threat to the environment and ecosystems (Wu et al. 2019). Recent research has shown MP buildup and abundance in the marine world, freshwater environments, and fish (Ashar et al. 2020). The overall amount of plastic garbage has been attempted to be measured by scientists. Most scientists acknowledge that there is no precise estimate of the amount of global plastic waste or debris that pollutes soil and bodies of water, or how much plastic waste enters the aquatic ecosystem, but they all agree that the amount of plastic waste continues to be significant based on production figures (Derraik 2002). Land-based sources of plastic waste account for 80% of the plastic debris found in the ocean globally, and rivers have the potential to be a major transport channel for all sizes of plastic debris. Plastic in the aquatic environment can also return to land during high tide or flooding occurrences, rather than being carried from land to sea by rivers (Ashar et al. 2020).

Plastic waste types were assessed in a variety of aquatic ecosystems in order to develop a worldwide system of waste plastic movement and development that may be used to help reduce plastic pollution in water surroundings. Packages and consumable goods seem to be the most common item types in rivers, whereas fisheries items were the most common in the ocean. Electronics, construction and renovation materials, and transportation plastics were all in short supply. The most polluting polymers in all conditions were polyethylene (PE) and polypropylene (PP). The most polymer diversity was found to have accumulated in oceanic and freshwater sediments. This suggests that plastic density, surface area, and scale have the most influence on plastic trash movement and accumulation trends. The majority of litter is expected to be kept in sediments or beaches, while only thick-walled, larger plastic garbage consisting of low-density polymers is carried from rivers to the ocean by tides (Schwarz et al. 2019).

Further, natural and synthetic polymers are used in a variety of applications, including furniture, household appliances, textiles, agriculture, packaging, healthcare, electronics, automotive, and energy. Because of its exceptional features like lightness, strength, low cost, and ease of manufacture, polymers are employed all over the world (Dogu et al. 2021). Additionally, according to Zhou et al. (2016), the polymer sector creates jobs and fosters global growth and competitiveness. In 2016, the plastics industry directly employed about 1.5 million workers and generated approximately 30 billion euros to a European economy

(Dogu et al. 2021; Plastics Europe 2017). Over the last 50 years, demand for plastics has steadily increased, and just about 8% of all crude oil used worldwide is currently utilized to make plastics and fuel plastic manufacturing processes. A more recent trend in the sector is the growth in the worldwide plastics output from 322 Mts during 2015 to 335 Mts during 2016 (Dogu et al. 2021; PlasticsEurope 2017). According to a report published by PlasticsEurope (2017), the main plastics manufactured in Europe are divided into four categories based on demand; polypropylene (PP) (19.3%) > low-density polyethylene (LDPE) (17.5%) > high-density polyethylene (HDPE) (12.3%) > poly(vinyl chloride) (PVC) (10%) > polyurethanes (PUR) (7.5%) > poly(ethylene terephthalate) (PET) (7.4%) > polystyrene (PS) (6.7% wt). In another report published by PlasticsEurope (2017), the main plastics manufactured in Europe are divided into four categories based on demand. With 16% of global demand, the construction industry is the second-largest end-use sector. Due to its hardness and durability, PVC is a big contributor to the construction sector, since it's utilized for door and window frames, as well as subterranean pipelines. Synthetic textiles are the world's third-largest end-use market, with 15% of global demand. Synthetic fibers are used in a variety of items, including rope, carpet, and fabric, as well as a wide range of specialty textile applications. For example, bullet-proof body armor is made from Kevlar[®], a lightweight, heat-resistant, and durable synthetic fiber. One of the world's top policy priorities is the transition to a sustainable and circular economy. Waste management of solid plastic waste (SPW) is an essential part of the circular economy solution (Dogu et al. 2021).

11.2.2 Plastic Transportation in Aquatic Environments

The features of plastic trash, along with local environmental conditions, combine to create vertical and horizontal movement of plastics in the water. Three aspects determine the vertical migration of plastics in aquatic system; particle size, density, and polymer surface area (Chubarenko et al. 2016; Kowalski et al. 2016). The density of a polymer could be greater or lesser than that of the water (ocean density 1.02–1.04 g/cm vs. freshwater density 1.00 g/cm). Quite a few polymers float in water, whereas others sink, and yet another has a central density. Local environmental conditions have a greater influence on their sedimentation patterns. PS is available in two types; foamed (e.g. expanded PS) and non-foamed (e.g. PS), with the density of the polymer being difficult to identify because it was not always indicated explicitly in research. The polymer contents in the epipelagic divisions show the impact of density. Low density polymers (PE and PP) were the most common, with high-density polymers showing up seldom or not at all. Rivers have a greater proportion of high-density polymers than those in the ocean; the fact that rivers have stronger currents can explain this. As a result of the increased turbulence, high-density polymers stay buoyant and could be delivered horizontally rather than vertically. Vertical transport can take over when ocean currents are weak, explaining the disparity in high-density polymer involvement. Investigations have indicated that the impact of high density polymers reduces as one moves farther from coastal locations (Schwarz et al. 2019). Due to increasing turbulence, low-density polymers, on the other hand, are forced to settle. Water turbulence is caused by wind, waves, and/or tides, which results in increased vertical movement in the water column. This is especially true for smaller macroplastics and MPs, which have lesser terminal rising velocities owed to their increased surface area (Kukulka et al. 2012;

Lebreton et al. 2012). Low-density polymer micro- and nanoplastics (<100 nm) heteroaggregate with particulate matter during mixing, leading in deposition in the top layer that may surpass 50% of entire particles in some models (Besseling et al. 2017). Whereas these particulates could be resuspended, they also have the potential to become deposited in deeper strata. Vertical flow, or sedimentation, is influenced through environmental influences. Higher salinity and lower ocean temperatures, for example, cause a surge in high-density polymers in the surrounding water. In short, the polymer form has a significant impact on the composition of plastics in an existing compartment, with variances in content in the same part possibly clarified via environmental circumstances such as salt and temperature (Kowalski et al. 2016; Schwarz et al. 2019).

11.3   An Introduction to Waste Management

Considerable shifts toward a more sustainable culture are expected in the waste management industry. For coming generations, it will be even more important to take a much more comprehensive approach to sustainable practices. Many businesses are now considering the recyclability and functionality of their plastic packaging. Recycling design is a major topic among designers. For a circular economy to succeed, communication throughout the distribution network is critical. Recycling, composting, anaerobic digesters, landfill, and incineration are the final options for plastic items in the waste management industry. In an ideal world, the 4Rs method is used, which entails reducing, reusing, recycling, and eventually recovering energy, with landfill being the least environmentally friendly alternative. Compostable biopolymer solutions have become extremely attractive. The benefits and drawbacks of various circumstances are discussed in this section (Letcher 2020).

11.3.1   Plastic Waste Treatment

The most frequent way to deal with plastic waste is to return it to the plastic manufacturing line's heating cycle in order to increase output and lower overall production costs. The secondary method is to mechanically re-extrude, progress, and transform plastic wastes to fresh plastic items that are mixed with virgin polymers; and the tertiary method is to chemically or thermochemically change the polymer matrix of plastic to use as a monomer source in manufacturing recycling loops. Although the tertiary method is one of the most common methods of recycling plastic, most plastics could not be recycled. For example, thermoplastics and thermosetting polymers are two different forms of plastics; the others comprise polymers that cross-link to establish a permanent chemical connection and cannot be remelted into fresh material, no matter how much heat is applied. Moreover, in terms of enhancing the quality of the polymers produced, recycling plastics is inefficient and ineffective (Chia et al. 2020; Post et al. 2020; Vijayakumar and Sebastian 2018). The 4Rs approach to waste management is based on this thinking: Minimize, reuse, recycle, and recover are the best management strategies, with landfill being the minimum desirable. It is also feasible for such a similar polymer to flow over several phases, such as creating a reusable container that is gathered and recycled into something such as a lasting application that is then reused for energy when it's no longer useful (Hopewell et al. 2009).

Landfills are the traditional method of waste disposal, but landfill space has become rare in certain countries. A really good landfill location causes minimal instant environmental damage apart from the impacts of collecting and transporting, however there are long-term threats of pollution of groundwater and soils through certain additives, and waste byproducts that break down in plastics that can become persistent organic pollutants (Oehlmann et al. 2009; Teuten et al. 2009).

While incinerating plastic trash avoids the need for landfilling, dangerous substances may be emitted into the atmosphere. Because of the perceived pollution danger, plastic burning is less widespread as a waste-management approach than landfill and mechanical recycling. Further, reduced waste volumes are achieved by decreasing the quantity of packing used per item. Due to cost considerations, many manufacturers would employ the bare minimum of material available for a specific purpose. However, this idea must be balanced in contrast to esthetics, usability, and promotion benefits, all of which might contribute to needless packing, along with the impact of present tooling and manufacturing methods that could lead to too much packaging of specific things.

In addition, because of the huge number of packs and containers used for branding and promotion, direct take-back and restocking is challenging. While take-back and refilling programs for both PET and glass bottles exist in several European nations, they are often considered as a specialized practice for small enterprises rather than a viable large-scale solution for reducing packaging waste. Plastics used in product transportation have much re-use potential, and specific plastic elements in high-value customer items such as automobiles and electronics also have much re-use or re-manufacturing potential. The re-use of containers in transporting goods is a good example of this on a large scale. The terminology for recycling of plastics is unclear and occasionally misleading due to the extensive diversity of recovery and recycling processes; secondary, tertiary, and quaternary are the four varieties. The most important recycling is referred to as closed-loop recycling; whereas subordinate recycling is stated as degrading. Chemical recycling refers to the process of depolymerizing a polymer to their chemical elements. Quaternary recycling is as well-known as energy recovery; energy from trash. Another type of tertiary recycling, often known as organic or biological recycling, involves composting biodegradable plastics. Furthermore, degradable and compostable plastics should be correctly branded and utilized in a pathway which counterpart rather than undermine waste controlling plans (Hopewell et al. 2009; Song et al. 2009).

11.4 Plastic Recycling Systems

Plastics are recycled in a range of methods, with the ease of recycling varying according to the polymer, package design, and product type. Multi-layer and multi-component packaging, for example, is more difficult and expensive to recycle than rigid polymer containers. Thermoplastics with a high mechanical recycling capacity include PET, PE, and PP. Unsaturated polyester and epoxy resin are thermosetting polymers that cannot be physically recycled, but they can be repurposed as filler materials after being reduced to small particles (Rebeiz and Craft 1995). Since thermoset polymers are lastingly cross-linked throughout the manufacturing process, they cannot be re-melted. Because of

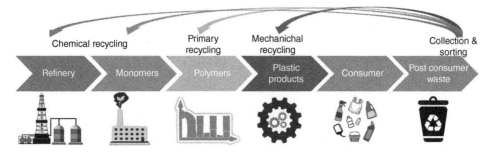

Figure 11.1 For the recycling of SPW and their related items, various processing techniques are used (Dogu et al. 2021).

contamination across polymer types, multi-layer/multi-component materials are rarely recycled. The post-consumer recycling process includes procedures such as collection, processing, washing, size reduction, separation, and minimizing contamination by incompatible polymers (Hopewell et al. 2009).

In global waste management systems, plastic recycling is done through four main methods. The numerous types of recycling available are primary, secondary, tertiary, and quaternary recycling (Rahimi and García 2017; Zhou et al. 2016). The remaining noticeable substitute to these procedures is the landfilling of SPW. The many processing procedures that may be utilized to recycle SPW and related items are depicted in Figure 11.1. Mechanical recycling is the mechanical recovery of plastics, whereas primary recycling is the recycling of pre-consumer waste. The measures in the next sections are usually followed throughout the recycling process. Recovery, planning, primary recycling, energy recovery, and chemical and mechanical recycling of SPWs, are examples of these processes.

11.4.1 Recovery

The recovery stage prioritizes collecting and logistical operations to gather plastic garbage. The reject step has gotten scant attention in the literature, which is a crucial topic to emphasize. The study stressed the usage of recycled materials as a benefit rather than a subject of conflict. On the other hand, one theory at this point is that dispersed recycling will make supply chains shorter and simpler, with a focus on lowering transportation effects (Despeisse et al. 2017; Garmulewicz et al. 2018; Sanchez et al. 2020).

11.4.2 Preparation

To render waste materials suitable for use as recycled feedstock, the preparation stage includes processes including identification, sorting, and size reduction (Sanchez et al. 2020).

11.4.3 Primary Recycling

Primary recycling is the process of reusing a product without modifying it. In the context of recycling of plastics, this is also known as re-extrusion because it relates to the reusing of pre-consumer leftovers (Zhou et al. 2016). During the polymer processing process or when the polymers are turned into the end product, pre-consumer wastes are produced. A substantial amount of today's recycling procedures is primary recycling, which includes a clean, single-use plastic input (Rahimi and García 2017). Pre-consumer commodities are typically well-identified, clean, and uniform because they are made from a single feedstock. The advantage of primary recycling of industrial process scrap is that it eliminates the requirement for sorting and transportation. It is well-known in the industry as a result of these factors (Al-Salem et al. 2009).

Primary recycling, on the other hand, has a number of disadvantages that limit its application outside of the recycling business. Primary recycling's only drawback is that it requires near-perfect input in the form of single-use, uncontaminated plastics (Al-Salem et al. 2009). As a result, primary recycling is claimed to be unsuitable for SPW obtained from municipal solid waste (MSW) because dirty mixed plastics cannot be processed. As a result, the technique's application range is severely decreased. Second, with primary recycling, end-stability products are a prevalent problem. Continuously reprocessing recycled polymers, it has been claimed, has a considerable impact on the mechanical qualities of the post-recycled manufactured goods. This decreases the overall quantity of reusing cycles by a large amount, posing a significant long-term hurdle.

11.4.4 Energy Recovery

Energy recovery is the technique in which plastics are burned and the energy released in the form of heat is utilized for a variety of purposes. Incineration is another name for this procedure. This method has been widely used since the dawn of the plastics age, however the poisonous gases emitted while incineration, that contribute to global warming, have made it less desirable. Incineration minimizes the quantity of garbage released in landfills while simultaneously producing electricity (Singh et al. 2017). Additional benefits of incineration technique include plants within city boundaries, energy output, and constant feed resulting in increased yield. Costly operation, high maintenance costs, unviable outcomes for components with enhanced moisture and chlorinated chemicals, and high ash content are other constraints of energy recovery (Silva et al. 2019). Plastics release harmful and noxious dioxins when they are burned, which cannot be discharged directly into atmosphere (Jubinville et al. 2020). As a result, environmental rules should be adhered to, contamination protection measures should be implemented, and the entire process must all be monitored closely. The collected plastic garbage is processed before being delivered to an incinerator to be burned in a standard energy process of recovery. The ash that is gathered in this stage is disposited in landfills. The heat recovered is utilized to generate electricity from the gas generated, while combustion is cooled to remove air pollutants. Toxic gases are treated (SO_x, NO_x) before being discharged into the environment by an emission stack. The plant's leftovers are disposed of in landfills (Jubinville et al. 2020).

The energy recovery route for SPW incineration is also known as quaternary recycling (Van Caneghem et al. 2012). At first glance, secondary fuels such as plastic waste appear to be a viable and inexpensive energy source. It should be noted, however, that the incineration of plastics necessitates the implementation of expensive emission control measures. Plastic incineration often poses significant public health and environmental risks. Based on the shape and content of the plastic trash, incineration might produce harmful volatile organic chemicals such as dioxins, halogenated gases, polychlorinated dibenzofurans, polyaromatic hydrocarbons, cyano compounds, furans, and polychlorinated dibenzo-p-dioxins (Vinu et al. 2016).

11.4.5 Mechanical Recycling

An additional prominent waste managing strategy is mechanical recycling, which includes physically reusing SPW into secondary raw materials and items (Yin et al. 2015). This approach is one of the most regularly utilized recycling ways worldwide, and it provides a suitable beginning point for recycling plastics because it requires very simple technology. The inevitability of a sequence of treatments and planning measures is one of mechanical recycling's major drawbacks (Yin et al. 2015). Furthermore, mechanical recycling is only cost-effective when no expensive separation stages are necessary. It is indeed worth noting that with SPW, total separation of separate components is rarely done due to the time and cost involved. As a result, mechanical recycling's frequency and application areas are clearly constrained. The following are the techniques used in the mechanical recycling process.

11.4.5.1 Sorting/Separating

Sorting is the start of all recycling procedures (Pacheco et al. 2012). The size, density, form, color, and chemical makeup of the polymers are all factors in the sorting process (Mastellone 2019). The utilization of diverse technologies such as x-ray, infrared, and fluorescence, is used to develop plastic sorting processes. Sorting plastics thoroughly guarantees that pure plastic types are obtained and that impurities are kept out of the recycling stream. Sorting has the goal of removing those plastics that cannot be processed. Sorting produces high-quality polymers that can be reused (Hahladakis and Iacovidou 2019).

The region of application of plastic products should be kept in mind during processing; for example, to prevent contamination, plastics used for food and non-food packaging applications should be treated separately. Also, don't mix biodegradable and non-biodegradable plastics. Plastic sorting that is effective improves the reusability of plastics while lowering environmental dangers. When diverse chemical compositions of plastics are combined during recycling, different phases can develop. As a result, the product's quality may suffer. Different processing needs, as well as the plastic's heterogeneity, have an impact on this (Khangale et al. 2021).

11.4.5.2 Electrostatic Separation

Plastics are charged with opposing polarities via a particle-to-particle charging process in the electrostatic separation procedure. The formation of positively and negatively charged particles occurs when particles collide with one another (Ragaert et al. 2017). The tribo-charging technique is used once the particles have been charged; this includes sorting particles with a diameter of approximately 5 mm using electric forces (Chandrasekaran and Sharma 2019; Khangale et al. 2021). Triboelectrostatic separation, corona discharge, and electrostatic induction are some of the other electrostatic separation techniques. Because of the inherent low cost of these methods, electrostatic separation appears to be quite reliable. Simultaneously, the approach produces no secondary pollutants.

11.4.5.3   Manual Sorting

Based on the type of plastics involved, manual sorting may necessitate the services of a skilled sorter. The type of plastics that will be made with the sorted ingredients could possibly explain the high sorting proficiency. The sorting approach appears to be incredibly efficient, but it may come at a great cost. This procedure cleans the plastics of stray items and needless packing. Manual sorting is time-consuming, and it is not viable to sort large numbers for a low cost when compared to automated sorting systems (Richard et al. 2011).

11.4.5.4   Sink Float Method

Submerging plastics in water is required for the sink-float method or the friction cleaning process. Centrifugal force or gravitational force are used to separate the particles. The sink-float technique can help identify plastic types by measuring hydrogen ion concentration (pH), froth content, wetting agents, and depressants. The approach, however, is mostly based on plastic density (Chandrasekaran and Sharma 2019). PE, HDPE, and LDPE are among the polyolefin plastics that are most suited to the sink-float method (Khangale et al. 2021). Certain plastics float, whereas others sink during this procedure. Plastics such as PP and PE float, while PET, PS, and PVC sink. When sifting crushed flakes, however, this method is more efficient.

11.4.5.5   Plastic Identification

Plastic identification methods mentioned in the literature include Fourier transform infrared spectroscopy (FTIR), differential scanning calorimetry (DSC), image recognition, and optical color recognition (de Römph and Van Calster 2018). According to Ragaert et al. (2017), companies would benefit from performing laboratory analysis focused at correctly detecting plastics. The plastic source should be examined before sourcing plastics for recycling because it can alter the recyclability of plastics. Some strategies for identifying and sorting can be confusing; as a result, more description of each method's notion cannot be considered excessive for this investigation.

11.4.5.6   Shredding

Shredding is used to reduce the volume of polymers and acquire the suitable size for subsequent processing. Non-plastic materials can also be easily separated using this method. Shredding plastics is done with a rotor knife and a fixed counter knife, as well as water. Water is needed since the heat from the blades may melt and weaken the plastic. If polymers are to be transformed into pellets, this procedure is particularly crucial (Khangale et al. 2021).

11.4.5.7   Agglomeration

Because delicate plastic agglomerates are readily broken down, the agglomeration process should be done cautiously and accurately. Plastics are heated, molded together, and further shredded during agglomeration. Heat exposure may cause changes in the mechanical characteristics of polymers. Mechanical recycling may be limited due to the many polymer combinations detected during mixing. The use of thermoplastics in secondary recycling, on the other hand, preserves the purity of the plastics because the technology is impervious to contaminants.

11.4.5.8 Washing/Cleaning

Drum screens are used to keep sediments and dust off the plastics while they are being washed. Manual separation is then utilized to remove non-recyclable materials such as wood, aluminum, and plastics. To eliminate all micro and macro pollutants, a washing procedure is required. After the washing procedure, all plastics should be dried using thermal drying techniques and washed again before being dried completely (Briassoulis et al. 2013; Khangale et al. 2021).

11.4.6 Chemical Recycling of Solid Plastic Pollutants

Chemical and tertiary recycling are synonymous terms; processes that break down end-of-life polymers into smaller compounds are included (Sasse and Emig 1998). After that, these elements can be employed to create new petrochemicals and plastics. These methods are regarded as chemical because the chemical structure of the polymer is transformed in the course of the process (Al-Salem et al. 2009). As previously mentioned, tertiary recycling targets to alleviate the shortcomings of primary and secondary recycling technologies. A number of activities are used to detect, segregate, and sort waste materials in primary and secondary recycling methods (Dogu et al. 2021)

Chemical recycling is a process that involves breaking down polymers by the reactivity of various chemical or thermochemical agents, and only thermosetting polymers are acceptable. Chemical recycling refers to the use of chemicals to depolymerize polymers and produce monomers or oligomers as feedstock materials. Chemical recycling is also known as compound reuse, which is one of the many advantages of chemical recycling. Polymers are broken down into smaller molecules known as monomers and oligomers, which are chemical intermediates. These monomers are utilized as raw ingredients in the synthesis of polymers, which are used as starting materials for recycled plastics. Chemically regenerated polymers have a number of drawbacks, including expensive raw material costs, significant capital investment, and a wide scale of operation. It also necessitates sophisticated equipment and consumes a significant amount of energy. Chemical recycling can be used to create high-strength fiber. Chemical recycling is not acceptable for plastics used in the manufacture of electronic gadgets because they contain chemical retardants (Khangale et al. 2021).

11.4.7 Reuse and Re-stabilization

For abandoned or discarded phytochemical constituents, reuse is a widespread practice and the preferable option. Because polymer goods are already prepared for their intended purpose, they are often reprocessed without affecting their chemical composition. Recycling polymeric materials is more costly than reuse because separating is difficult, notably as that the design complexity of such product increases (Niaounakis 2013). Direct material recovery while molding is small-scale mixing with the other polymers to bring in new products, sometimes with the addition of compatibilizers as well as other chemicals such as antioxidants or reinforcements. Finally, some of the original qualities of the polymer may be recovered by modifying the structure of the polymer (Niaounakis 2013).

Recuperation is a very well used industry procedure that involves repurposing waste from the manufacturing processes for items that fall short of predetermined quality criteria. This procedure involves grinding and combining scrap with virgin material, which is then fed directly into the processing machinery. Melt lumps, mold sprues, damaged pieces, and waste materials from the molding process may all be sources of scrap (Niaounakis 2013). In other situations, the plastic trash should be reduced to a flake-like component that is heat-treated into very compact and available granules. It's essential to remember to not heat the plastic past its melting point, as this will cause it to break down and render the industrial waste useless. Another crucial thing to consider is ensuring that the recovered granules are from the same grade as that of the virgin material in order to maintain quality of the product and management. The patent by Erema Engineering (WO2009100473) outlines a procedure in which an injection molding machine is supplied with a specific grinder, a die-face blade, and a cooling method that enables the regenerated granules to be re-introduced directly into the process (Hackl et al. 2013; Niaounakis 2013). When reprocessing polymeric material from waste streams, the qualities of the reprocessed material usually deteriorates when matched with those of the virgin material, owing to chemical structural changes. Chain scission reactions, hydroxyl radicals, and other thermo-degradative agents such as oxygen and contaminants are all responsible for changes in chemical structure. Re-stabilization is a technique for preventing the degradation of manu-factured parts, and aids in the long-term stability of reprocessed materials by protecting them against thermo-mechanical deterioration mechanisms. Re-stabilization, contrary to popular belief, does not restore lost characteristics; rather, it aids in the prevention of addi-tional property loss due to oxygen moiety and moisture existing in the structure. Hampered phenols, compatibilizers, hindered amine stabilizers, carbodiimide, and other agents are employed in re-stabilization. Blending recycled polymers with virgin materials is the most popular and widely used process in the business, with the goal of improving the qualities of a single waste stream or municipal garbage. Numerous research on the impact of com-bining recycled polyolefin material with virgin polyolefin material have been conducted during the last few decades. In most situations, recyclate-polymer blends are treated with a compatibilizer such as maleic anhydride grafted polypropylene (MAPP) or polyethylene-grafted-maleic anhydride (MAPE), as well as additives such as light stabilizers and chain extenders (Jubinville et al. 2020).

11.5 Latest Industry Trends and a Future Perspective

The ability to successfully recycle mixed plastic trash is another significant hurdle for the plastics recycling division. By expanding post-consumer plastic package collections to accommodate a broader range of items and pack sizes, it is conceivable to reprocess a higher percentage of the plastic waste stream. As a result, strengthening laws to encourage industry to apply environmental design standards will have a big influence on recycling efficiency, growing the amount of packing that could be gathered and diverted from waste at a reasonable cost (Shaxson 2009).

Plastic is produced in amounts of 300 million tons each year, with 8 million tons ending up in the oceans. In 30 years, the amount of plastic in the oceans is expected to surpass the amount of fish if this current trend continues. The widespread expectation that chemists would develop biodegradable plastics for mass production proved illusory, and the few examples available are economically irrelevant. Recycling, which involves turning plastic waste into useful products and thereby reducing the amount of plastic litter, is one partial solution to the problem. When two non-miscible polymers are melt-mixed, extruded, cold drawn, and then compression- or injection-molded to form polymer-polymer composites, microfibrillar composites (MFCs) are investigated as a promising technique for polymer recycling. At the same time, plastic recycling may not address the worldwide problem of plastics' harmful impact on the environment; it simply delays it. Recycling, on the other hand, helps to reduce the quantity of litter produced by plastics. The actual solution to the problem would entail the involvement of all members of society (Fakirov 2021).

The circular economy has surpassed the previous paradigm of create, use, and disposal as a preferred method of dealing with trash. This new create, use, reuse, and recycle approach is currently informing businesses with the goal of being more sustainable. Three principles guide the circular economy: eliminate waste and pollution, maintain products and resources in use, and regenerate a natural system. To combat single-use packaging, the Ellen MacArthur Foundation and Waste and Resources Action Programme (WRAP) UK have collaborated. The goal of the study is to develop materials that are reusable, recyclable, or biodegradable, as well as to boost garbage collection, stimulate demands for secondary markets for recyclable material, and involve the wider populace in the circular economy. In July 2018, the European Union updated its waste-related legislation. In the waste management and recycling industries, the circular economy package has established goals. The following points on plastic and trash management are highlighted in the updated document. The targets included are some of the objectives.

- Adding separate collections for hazardous household garbage (by end 2022), biowaste (by end 2023), and textiles (by end 2025);
- Reducing the quantity of waste going to landfill to no more than 10% by 2035;
- An EU objective of recycling 65% of packaging waste by 2025 and 70% by 2030;
- Recycling targets of 55% for plastic packaging;
- Minimize standards for producer responsibility schemes;
- Increase preventative actions to combat food waste and marine litter.

Companies should plan for such conclusion of a life cycle of the product and engage with customers throughout the distribution chain. Better policies are increasing demand for secondary material markets, making high-quality recycled materials more easily accessible. Materials are used more effectively and sustainably if the concepts of the circular economy are considered. Because there is presently no legislation requiring the use of recycled materials, the bulk of plastic packaging in the UK is composed of virgin plastic. We are producing higher carbon emissions as a result, even as an industry we are contributing to climate change in a more detrimental manner than if we used recyclable plastic and less energy and water. Introducing a charge is one way to encourage people to utilize recycled materials. The 2018 Budget unveiled a new tax which would take effect in April 2022 and apply to packaging that contains less than 30% recycled plastic. As a result, the use of recycled plastic is encouraged across the supply chain. The tax credit will allow for further investment in the plastics industry to solve waste challenges. The carrier bag 5p tax in 2015 was an illustration of how customers react to rewards. As just a result, 15.6 billion bags have been removed from the market. Customers choose to recycle bags rather than purchase new ones, which is a big step that will influence how the government considers policy recommendations. Consumers who buy a drink in a single-use bottle are charged a deposit, which is recovered whenever the container is restored. Deposit return schemes (DRSs) have shown to be effective in other European countries, having existing recycling rates exceeding 95%. According to reports, the UK recycles 70% of plastic bottles, enabling 4 million bottles to be discarded instead of recycled (Letcher 2020). The DRS will most likely be run by reverse vending machines and stores, but the specifics are still being worked out. While this will assist in meeting recycling goals, exact volumes and uptake are difficult to forecast. The main concerns for plastics recyclers are lowering costs and restoring material quality to the point where it can be reused in high-value applications such as food-grade and nonfood packaging. This will necessitate financial commitment as well as the support of brand owners (Letcher 2020).

Despite the fact that many plastic recycling methods concentrate on primary and secondary recycling, there is a significant gap between the amount of plastic trash generated and the quantity recycled. In order to address the challenges related with the vast volumes of plastic trash being created, we must now turn to additional procedures that may be employed in conjunction with present operations. Several scientific reports have been conducted that define general methodologies for the chemical recycling of hydrolyzable polymers such as PET, along with nonhydrolyzable polymers such polyolefins, as described. According to these findings, chemical transformation of waste plastics into value-added chemicals could be a useful way to enhance current recycling procedures. Commercialization of chemical recycling methods for plastic waste reveals more complex difficulties of separation and purification connected with real-world waste streams, ranging from combined waste plastic to multicomponent metal–plastic parts to full spectrum municipal waste. Considering the current problems complexities, significant yields of desired products have indeed been produced by carefully selecting catalysts, solvents, temperature stages, residence periods, and beginning feedstock. Continued emphasis on catalysts is needed to enhance total efficiencies and reduced temperature in recycling, thus lowering energy needs. Chemical recycling of plastic trash is now in its infancy, but it has the potential to become one of the most important procedures for effectively reducing waste in landfills (Thiounn and Smith 2020). Shi et al. (2020), present a monomer design method which is based on a bridging bicyclic thiolactone which results in stereo-disordered to entirely stereo-ordered polythiolactones with good crystallinity and chemical recyclability. These polythioesters defy the previously noted tradeoffs by providing a distinct combination of desirable properties, such as intrinsic tacticity-independent crystallinity and chemical recyclability, tunable tacticities from stereo-disorder to perfect stereoregularity, and combined high-performance properties such as high thermal stability and crystallinity, as well as high mechanical strength, ductility, and strength.

Failing to solve plastics' end-of-life challenges today has not only hastened the degradation of finite resources, but it has also resulted in significant global plastics contamination and a significant loss of energy and material value in the global economy. To resolve this worldwide problem, the development of next-generation polymers must take into account afterlife issues and develop closed-loop lifecycles in the direction of a circular economy. In this context, the advancement of chemically reusable polymers which can be depolymerized back to their monomer building blocks with high selectivity and purity for virgin-quality polymer reproduction represents a circular economic way to address these serious environmental and economic issues (Shi et al. 2020). For instance, ring-opening polymerization (ROP) of unstrained Gamma-butyrolactone (GBL) produces polyester poly(GBL), which may be depolymerized back to GBL in quantitative purity and yield with a low energy input (Hong and Chen 2016). However, the performance attributes of poly(GBL) are insufficient for most applications. To resolve this depolymerization/performance trade-off, ring-fused bicyclic GBL structural variants were developed to improve monomer polymerizability, polymer thermal stability, and crystallinity without compromising chemical recyclability, However, the crystalline materials that result have high melting transition temperatures (T_m) and are mechanically brittle, necessitating the addition of flexible copolymers to provide practical ductility. Furthermore, either stereocomplexation of preformed enantiomeric polymers from distinct pools of enantiopure monomers or elaborate stereoselective polymerization of the racemic monomer pool is required to obtain these crystalline materials (Shi et al. 2020).

11.6 Conclusions

The production of waste plastics is rapidly increasing, posing major environmental challenges. Furthermore, waste plastic contains useful chemicals and have a high calorific value. Despite significant efforts to advance waste plastics reduction, reuse, and recycling, emerging technologies that produce high-value goods have yet to be commercialized. Moreover, plastics have outgrown the majority of man-made products and have long been scrutinized by environmentalists. Nonetheless, there is a lack of reliable global information, especially concerning their end-of-life fate. For effective plastic waste management, knowing the amount and composition of plastic waste produced is critical. This information will aid local authorities, policymakers, and the public and private sectors in understanding the complexities of plastic waste generation, and identifying the potential of plastic waste recycling, which will not only minimize the burden of landfilling, but also increase revenue generation by formalizing informal recycling.

References

Al-Salem, S.M., Lettieri, P., and Baeyens, J. (2009). *Recycling and recovery routes of plastic solid waste (PSW): a review. Waste Management* 29 (10): 2625–2643.

Ashar, M., Fraser, M.A., Li, J. et al. (2020). *Interaction between microbial communities and various plastic types under different aquatic systems. Marine Environmental Research* 162: 105151.

Besseling, E., Quik, J.T., Sun, M., and Koelmans, A.A. (2017). *Fate of nano-and microplastic in freshwater systems: a modeling study. Environmental Pollution* 220: 540–548.

Briassoulis, D., Hiskakis, M., and Babou, E. (2013). *Technical specifications for mechanical recycling of agricultural plastic waste. Waste Management* 33 (6): 1516–1530.

Browning, S., Beymer-Farris, B., and Seay, J.R. (2021). *Addressing the challenges associated with plastic waste disposal and management in developing countries. Current Opinion in Chemical Engineering* 32: 100682.

Bucci, K., Tulio, M., and Rochman, C.M. (2020). *What is known and unknown about the effects of plastic pollution: a meta-analysis and systematic review. Ecological Applications* 30 (2): e02044.

Chandrasekaran, S.R. and Sharma, B.K. (2019). *From waste to resources: how to integrate recycling into the production cycle of plastics.* In: *Plastics to Energy* (ed. S. Al-Salem), 345–364. William Andrew Publishing.

Chia, W.Y., Tang, D.Y.Y., Khoo, K.S. et al. (2020). *Nature's fight against plastic pollution: algae for plastic biodegradation and bioplastics production. Environmental Science and Ecotechnology* 4: 100065.

Chiba, S., Saito, H., Fletcher, R. et al. (2018). *Human footprint in the abyss: 30 year records of deep-sea plastic debris. Marine Policy* 96: 204–212.

Chubarenko, I., Bagaev, A., Zobkov, M., and Esiukova, E. (2016). *On some physical and dynamical properties of microplastic particles in marine environment. Marine Pollution Bulletin* 108 (1–2): 105–112.

Cottafava, D., Costamagna, M., Baricco, M. et al. (2021). *Assessment of the environmental break-even point for deposit return systems through an LCA analysis of single-use and reusable cups. Sustainable Production and Consumption* 27: 228–241.

Derraik, J.G. (2002). *The pollution of the marine environment by plastic debris: a review. Marine Pollution Bulletin* 44 (9): 842–852.

Despeisse, M., Baumers, M., Brown, P. et al. (2017). *Unlocking value for a circular economy through 3D printing: a research agenda. Technological Forecasting and Social Change* 115: 75–84.

Dogu, O., Pelucchi, M., Van de Vijver, R. et al. (2021). *The chemistry of chemical recycling of solid plastic waste via pyrolysis and gasification: state-of-the-art, challenges, and future directions. Progress in Energy and Combustion Science* 84: 100901.

Fakirov, S. (2021). A new approach to plastic recycling via the concept of microfibrillar composites. *Advanced Industrial and Engineering Polymer Research* 4: 187–198.

Garmulewicz, A., Holweg, M., Veldhuis, H., and Yang, A. (2018). *Disruptive technology as an enabler of the circular economy: what potential does 3D printing hold? California Management Review* 60 (3): 112–132.

Hackl, M., Feichtinger, K. and Wendelin, G. (2013). Method and device for injection molding plastic material. In: *EREMA Engineering Recycling Maschinen und Anlagen GesmbH*, U.S. Patent 8,419,997.

Hahladakis, J.N. and Iacovidou, E. (2019). *An overview of the challenges and trade-offs in closing the loop of post-consumer plastic waste (PCPW): focus on recycling. Journal of Hazardous Materials* 380: 120887.

Hong, M. and Chen, E.Y.X. (2016). *Completely recyclable biopolymers with linear and cyclic topologies via ring-opening polymerization of γ-butyrolactone. Nature Chemistry* 8 (1): 42–49.

Hopewell, J., Dvorak, R., and Kosior, E. (2009). *Plastics recycling: challenges and opportunities. Philosophical Transactions of the Royal Society B: Biological Sciences* 364 (1526): 2115–2126.

Jiang, J.Q. (2018). *Occurrence of microplastics and its pollution in the environment: a review. Sustainable Production and Consumption* 13: 16–23.

Jubinville, D., Esmizadeh, E., Saikrishnan, S. et al. (2020). *A comprehensive review of global production and recycling methods of polyolefin (PO) based products and their post-recycling applications. Sustainable Materials and Technologies* 25: e00188.

Khangale, U.B., Ozor, P.A., and Mbohwa, C. (2021). *A review of recent trends and status of plastics recycling in industries. Engineering and Applied Science Research* 48 (3): 340–350.

Kowalski, N., Reichardt, A.M., and Waniek, J.J. (2016). *Sinking rates of microplastics and potential implications of their alteration by physical, biological, and chemical factors. Marine Pollution Bulletin* 109 (1): 310–319.

Kukulka, T., Proskurowski, G., Morét-Ferguson, S. et al. (2012). *The effect of wind mixing on the vertical distribution of buoyant plastic debris. Geophysical Research Letters* 39 (7).

Lebreton, L.M., Greer, S.D., and Borrero, J.C. (2012). *Numerical modelling of floating debris in the world's oceans. Marine Pollution Bulletin* 64 (3): 653–661.

Letcher, T.M. (2020). *Plastic Waste and Recycling: Environmental Impact, Societal Issues, Prevention, and Solutions*. Academic Press.

Mastellone, M.L. (2019). *A feasibility assessment of an integrated plastic waste system adopting mechanical and thermochemical conversion processes. Resources, Conservation & Recycling: X* 4: 100017.

Niaounakis, M. (2013). *Biopolymers: Reuse, Recycling, and Disposal*. William Andrew Publishing.

Oehlmann, J., Schulte-Oehlmann, U., Kloas, W. et al. (2009). *A critical analysis of the biological impacts of plasticizers on wildlife. Philosophical Transactions of the Royal Society B: Biological Sciences* 364 (1526): 2047–2062.

Pacheco, E.B., Ronchetti, L.M., and Masanet, E. (2012). *An overview of plastic recycling in Rio de Janeiro. Resources, Conservation and Recycling* 60: 140–146.

PlasticsEurope (2018). Plastics – the facts 2018: an analysis of european plastics production, demand and waste data. https://www.plasticseurope.org/application/files/6315/4510/9658/Plastics_the_facts_2018_AF_web.pdf (accessed 1 September 2019).

PlasticsEurope (2019). Plastics – the facts 2019. an analysis of european plastics production, demand and waste data. https://www.plasticseurope.org/application/files/9715/7129/9584/FINAL_web_version_Plastics_the_facts2019_14102019.pdf (accessed 18 May 2020).

PlasticsEurope – AISBL (2017). Plastics – the facts 2017. An Analysis of European Plastics Production, Demand and Waste Data. On-line.

Post, W., Susa, A., Blaauw, R. et al. (2020). *A review on the potential and limitations of recyclable thermosets for structural applications. Polymer Reviews* 60 (2): 359–388.

Ragaert, K., Delva, L., and Van Geem, K. (2017). *Mechanical and chemical recycling of solid plastic waste. Waste Management* 69: 24–58.

Ragusa, A., Svelato, A., Santacroce, C. et al. (2021). *Plasticenta: first evidence of microplastics in human placenta. Environment International* 146: 106274.

Rahimi, A. and García, J.M. (2017). *Chemical recycling of waste plastics for new materials production. Nature Reviews Chemistry* 1 (6): 1–11.

Rebeiz, K.S. and Craft, A.P. (1995). *Plastic waste management in construction: technological and institutional issues. Resources, Conservation and Recycling* 15 (3–4): 245–257.

Richard, G.M., Mario, M., Javier, T., and Susana, T. (2011). *Optimization of the recovery of plastics for recycling by density media separation cyclones. Resources, Conservation and Recycling* 55 (4): 472–482.

de Römph, T.J. and Van Calster, G. (2018). *REACH in a circular economy: the obstacles for plastics recyclers and regulators. Review of European, Comparative & International Environmental Law* 27 (3): 267–277.

Ryberg, M.W., Hauschild, M.Z., Wang, F. et al. (2019). *Global environmental losses of plastics across their value chains. Resources, Conservation and Recycling* 151: 104459.

Sanchez, F.A.C., Boudaoud, H., Camargo, M., and Pearce, J.M. (2020). *Plastic recycling in additive manufacturing: a systematic literature review and opportunities for the circular economy. Journal of Cleaner Production* 264: 121602.

Sasse, F. and Emig, G. (1998). *Chemical recycling of polymer materials. Chemical Engineering & Technology: Industrial Chemistry-Plant Equipment-Process Engineering-Biotechnology* 21 (10): 777–789.

Schwarz, A.E., Ligthart, T.N., Boukris, E., and Van Harmelen, T. (2019). *Sources, transport, and accumulation of different types of plastic litter in aquatic environments: a review study. Marine Pollution Bulletin* 143: 92–100.

Shaxson, L. (2009). *Structuring policy problems for plastics, the environment and human health: reflections from the UK. Philosophical Transactions of the Royal Society of London. Series B Biological Sciences* 364: 2141–2151.

Shi, C., McGraw, M.L., Li, Z.C. et al. (2020). *High-performance pan-tactic polythioesters with intrinsic crystallinity and chemical recyclability. Science Advances* 6 (34): eabc0495.

Silva, L.J.D.V.B.D., Santos, I.F., Mensah, J.H.R., and Gonçalves, A. (2019). *Incineration of municipal solid waste in Brazil: an analysis of the economically viable energy potential. Renewable Energy* 149: 1386–1394.

Singh, N., Hui, D., Singh, R. et al. (2017). *Recycling of plastic solid waste: a state of art review and future applications. Composites Part B: Engineering* 115: 409–422.

Song, J.H., Murphy, R.J., Narayan, R., and Davies, G.B.H. (2009). *Biodegradable and compostable alternatives to conventional plastics. Philosophical Transactions of the Royal Society B: Biological Sciences* 364 (1526): 2127–2139.

Teuten, E.L., Saquing, J.M., Knappe, D.R. et al. (2009). *Transport and release of chemicals from plastics to the environment and to wildlife. Philosophical Transactions of the Royal Society B: Biological Sciences* 364 (1526): 2027–2045.

Thiounn, T. and Smith, R.C. (2020). *Advances and approaches for chemical recycling of plastic waste. Journal of Polymer Science* 58 (10): 1347–1364.

Thompson, R.C., Swan, S., Moore, C.J., and vom Saal, F.S. (2009). *Our plastic age. Philosophical Transactions of The Royal Society of London. Series B, Biological Sciences* 364: 1973–1976.

Van Caneghem, J., Brems, A., Lievens, P. et al. (2012). *Fluidized bed waste incinerators: design, operational and environmental issues. Progress in Energy and Combustion Science* 38 (4): 551–582.

Vijayakumar, A. and Sebastian, J. (2018). Pyrolysis process to produce fuel from different types of plastic–a review. In: *IOP Conference Series: Materials Science and Engineering*, vol. 396, No. 1, 012062. IOP Publishing.

Vinu, R., Ojha, D.K., and Nair, V. (2016). Polymer pyrolysis for resource recovery. In: *Elsevier Reference Module in Chemistry, Molecular Sciences and Chemical Engineering* (ed. J. Reedijk). Waltham, MA: Elsevier.

Wu, P., Huang, J., Zheng, Y. et al. (2019). *Environmental occurrences, fate, and impacts of microplastics. Ecotoxicology and Environmental Safety* 184: 109612.

Yin, S., Tuladhar, R., Shi, F. et al. (2015). *Mechanical reprocessing of polyolefin waste: a review. Polymer Engineering & Science* 55 (12): 2899–2909.

Zhou, X., Broadbelt, L.J., and Vinu, R. (2016). *Mechanistic understanding of thermochemical conversion of polymers and lignocellulosic biomass. Advances in Chemical Engineering* 49: 95–198.

12

Microbial Degradation of Micro-Plastics

Pooja Sharma, Sophayo Mahongnao, and Sarita Nanda

Department of Biochemistry, Daulat Ram College, University of Delhi, New Delhi, India

12.1 Introduction

Since 1950, the worldwide plastic production has increased tremendously. It was found to be 359 million tons in the year 2018 and continues to grow. According to Zhu and Wang (2020), 26 billion metric tons of plastic waste will have been generated by 2050; half of which will be contributing to the waste disposal sites. This could be attributed to the fact that plastic is indeed a revolutionary chemical product for its properties of elasticity, non-corrosiveness, and water-resistance. Thus, plastics find essential and irreplaceable functions in almost every type of manufacturing industry as well as in our day-to-day life. The source of contamination is multidirectional such as urban, agricultural, and indiscriminate effluent discharge from industries and wastewater treatment plants (WWTPs). They generally end up accumulating in water bodies such as lakes, rivers, estuaries, and oceans (Li et al. 2016). These negatively affect the life of marine inhabitants; some lose their lives when becoming entangled in plastics due to suffocation and physiological stress, while others risk their lives by ingesting the MPs, which leads to health and reproductive complications (Alimba and Faggio 2019). However, air and soil ecosystems are not far from also becoming contaminated. The land becomes tainted with the sludge coming from the WWTPs as organic fertility boosters and with the leachates from the surrounding landfills. Although plastics are non-biodegradable they may become reduced to smaller particles of varying sizes such as meso-, micro-, and nano-plastics. MPs are the plastic particles which fall in the size range of about 5–100 nm, and may be formed due to the mechanical shredding, photodegradation, and bioremediation of plastic pieces (Yuan et al. 2020). There are several environmental, physical, and chemical properties of these polymers that determine their respective rate of degradation. Based on their reluctancy toward microbial deterioration, the polymerase is either biodegradable or non-biodegradable. The commonly used biodegradable plastics include

Plastic and Microplastic in the Environment: Management and Health Risks, First Edition.
Edited by Arif Ahamad, Pardeep Singh, and Dhanesh Tiwary.
© 2022 John Wiley & Sons Ltd. Published 2022 by John Wiley & Sons Ltd.

polyhydroxyalkanoate (PHA), polyhydroxybutyrate (PHB), polylactic acid (PLA), polycaprolactone (PCL) and polybutylene succinate (PBS); while polyethylene (PE), polypropylene (PP), polystyrene (PS), polyethylene terephthalate (PET), polyurethane (PUR), polyvinyl chloride (PVC), high-density polyethylene (HDPE), and low-density polyethylene (LDPE) are commonly used non-biodegradable plastic polymers for various purposes (Ahmed et al. 2018).

The reluctance toward biodegradation is one of the major problems associated with management of polymeric waste. Having biodegradable plastic does not guarantee their escape from the environment, as they persist in the nature for a long time. Over time, they become fragmented, adding to their easy dispersion and involvement with the flora and fauna. Micro- and nanoplastics, due to their smaller sizes, possess a greater threat as they have been found interfering with the food webs and are redistributed (Von Moos et al. 2012). Plastics have been found accumulating on persistent organic pollutants, metals, and pathogens, distributing these toxic elements to different habitats and their inhabitants. These substances along with amino acids precondition the polymeric surface for initiating colonization by microorganisms and support their growth and biofilm maturation (Liu et al. 2021). Once attached to the plastic surface, microbes degrade the polymer by fragmentation, destabilizing its chemical structure and utilizing them for energy production using various exozymes and endozymes.

The increasing contamination levels and their associated environmental issues have become the current topic of interest for research. There is very limited knowledge about the plastic removal strategies; although some advanced physical technologies such as the rapid sand filter, membrane bioreactor, and dissolved air floatation have been tested by treating wastewater MP contamination. However, these final-stage treatments are still inadequate as the final effluents have been found to contain MPs of smaller sizes (<20 µm) and fibrous shapes (Poerio et al. 2019). Current research focuses on bioremediation by exploiting those microbes having the potential to degrade plastic polymers. Microbes, being cosmopolitan, are a very essential part of Earth's biodiversity, and targeting them for plastic degradation may be more advantageous since it will generate no adverse side effects. It would be of immense interest to further our research on the various microbial–plastic and microbial–microbial interactions responsible for plastic colonization leading to biofilm formation that further attract other species. A better understanding of the basis for these interactions and polymer degradation mechanisms such as the utilization of plastics as carbon sources, and microbial enzyme-based catalysis may help us deduce the ways of enhancing the bioremediation process. There is a need to explore if we can combine the microbe-based remediating technologies with our known physical technologies for remediating MPs for a more effective result.

12.2 Plastic Categorization Based on Biodegradability

Plastics can be categorized as biodegradable and non-biodegradable based on their potential to be degraded by living organisms. The various types of plastic polymers along with their constituent monomeric units are summarized in Table 12.1. The flowchart showing different types of plastics based on degradability is given in Figure 12.1.

Table 12.1 Types of the plastic polymer along with constituent monomeric unit (Vroman and Tighzert 2009).

Biodegradable plastics	Monomer unit	Non-biodegradable plastics	Monomer unit
Poly α-hydroxy acids Ex: Polylactic acids	$(-O-CH(CH_3)-CO-)_n$	Polyethylene	$(-C_2H_4-)_n$
Poly β-hydroxy acids Ex: Polyhydroxybutyrate -co-hydroxyvalerate	$[-COCH_2CH(CH_3)O]_m[COCH_2CH(C_2H_5)O-]_n$	Polystyrene	$(-C_8H_8-)_n$
Poly ω-hydroxy acids Ex: Polycaprolactone	$(-O-(CH_2)_5-CO-)_n$	Polyvinylchloride	$(-CH_2CHCl-)_n$
Polyalkene decarboxylate Ex: Polyethylene succinate	$(-O-(CH_2)_2-O-CO-(CH_2)_2-CO-)_n$	Polypropylene	$(-C_3H_6-)_n$
Polyester amide Ex: BAK1095	$[-NH-(CH_2)_5-CO-]+[-OH-(CH_2)_4-OH-] +$ $[-COOH-(CH2-)]$	Polyethylene terephthalate	$[-(CO)C_6H_4 (CO_2CH_2CH_2O)-]_n$
Starch based polymer Ex: Starch/ polyethylene	Starch $+ (-C_2H_4-)_n$	Polyurethane	$[-NH-(CH2)6-NH-CO-O-(CH2)4-O-CO-]$
Cellulose acetate Ex: BIOCETA or Cellulose diacetate		Polyester	$-O-(CH2)6-O-CO-(CH2)4-CO-$

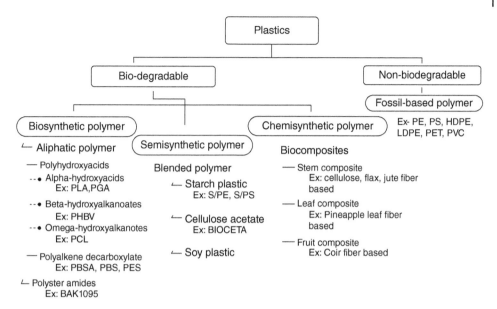

Figure 12.1 Types of plastic polymers based on their reluctance toward the feasibility of biodegradation process. The biodegradable polymers are of various types depending on the structural modification introduced in order to enhance their biodegradation.

12.2.1 Non-biodegradable Plastics

Non-biodegradable plastics are the plastic polymers that are not easily degraded by microbes or have slower rate of degradation; these are the fossil-based polymers that are derivatives of hydrocarbons and petroleum. They are formed by extensive repetition of monomeric units, thus having high molecular weight, which is one of the physical hindrances to their biodegradation (Ahmed et al. 2018). These are the most commercial forms of plastic; for example, polyethylene and polycaprolactone used for packaging and plastic shopping bags, which are a major contributor of environmental contamination.

12.2.2 Biodegradable Plastics

Biodegradable plastics can be easily degraded by microorganisms such as bacteria and fungi. These are the man made alternatives that have served the purpose of plastic essentiality and protection from plastic pollution by replacing the fossil-based polymers in the environment. They are further classified as biosynthetic polymers, blended polymers, and bio-composites. Biosynthetic polymers are the neat polymers that will easily biodegrade, as they are derived from renewable sources, while the other two types are the polymers that will be biodegraded if they exhibit chemical homology with a natural compound.

12.2.3 Biosynthetic Plastics

Aliphatic polyesters are a type of biosynthetic polymer, depending on the type of constituent monomeric unit, aliphatic polyesters are of two types; the polyhydroxy acids where the monomeric unit is hydroxyl acid, and polyalkene dicarboxylate which are diols and dicarboxylic acids. Further, these polyhydroxy acids can be categorized into the following types based on the position of the hydroxyl group with respect to the carboxylic acid.

- Poly(α-hydroxy acids), such as polylactic acid (PLA) or poly glycolic acids (PGA), which are formed by condensation polymerization of specific hydroxyl acids. These can also be derived from natural sources such as corn and sugar beets.
- Poly(β-hydroxyalkanoates) are synthesized by microbial fermentation utilizing glucose and propionic acid as the carbon sources, for example, the copolymer PHBV (polyhydroxybutyrate-co-valerate), which is marketed under the commercial name as Biopol.
- Poly(ω-hydroxyalkanoate) is a condensation product of aromatic cyclic compound lactone; such as polycaprolactone (PCL). These types of polymers have been found to be degraded by lipase enzyme. These have strength and rigidity properties lying between high and low density polyethylene.
- Poly(alkylene dicarboxylate) are formed by condensation of glycols and aliphatic dicarboxylic acids (such as succinic acid). These are marketed under the tradename of Bionolle with three different series; polybutylene succinate (PBS), polyethylene succinate (PES), and polybutylene succinate co-butylene adipate (PBSA). They have tensile strength comparable to polyethylene or polypropylene (Vroman and Tighzert 2009).
- Polyester amide is another variant of biodegradable polymer, derived from aliphatic polyesters and alpha amino acids such as glycine and alanine. Glycine-derived polymers are resistant to proteolytic degradation by chymotrypsin and elastase enzymes, whereas addition of phenylalanine to these polymers may enhance the degradation process by the same enzymes. However, papain can degrade all types of polyester amides. BAK 1095 resin is a known example of a polyester amide which undergoes complete biodegradation, releasing CO_2, H_2O, and biomass (Poliakoff et al. 2002).

12.2.4 Blended Plastics

Blended polymers refer to the types of polymers that are synthesized by blending naturally degradable materials such as starch with other plastic polymers. Starch is a plant-based polymer constituted by linear chain amylose and branch chain amylopectin. Natural sources of starch include corn, potato, wheat, and rice. Being the most economical and ecological polymer, most of its production is used to meet the non-food industrial demands.

To overcome the shortcomings of this category of polymer, such as the moisture content, durability, and high degradability, structural modifications such as alteration in the constituents' proportions and blending with other polyesters is carried out, for example, Thermoplastic Starches (TPS). Blends of TPS and aliphatic polyesters such as PVC or PCL are marketed under the tradename Mater Bi, and have similar properties to those of PE and PS, as well as being easily recyclable and degradable. These are used in a wide range of applications such as packaging, disposable items, personal care, and hygiene. Cellulose acetate is another type of blended polymer formed by blending acetic anhydride and cotton lint or wood pulp which may be obtained from recycled paper and sugarcane. However, they are reluctant to bio-degradation due to the presence of ether linkages. BIOCETA and

Enviro Plastic Zare are the commercial polymers used in packaging and industrial markets. Soy plastics are also blended polymers, although they are more sensitive to water. These are blended with polyphosphate fillers in order to reduce their water reactivity. A reaction product of soya bean, polysaccharide polymer, reducing agent, and additives exhibits higher tensile strength and biodegradation. Soy plastics have been used by the Ford Company in the manufacturing of automobile parts (Ly et al. 1998).

12.2.5 Biocomposite Polymers

Biocomposite polymers are a type of biodegradable polymers composed of a matrix and a reinforcing fiber. A matrix is generally comprised of plastic polymer and fiber; in the case of bio-composites, they are derived from natural sources. These composites are also called 'green' polymers as they are formed of fiber derived from plant parts such as the stem (bast fiber), leaf (pineapple leaves), and fruit (coconut coir) fiber. There are three major types of stem-based composites: cellulose fiber-based; flax-, hemp-, ramie-based; and jute fiber-based biocomposites. Cellulose fiber with polyhydroxybutyrate (PHB) matrix produces a composite with high stiffness comparable to polypropylene (PP) and polystyrene (PS). Jute-based composites have tensile strength and stiffness as that of glass fiber, so they are used as thermo-set and thermoplastics. Surface modifications of jute fiber such as de-waxing, alkali treatment, cyanoethylation, and grafting are employed to decrease water absorption properties and enhance mechanical properties such as hydrophobicity and fiber-matrix adhesions. Alkali treatment specifically removes impurities and increases roughness of the surface. The biocomposites exhibit a certain degree of flame-resistance making them useful for paneling in cars, railways, or aircrafts reinforced in the biodegradable polymer matrix, thus making them the most environment-oriented polymers (Mohanty et al. 2000). Thus, modifications of the molecular structures of polymers such as the incorporation of starch, along with altering functional groups or introducing electrophilic atoms such as oxygen, makes it easier for the microbes to access and attack them. Fragmentation of larger polymers leads to formation of smaller plastic particles called microplastics (MPs).

12.3 Microplastics Cycling into the Environment

MPs, the new trend in pollution, have become ubiquitous in the environment including the soil, water, and air, even reaching the isolated ecosystems of the Arctic. There are two variants to these types of plastics called primary and secondary microplastics (Figure 12.2). The primary type refers to the MP particles manufactured in daily-use products such as cosmetics, toothpastes, etc., while the secondary types are derivatives of larger plastics that have undergone several stages of deterioration. Thus, their entry into the ecosystem is multidirectional, such as from households, garbage landfills, industrial sectors, and WWTPs (Alimba and Faggio 2019). With respect to the marine environment, the contamination source is the waste disposal from cargo and passenger ships, as was first reported by Carpenter and Smith (1972). In the marine environment, MPs interfere with the feeding habits of marine species from algae, zooplanktons, small invertebrates, and small and large fishes, thus making their way into the food web (Picó and Barceló 2019). They even

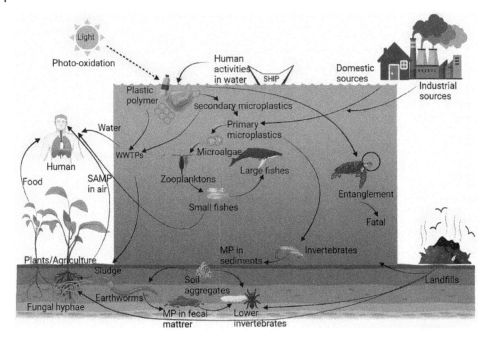

Figure 12.2 Plastic and plastic-derived products make their way into the environment from a wide array of sources such as from households to industries and from landfills to tourism. The deteriorated smaller particles interfere with marine or land food chains, bioaccumulating in higher organisms and returning to nature, completing the cycle.

become fatal to marine species as they cause respiratory choking when becoming entangled with them, or they may cause reproductive abnormalities (Gilardi et al. 2010). From water bodies, MPs find their way to agricultural lands, as the water is used for irrigation purposes, and are spread as natural fertilizers from the WWTPs in the form of sludge. Chen et al. (2020), showed that over time, the subsequent application of sludge in the fields leads to increase in the plastic contamination level in that soil, by experimenting with different cultivable fields that undergo varied application of sludge, indicating the median microplastic content in soil to be of 320–12560 items per kilogram of dry weight of the soil. These polymers accumulating in soil are constantly affecting the local flora–fauna interactions as well as soil microbiota. The polymer particles are redistributed by soil organisms like earthworms, arthropods, ticks, and mites. The earthworms may feed on the polymer particles causing its death, or may be excreted in feces that are fed upon by lower invertebrates in soil. MPs together with soil form soil aggregates that interfere with the physical property of soil such as its porosity, affecting soil microbial community, nutrient cycling, and thereby soil fertility.

Although there is a lack of evidence to establish a direct link for transportation of MPs from soil to plants, a possible mechanism suggests the indigenous fungal mycelia may act as a carrier by transporting the pollutant-degrading microbes. The MPs following this cycle may return to humans in the form of contaminated food and water (Guo et al. 2020). From the surface of the soil, these plastic particles are carried away, contaminating the atmosphere. The atmospheric contamination was first studied by Dris (2015), who reported that lighter densities of these suspended atmospheric microplastic particles (SAMP) aid in their dispersion into the environment. MPs being suspended in air may lead to serious respiratory issues upon inhalation (Dris et al. 2016). The SAMP are comprised of microfibers, micro-fragments, and granules. A comparative study showed that MP contamination in indoor air is higher when compared to outdoor air; with the indoor deposition rate of 1586 to 11130 fibers/m^2 per day (Dris et al. 2017). From air these are carried to soil and water bodies by the wind. As reported by Tiwari et al. (2019) the accumulation of MPs in sand is found to be 220 particles per kg of dry sand along the Indian coast. The contamination is not confined to the marine ecosystems but has also reached the freshwater ecosystem. The backwater of the Xiangxi River in China was reported to have $3.42 \times 10^7/km^2$ of MPs (Zhang et al. 2017a). Thus, there is a strong need to focus our research on the knowledge gap that we face, while linking MP contamination reaching different parts of our environment and the levels of impact they have on the living organisms of a particular ecosystem, and finally making their way to the humans, bioaccumulating, and becoming a persistent part of this very nature.

12.4 Microorganisms and Interactions with Microplastics

With the widespread accumulation of MPs, their chances of interaction with another cosmopolitan community of microbes have increased. These plastics are the newly discovered breeding grounds where the microbes to flourish. Being fragmented, non-biodegradable, and persistent particles, MPs have more chances of incorporation into biological food chains, thereby leading to biomagnification at the highest trophic level (Eriksson and Burton 2003). The microbes commonly found inhabiting plastics include algae, diatoms, fungi, and bacteria (Mammo et al. 2020). The various modes of interactions involve colonization, chemical adsorption, ingestion, and degradation. After colonization, genetic machinery is influenced by the expression of surface attachment and motility genes, thereby altering cellular behavior (Tuson and Weibel 2013). MPs not only attract microbes, they are also enriched with organic pollutants and metals that help maintain the interaction between microbes and plastics, as the organic pollutants keep a check on the nutrient availability, and the metals enhance cellular pathways such as electron transportation via respiration (Hara et al. 2017; Shen et al. 2019). Microbes deteriorate the plastic surfaces through biofilm formation by secreting extracellular polymeric substances. The biofilms provide support from mechanical shearing and protection from predators. Biofilm also determines the surrounding microbial community due to leaching of dissolved organic carbon (DOC), thus increasing the marine DOC pool influencing the food web (Romera-Castillo et al. 2018). Proteobacteria is the most commonly found bacterial phylum in the Plastisphere accompanied by Firmicutes in freshwater, while cyanobacteria colonize profoundly in the marine environment. The accumulation of MPs on the surface of planktons and microalgal cell walls has been reported by Yang et al. (2014) and Bergami et al. (2017), respectively. These MPs become embedded into the caveolae of the phytoplanktonic cell wall such as that of *Skeletonema costatum* (Zhang et al. 2017b). There arises some chemical signaling in the form of dimethyl sulfide, an algal derivative, plus chemicals from phytoplankton, which create prey confusion leading to MP ingestion by marine animals (Procter et al. 2019). The ingestion of MPs by the microorganisms may take place via adsorption, surface feeding, or endocytosis (phagocytosis or pinocytosis). Since the MPs size lies within the range of food habits of zooplankton, their tendency to be ingested and become incorporated into the food chain is high (Cole et al. 2013). Botterell et al. (2018) reported that copepod, a zooplankton, feeds on MPs of sizes 5–816 μm through filter feeding. Among the microbes there are pathogens, which are found attached to the plastic surfaces, and thus plastics act as a carrier for their transmission, such as the human pathogens *Haemophilus, Aeromonas, Acinetobacter, Pseudomonas mendocina,* and pathogenic *Escherichia coli* strains; along with *Aeromonas salmonicida, Tenacibaculum* sp., *Phormidium* sp., and *Leptolyngbya,* which are pathogens of fish and shrimp (Gilardi et al. 2010). *Pseudomonas syringae,* a plant pathogen, has also been reported on the MP biofilms (Naik et al. 2019). The microbes that constitute the marine Plastisphere are known to play an important role in the biogeochemical cycles of oceans. As reported, the heterotrophic bacteria on the MPs are comparatively more active than the free-living microbes; thus reflecting the influence of MPs on biogeochemical cycles (Hutchins and Fu 2017; Jacquin et al. 2019). Even photosynthetic bacteria such as blue green algae are found to be overexposed on MPs that might give out more primary production in water bodies, thus referring to the less explored world of the microbial–microplastic interactions.

12.5 Factors Affecting Biodegradation of Microplastics

Biodegradation of MPs refers to the process of reducing a polymer into an oligomer and then a monomer, or breaking down the hydrocarbon backbone by the action of microorganisms (Hadad et al. 2005). It is the molecular structure, microbial diversity, and surrounding conditions of plastics that determine the rate of biodegradation. The environmental factors that may affect the degradation process include temperature, pH, air, moisture in air, sun, salinity, and stress (Gu 2003; Figure 12.3).

The biomass produced by the microbes undergoes the process of mineralization, producing carbon dioxide and water in the presence of air, and carbon dioxide and methane in absence of air (Ng et al. 2018). The pH influences the optimum enzymatic activity, as it has been observed that at a low pH the fungal ligninolytic activity was enhanced, helping the degradation as indicated by the maximum evolution of CO_2, one of the major parameters to check degradation. Regarding colonization of the surface, E. coli has been found to decrease the pH on the surface, thus creating a gradient of hydronium ions producing proton motive force, which effect the cellular energy production. The level of humidity affects the microbial growth and action of microbial enzymes on hydrolysis of polymers. Salinity plays an important role in determining the adhesion of the microbial community in both marine and fresh water.

Similarly, osmolality effects microbial–polymer interaction by influencing the genetic machinery, as the genes for pili development are suppressed while the biofilm formation genes are over expressed. Thermal degradation occurs on polymeric exposure to a temperature of over 200 °C, causing a break in the long chains at the impurity sites, such as the site where there are unsaturated bonds and head-to-head arrangements of units. Increase in temperature is accompanied by phase transitions increasing the hydrophobicity of the surface and enhancing the microbial surface adhesion (Tuson and Weibel 2013). UV radiation (290–400 nm) is

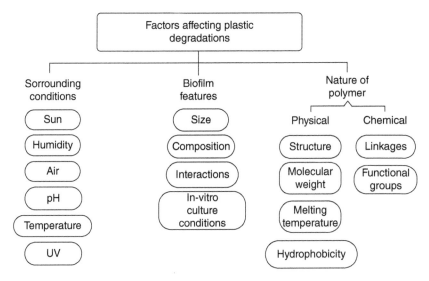

Figure 12.3 There are several factors affecting polymer degradation: (a) surrounding conditions includeing sun, humidity, air, pH, temperature, and UV radiation; (b) biofilm features including the type of microbes and their modes of interactions; and (c) nature of the polymer, which are it's physical and chemical properties.

well reported to initiate the polymer degradation, as they can cleave the C–C bond and generate free radicals creating the weak points for microbial attack. This radiation brings changes to the mechanical properties such as tensile strength along with molecular weight reduction. Photo-oxidation via UV pretreatment has been found to enhance the degradation process by making the plastic brittle and changing the physical features by introducing hydrophilic carbonyl group; for example, polyethylene that underwent UV treatment for 500 hours aided the degradation by *Penicillium simplicissimum* YK (Yamada-Onodera et al. 2001). Similarly, the microbial population diversity, its size, and the optimum culture condition needed to keep the microbial enzymes activated also affect the plastic degradation (Ahmed et al. 2018). The chemical and physical properties of the polymer such as the high order structures, linkages, molecular weight, and the melting point affect enzymatic activity, for example *Rhizopus delemar* lipase was highly active at low molecular weight poly e-caprolactone diol (Tokiwa et al. 2009). The hydrophobicity of plastic plays a crucial role in formation of biofilms following biodegradation. The **molecular structure** of the polymer like the crystallinity, modulus of elasticity and branching slow down the rate of degradation. Melting temperature also has the reciprocal relation with the degradation rate (Tokiwa and Calabia 2004).

12.6 Mechanisms of Microplastic Biodegradation

The inadequacy of commonly applied plastic disposal methods such as land filling, burning, and recycling has not only contributed to the ecological imbalance but have also paved the way for growing concerns and research-based approach toward the microbial methods of plastic degradations (Krueger et al. 2015). Biodegradation of plastic by microbes involves bioaccumulation of microbes on the plastic surface forming the biofilms; thus, biofilms are the microbial aggregates present on the surface of polymers. These microbial aggregations occur when the polymers are exposed to a specific environment such as soil, seawater or freshwater, or even the gut of specific organisms such as the worms and lower invertebrates that feed on these polymer particles, mistaking them as food due to their small sizes. Several factors may affect the formation of biofilms. As the Plastisphere community will be specific for a specific environment the exposed surrounding conditions such as the temperature, presence or absence of light, UV radiations, etc. will play a major role in the colonization of polymer surface by the pioneer microbial species. UV rays may initiate the process of fragmentation by forming cracks and crevices, and thus easing the process of microbial entry onto the large polymer surface (Yang et al. 2014). With respect to the formation of biofilm in the water, the level of nutrient enrichment, that is the abundance of nitrogen and phosphorus, will determine the microbial interactions with MPs. Therefore, the Plastisphere composition will vary from freshwater systems (less eutrophication) to lakes and oceans (more eutrophication). Plastic polymers have been found to accumulate potentially toxic elements such as copper, lead, and cadmium, as well as persistent organic pollutants such as polyaromatic hydrocarbons (PAHs), polychlorinated biphenyls (PCBs), and the insecticide dichlorodiphenyltrichloroethane (DDT) (Bakir et al. 2016). Few studies have considered these chemical attachments to mediate microbial–microplastic interactions, as these toxic chemicals attract pollutant-degrading microorganisms, and metals take part in electron transportation aiding the microbial utilization of polymer as a carbon source. Moreover, the polymers' surface properties such as texture and hydrophobicity act as the key determinants for the attraction of microbes and their attachment to the surface. Thus, all the factors discussed guidesthe process of biofilm formation, microbial attachment, and subsequent colonization leading to biofilm maturation (Rummel et al. 2017).

The subsequent interaction together with the exposed conditions initiates the degradation process as the microbes secrete exoenzymes such as polymerases that ease the process of fragmentation by breaking branching, hydrocarbon linkages, and incorporating electrophile, such as oxygen-forming alcohol or peroxyl group [(Elbanna et al. 2004), Figure 12.4]. This is followed by assimilation of these fragmented polymers or MPs into the cellular compartment of these microbes through the semi-permeable membranes, utilizing these for nutrition and then finally mineralization leading to the formation of CO_2 and H_2O under aerobic conditions and methane under anaerobic conditions. There are certain parameters that can be considered for determining polymer degradation such as the physical changes including reduction in the molecular weight, tensile strength, and change in topography, increase in the microbial density growing on the polymer surface, increase in the biological oxidation demand, and CO_2 evolution. Microbes bring changes to the functional groups of MPs and thereby alter the properties as well as their physical appearance. The most commonly involved microbes in the degradation process include bacteria, fungi, and algae (Shah et al. 2014, Table 12.2).

Bacteria being the most abundant microbes, form a major part of the Plastisphere. The horizontally floating polymer surface act as a storehouse for nutrients such as basic amino acids that can stimulate growth, metals and other organic pollutants are also present which bacteria use as metabolites and co-factors, metals (e.g. Fe and Mg) also act as electron acceptors during respiration. All these factors make polymer a suitable surface for adhesion. Bacteria generally use its external appendages such as pili and fimbria for the attachment, and these provide a channel for charge transfer.

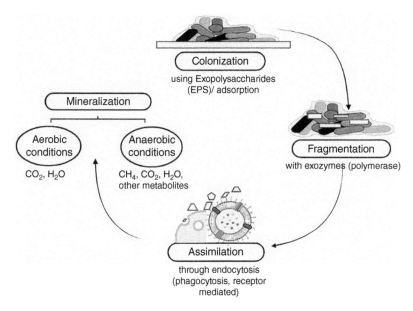

Figure 12.4 Steps involved in the biodegradation process: (a) colonization – attachment to the polymeric surface; (b) fragmentation – breaking polymers into oligomers and monomers; (c) assimilation – incorporating the smaller molecules inside; and (d) mineralization – utilizing MPs for energy production along with CO_2, H_2O, and other metabolites.

Table 12.2 List of various microorganisms bioremediating plastic polymers.

Microorganism	Source	Incubation conditions	Polymer	Mode of action	References
Actinomadura rubrobrunea	Compost	50 °C, pH = 7, several days	Cellulose acetate	Cellulose breaks glycosidic bond, esterase breaks ester bond	Degli-Innocenti et al. (2002)
Pseudomonas chlororaphis	Microbial consortium	37 °C, 4 days	Polyurethanes	Almost complete degradation using polyurethenase enzymes (esterase and lipase)	Howard et al. (2001a)
Aspergillus niger	Soil	25 °C, 70 days, 100 rpm shaking	Polyurethanes	Isocyanate group disappears	Filip (1979)

(Continued)

Table 12.2 (Continued)

Microorganism	Source	Incubation conditions	Polymer	Mode of action	References
Pestalotiopsis microspore E2712A	Plant stem	23 °C, 2 weeks, both in presence and absence of O_2	Polyurethanes	Polyurethanase (seriene hydrolase), enzyme breaking ester bond	Russell et al. (2011)
Bacillus, Burkholderia, Nocardiopsis, Cupriavidus, Micromycetes	Soil; polymer specimen	30°C 25°C	Poly-β hydroxy -alkanoates	Fragmentation, 98% degradation, Weight loss	Boyandin et al. (2013)
Brevibaccillus borstelensis	Soil	15 °C, 30 days	Polyethylene	Degrades CH_2 backbone, reduced CO group, weight loss by 30%	Hadad et al. (2005)
Penicillium simplicissimum YK	Soil and leaves	UV or HNO3, pretreated PE,3 months	Polyethylene	Functional group insertion	Yamada-Onodera et al. (2001)
*Enterobacter asburiae*YT1, *Bacillus sp*.YP1	Waxworm gut	30°C, 28 days	Polyethylene succinate	Hydrophobicity decreases, 50% tensile strength decrease	Anderson and Shive (2012)
Ideonella sakaiensis 201-F6	PET contaminated environmental samples	30°C, 6 weeks	Polyethylene terephthalate	Frequent pits development by PETase (extracellular hydrolytic enzyme)	Yoshida et al. (2016)
Rhizopu sarrhizus, Rhizopus delemar	Soil	30°C	Starch/polyester	Lipase degrades polyesters, Hydrolyse ester bonds	Tokiwa and Calabia (2007) Tokiwa and Suzuki (1977)
Aspergillus sp. ST-01	Soil	50 °C, 6 days	Polycaprolactone	Complete degradation	Sanchez et al. (2000)
Pseudomonas sp. AKS2	Kolkata Municipal Solid waste site	30 °C, 3 days	Polyethylene succinate	Changes surface topography	Tribedi and Sil (2013)
Bacillus sp. TT96	Soil	50 °C, 3–4 days	Polyethylene succinate	Weight loss	Tokiwa et al. (2009)

Table 12.2 (Continued)

Microorganism	Source	Incubation conditions	Polymer	Mode of action	References
Bacillus brevis	Soil	60 °C, 20 days	Polylactic acids	Weight loss by 20%	Yang et al. (2014)
Amycolatopsis sp., *B. licheniformis*	Soil	30 °C, 14 days	Polylactic acids	Weight. loss by 60%	
Aspergillustubingensis VRKPT1	PE waste disposal	30 °C, 12 weeks, 150 rpm	High density polyethylene	Weight loss	Sangeetha Devi et al. (2015)
Aspergillus flavus VRKPT2	Marine costal area	30 °C, 12 weeks, 150 rpm	High density polyethylene	Carbonyl index increased	
Exiguobacterium sp. strain YT2	Mealworm gut	22–24 °C, 28 days	Polystyrene	Hydrophobicity decrease Carbonyl polar bond formation	Yang et al. (2015)
Bacillus gottheilii	Mangrove ecosystems	40 days, 150 rpm, shaking	Polystyrene	Weight loss	Auta et al. (2017)
Scenedesmus dimorphus, Anabaena spiroides, Navicula pupula	Marine ecosystem	30 days	Low density polyethylene	8% weight loss	Vimal Kumar et al. (2017)

After adhesion, the cells secrete extra polymeric substances to protect themselves from water currents. There occurs quorum sensing by which the bacterial cells interact with each other via chemical signaling, which influences genetic makeup such as switching the motility and adhesion genes. The *Pseudoalteromonas, Pseudomonas, Acinetobacter, Alteromonas, Micrococcus, Bacillus,* and *Vibrio* are the common genera which constitute the biofilm (Singh and Sharma 2008). The gram-negative bacteria such as the *Pseudomonas* and *Vibrio* are profoundly found on the biofilm of the large polymers such as PE and PET, as they can overcome the hydrophobicity of these polymers and colonize them by modifying their lipopolysaccharide structure (Pruzzo et al. 2008; Krasowska and Sigler 2014).

Since the polymers are the highly stabilized structures due to uniform charge distribution, the bacteria uses depolymerases to convert the polymer into oligomers and monomers. The bacteria may employ the enzyme oxygenase to introduce an oxygen atom or two into the polymer-forming alcohol or peroxyl groups, destabilizing its structure and enhancing the degradation process (Vijaya and Mallikarjuna 2008). *Stenotrophomonas matlophilia* LB 2-3 strain has been found to decrease polylactic acid, tensile strength, and molecular weight of MPs (Jeon and Kim 2013). Out of the eight strains isolated from mangrove ecosystems in Malaysia that were tested for their potential to degrade UV treated polymers such as PE, PS, PET, and PP, only two—*Bacillus cereus* and *Bacillus gottheilii*—have been recorded to possess degradation ability by utilizing the MPs as the sole carbon sources. After 40 days of incubation, the weight loss percentage of PE, PS, and PET by *B. cereus* was found to be 1.6%, 7.4%, and 6.6%, respectively; whereas the weight loss percentages of PE, PS, PET, and PP by *B. gottheilii* is 6.6%, 5.8%, 3.0%, and 3.6%, respectively. The polymer sinks to the bottom of the flask within one week, indicating massive colonization by the microbes recorded by the optical density measurements. The FT-IR spectra showed the disappearance of the carbonyl group introduced by the ultraviolet radiations, along with the appearance of new functional groups such as hydroxyl and amino groups, and terminal double bonds in the reaction with *B. cereus*, whereas there was a single broad peak corresponding to the hydroxyl group in case of *B. gottheilii* (Auta et al. 2017). The *Enterobacter asburiae* YT1 (gram positive) and *Bacillus* sp. YP1 (gram negative) strains isolated from the gut of waxworms or Indian mealmoths (the larvae of *Plodia interpunctella*), have been found to chew/degrade PE. After an incubation period of 28 days, it was reported that the waxworms change their physical properties; the hydrophobicity decreased and tensile strength was reduced to 50% with the development of cracks and crevices, as evident from scanning microscopy and atomic force microscopy (AFM); chemical structure altered with incorporation of the carbonyl group exhibited by X-ray photoelectron spectroscopy (XPS) and FT-IR; and molecular weight was reduced with release of some degraded metabolites (Anderson and Shive 2012; Yang et al. 2014). Further research should be done to identify these metabolites and the metabolic pathways associated with their production. Yang et al. (2015) similarly investigated the biodegradation potential of gut bacteria of mealworm (the larvae of *Tenebrio molitor Linnaeus*) in degrading polystyrene. The isolated strain *Exiguobacterium* sp. strain YT2 was reported to degrade PS and could form biofilm and develop pits and cracks on the polymer surface within the incubation period of 28 days. Tribedi and Sil (2013) investigated the role of hydrophobicity in polymer degradation using *Pseudomonas* sp. AKS2, which was already a proven PES degrader. It was found that with the increase in hydrophobicity, the biofilm formation was enhanced and hence it increased the degradation. Bacterial nutrition conditions (carbon and nitrogen stress) were exploited to modulate the cell surface hydrophobicity with the glucose minimum concentration bringing out the maximum cell surface hydrophobicity. The commonly used biodegradable plastic PLA has been reportedly degraded by *Amycolatopsis* sp. and *Bacillus licheniformis*, isolated from the soil. *Pseudomonas chlororaphis* isolated from a microbial consortium obtained from the Naval Research Laboratory, Washington, D.C., has been found to degrade PUR. A clear zone of inhibition appeared overnight around the bacterial colonies. *P. chlororaphis* degrades PUR by secreting hydrolases enzyme, two of the PUase genes—PueA and PueB—have been cloned, and three of the PUase enzymes have been purified and characterized. The PueA gene produces an unclear zone of inhibition with endo-polyurethenase activity, and the PueB gene produces a clearer zone of inhibition with the exo-polyurethenase activity. These two genes work synergistically to degrade PUR polymer into oligomers, although there is not much information about their actual site of cleavage (Howard et al. 2001a). Giacomucci et al. (2019) studied five strains of bacteria, out of which the preliminary screening of about three months selected the two strains—*Pseudomonas citronellosis* DSM 50332 and *Bacillus flexus* DSM 1320—which have been found to degrade PVC. *P. citronellosis* reduces the weight of PVC by 1% after an incubation of 30 days (Giacomucci et al. 2019). *Pseudomonas* and *Bacillus* are the genera that are extensively involved in degradations of varied types of the polymers. Yoshida et al. (2016) investigated the PET degradation by *Ideonella sakaiensis*, isolated from PET debris-contaminated environmental samples such as the soil, sediments, and sludge from the recycling unit. It brought about complete degradation of PET in six weeks at 30 °C. The bacteria use small appendages to stick to the polymer surface and might also secrete some extracellular enzymes through them. Two genes have been identified from this strain; PETase and MHETase. PETase shares 51% homology with the hydrolase and so was designated as PET hydrolase or PETase. MHETase, homologous to tannase family, the known ester linkage hydrolases, was named so as it was found to degrade mono-hydroxyethyl-terephthalic acid to terephthalic acid (TPA) and ethyleneglycol (EG).

Even the eukaryotes such as fungi have worthy degradation potential as they too utilize MPs as a carbon source. They aid in formation of functional groups such as CO, COOH, and RCOOR, decreasing the polymer hydrophobicity. *Chytridiomycota* is the fungi extensively found in biofilm. Fungal strains such as *Mucor rouxii* NRRL 1835, *Aspergillus flavus*, and *P. simplicissimum* YK have been reported to degrade PE. The strain *P. simplicissimum* YK was obtained from polythene by Yamada et al. (2001), who reported its potential to degrade polythene. It was observed that the strain grew well on the polythene-containing media along with Triton X-100, which is a surface-active substance, thus indicating the polymer utilization by fungi for its growth. They also found that the UV-pretreated polymer (in this case treatment was given for 500 hours) hastens the degradation process as it introduces polar groups such as the carbonyl group, compared to the untreated polymer without any functional group insertions. Additionally, the hyphae are more efficient in degrading polymer than the spores as the hyphae reduced the weight of the polymer within one month, whereas the spores could cause little or no degradation, even in six months of cultivation. The reduction in molecular weight was assessed from the High Temperature-Gel Permeation Chromatography (HT-GPC) spectra, and the appearance or disappearance of the functional groups was observed using FT-IR spectra. The FT-IR spectra for polythene after treatment with *P. simplicissimum* YK showed a decrease in the absorbance of the band corresponding to the C=C double bond (Yamada-Onodera et al. 2001). White rot fungi that degrades lignin, a widely occurring non-biodegradable polymer in plant wood, has been studied for its potential to degrade plastic polymer. Iiyoshi et al. (1998) reported that manganese peroxidase is involved in polyethylene degradation by lignin-degrading fungi such as IZU-154, *Phanerochaete chrysosporium*, and *Trametes versicolor*. Another enzyme secreted by this category of fungi is laccases, which causes oxidation of both aromatic and nonaromatic polymers. Lee et al. (1991) compared the degradation potential of this fungi with the bacterium strain *Streptomyces viridosporus* T7A, which has been found to reduce the tensile strength of biodegradable plastic by 50% and molecular weight by 68%; although the fungal strain was found to be efficient at decreasing plastic elongation. Russell et al. (2011) tested several endophytic fungi isolated from the plant samples at Amazonian rainforest for their potential to remediate PUR revealing that the *Pestelotiopsis microspora* has the highest degrading activity among them all, and it was double the degradation activity of *P. chrysosporium*, used as the positive control for PUR degradation. After a 16-day time period the translucent culture indicated the use of polymer as a sole carbon source by the strain, while the *P. chrysosporium* reached half of the clearance zone in 15 days. The growth from the bottom of the liquid culture flask instead of the top confirmed the anaerobic degradation, further analysis showed that *P. microspora* E2712A strain was active both in aerobic and anaerobic conditions. The fungal strain has the ability to break the ester bond, as evident from the infrared spectra analysis in which the ester (COO) absorbance peak was absent. Screening of *Pestalotiopsis microspore* revealed that serine hydrolases is the enzyme involved in the PUR degradation (Howard et al. 2001b; Russell et al. 2011). Enzymes such as esterases and lipases from *R. delemar*, *R. arrhizus*, and *Candida cyclindracea* have been found to degrade complex polymers such as PCL and PEA. The polymer hydrolyzing activity of these enzymes was determined by measuring the total organic carbon (TOC) content, available in the reaction mixture containing the enzymes and the polymer incubated at 30 °C using a TOC analyzer. On hydrolysis, the reaction mixture has PEA polymer dissociated into its constituent parts; adipic acid and ethylene glycol, as well as the PEA oligomers highlighting the esterase (breaking of ester bonds) activity of the *R. delemar*'s lipase (Tokiwa and Suzuki 1977; Tokiwa and Calabia 2007). Degradation of the untreated (no UV or thermal treatments) and unmodified (no insertion of functional groups or pro-oxidant groups) HDPE has been reported by the fungal strains *Aspergillus tubingensis* VRKPT1 and *A. flavus* VRKPT2, isolated from a PE waste dumping site near the marine coastal area of Tuticorin, India. FT-IR spectra showed that these two strains have caused the polymer weight reduction of about 6.02 ± 0.2 and $8.51 \pm 0.1\%$, respectively (Sangeetha Devi et al. 2015). The scanning electron microscopy analysis concluded that these strains change the polymer surface topography, as there were crack and pit development indicating the secretion of some extracellular enzymes by these strains. Further research could be carried out to explore the molecular structure and function of these enzymes.

Algae have also been found to interact with MPs, but the evidence to state their involvement in biodegradation of plastic are few and the mechanism of degradation is not well established. The common algae accumulating on the plastic surfaces involve blue green algae such as *Oscillatoria, Nostoc, Anabaena, Spirogyra*, and *Chlorella* (Sarmah and Rout 2018). *S. costatum* has been reported to be able to adsorb the MPs into caveolae present on their cell walls. The dimethyl sulfide secreted by algae chemo attracts marine animal such as zooplankton toward MPs, thus establishing their role in distribution of MPs through food chain. Vimal Kumar et al. (2017) reported that *Scenedesmus dimorphus, Anabaena spiroides*, and *Navicula pupula* possess the ability to degrade both high-density and low-density polyethylene with blue green algae, degrading 8% of LDPE after one month's incubation. However, their degradation rate as compared to bacteria and fungi are quite low, attributing to their photosynthetic mode of respiration (Dineshbabu et al. 2020). In a recent study, Hoffmann et al. (2020) showed that *Phaecodactylum tricornatum* has a huge potential for serving as a genetic host to synthesize PETase enzyme from *I. sakaiensis* used for PET degradation.

12.7 Conclusion and Future Perspectives

Today we are surrounded by products that are wholly or partially made of plastic. They have become an essential part of our lives, which accounts for their ever-increasing production. Due to their recalcitrant nature and inadequate methods of disposal, they continue to contaminate different areas of the environment, be it land, water, or air. Plastics influence the flora and fauna of every environment, as they have become a part of the food chain, and are thus redistributed and bioaccumulating. Recent scenarios are such that microorganisms seem promising as a natural means of remediating plastic polymers; of which bacteria, fungi, and algae are the most prominent polymer degraders. The commonly found genera on biofilm mainly consists of *pseudomonas, bacillus, vibrio, aspergillus, penicillium*, and blue green algae; which secrete several enzymes such as depolymerases, oxygenase, esterase, chitinase, and lipase which are crucial for the degradation process. Since employing several physical technologies such as bioreactors and membrane filters in WWTPs, we fear it is not possible to largely counteract the MP contamination. Currently, research is being carried out to deduce the ways by which the maximum potential of these microbes may be harnessed and to see if a combination of physical and biological methods may prove effective. Even in-vitro experiments may be carried out utilizing genetic engineering as a promising tool to isolate and amplify the genes that code for a particular type of polymer degrader enzyme. Research in fresh water system contamination could prove beneficial in understanding plastics' level of entrainment, their entry source, effect on fresh water flora and fauna, and the end paths they are destined to for further contamination. In order to have a better insight into microbial–microplastic interactions, the geographical and seasonal variation of the effects on formation and colonization of biofilms could be researched upon simultaneously with the study of changes in biofilm composition on transiting from one part of the environment to the other. It would be of more scientific interest to look into the role of biofilms in the nutrient cycling and marine biogeochemical cycles, as they have many chemical depositions along with the microbes interacting with them. There is certainly less clarity about the abundance of phototrophic microbes such as the blue green algae on the biofilm compared to the water surrounding MPs that could prove to be influencing the primary production in oceans, indicating the polymers acting as the autotrophic hotspots in seawater. More research may be sought in the areas concerning the mechanisms in use by the various microbes for the degradation of plastic and metabolizing them as the carbon source.

References

Ahmed, T., Shahid, M., Azeem, F. et al. (2018). *Biodegradation of plastics: current scenario and future prospects for environmental safety. Environmental Science and Pollution Research International* 25 (8): 7287–7298. https://doi.org/10.1007/s11356-018-1234-9.

Alimba, C.G. and Faggio, C. (2019). *Microplastics in the marine environment: current trends in environmental pollution and mechanisms of toxicological profile. Environmental Toxicology and Pharmacology* 68: 61–74. https://doi.org/10.1016/j.etap.2019.03.001.

Anderson, J.M. and Shive, M.S. (2012). *Biodegradation and biocompatibility of PLA and PLGA microspheres. Advanced Drug Delivery Reviews* 64: 72–82. https://doi.org/10.1016/j.addr.2012.09.004.

Auta, H.S., Emenike, C.U., and Fauziah, S.H. (2017). *Screening of Bacillus strains isolated from mangrove ecosystems in Peninsular Malaysia for microplastic degradation. Environmental Pollution* 231: 1552–1559. https://doi.org/10.1016/j.envpol.2017.09.043.

Bakir, A., O'Connor, I.A., Rowland, S.J. et al. (2016). *Relative importance of microplastics as a pathway for the transfer of hydrophobic organic chemicals to marine life. Environmental Pollution* 219: 56–65. https://doi.org/10.1016/j.envpol.2016.09.046.

Bergami, E., Pugnalini, S., Vannuccini, M.L. et al. (2017). *Long-term toxicity of surface-charged polystyrene nanoplastics to marine planktonic species* Dunaliellatertiolecta *and* Artemia franciscana. *Aquatic Toxicology* 189: 159–169. https://doi.org/10.1016/j.aquatox.2017.06.008.

Botterell, Z.L.R., Beaumont, N., Dorrington, T. et al. (2018). *Bioavailability and effects of microplastics on marine zooplankton: a review. Environmental Pollution* 245: 98–110. https://doi.org/10.1016/j.envpol.2018.10.065.

Boyandin, A.N., Prudnikova, S.V., Karpov, V.A. et al. (2013). *Microbial degradation of polyhydroxyalkanoates in tropical soils. International Biodeterioration & Biodegradation* 83: 77–84. https://doi.org/10.1016/j.ibiod.2013.04.014.

Carpenter, E.J. and Smith, K.L. (1972). *Plastics on the Sargasso Sea surface. Science* 175 (4027): 1240–1241. https://doi.org/10.1126/science.175.4027.1240.

Chen, Y., Leng, Y., Liu, X., and Wang, J. (2020). *Microplastic pollution in vegetable farmlands of suburb Wuhan, Central China. Environmental Pollution* 257: 113449. https://doi.org/10.1016/j.envpol.2019.113449.

Cole, M., Lindeque, P., Fileman, E. et al. (2013). *Microplastic ingestion by zooplankton. Environmental Science & Technology* 47 (12): 6646–6655. https://doi.org/10.1021/es400663f.

Degli-Innocenti, F., Goglino, G., Bellia, G. et al. (2002). *Isolation and characterization of thermophilic microorganisms able to grow on cellulose acetate*. In: *Microbiology of Composting* (eds. H. Insam, N. Ridech and S. JKlammer), 273–286. Springer http://doi.org/10.1007/978-3-662-08724-4, https://doi.org/10.1007/978-3-662-08724-4_23.

Dineshbabu, G., Uma, V.S., Mathimani, T. et al. (2020). *Elevated CO2 impact on growth and lipid of marine cyanobacterium Phormidiumvalderianum BDU 20041; towards microalgal carbon sequestration. Biocatalysis and Agricultural Biotechnology* 25: 101606. https://doi.org/10.1016/j.bcab.2020.101606.

Dris, R., Gasperi, J., Rocher, V. et al. (2015). *Microplastic contamination in an urban area: a case study in greater Paris. Environmental Chemistry* 12 (5): 592. https://doi.org/10.1071/en14167.

Dris, R., Gasperi, J., Saad, M. et al. (2016). *Synthetic fibers in atmospheric fallout: a source of microplastics in the environment? Marine Pollution Bulletin* 104 (1–2): 290–293. https://doi.org/10.1016/j.marpolbul.2016.01.006.

Dris, R., Gasperi, J., Mirande, C. et al. (2017). A first overview of textile fibers, including microplastics, in indoor and outdoor environments. *Environmental Pollution* 221: 453–458. https://doi.org/10.1016/j.envpol.2016.12.

Elbanna, K., Lütke-Eversloh, T., Jendrossek, D. et al. (2004). *Studies on the biodegradability of polythioester copolymers and homopolymers by polyhydroxyalkanoate(PHA)-degrading bacteria and PHA depolymerases. Archives of Microbiology* 182 (2–3): 212–225. https://doi.org/10.1007/s00203-004-0715-z.

Eriksson, C. and Burton, H. (2003). *Origins and biological accumulation of small plastic particles in fur seals from Macquarie Island. Ambio: A Journal of the Human Environment* 32 (6): 380–384. https://doi.org/10.1579/0044-7447-32.6.380.

Filip, Z. (1979). *Polyurethane as the sole nutrient source for* Aspergillus niger *and* Cladosporium herbarum. *European Journal of Applied Microbiology and Biotechnology* 7 (3): 277–280. https://doi.org/10.1007/bf00498022.

Giacomucci, L., Raddadi, N., Soccio, M. et al. (2019). *Polyvinyl chloride biodegradation by* Pseudomonas citronellolis *and* Bacillus flexus. *New Biotechnology* 52: 35–41. https://doi.org/10.1016/j.nbt.2019.04.005.

Gilardi, K.V., Carlson-Bremer, D., June, J.A. et al. (2010). *Marine species mortality in derelict fishing nets in Puget Sound, WA and the cost/benefits of derelict net removal. Marine Pollution Bulletin* 60 (3): 376–382. https://doi.org/10.1016/j.marpolbul.2009.10.016.

Gu, J.-D. (2003). *Microbiological deterioration and degradation of synthetic polymeric materials: recent research advances. International Biodeterioration & Biodegradation* 52 (2): 69–91. https://doi.org/10.1016/s0964-8305(02)00177-4.

Guo, J.J., Huang, X.P., Xiang, L. et al. (2020). *Source, migration and toxicology of microplastics in soil. Environment International* 137: 105263. https://doi.org/10.1016/j.envint. 2019.105263.

Hadad, D., Geresh, S., and Sivan, A. (2005). *Biodegradation of polyethylene by the thermophilic bacterium* Brevibacillusborstelensis. *Journal of Applied Microbiology* 98 (5): 1093–1100. https://doi.org/10.1111/j.1365-2672.2005.02553.x.

Hara, T., Takeda, T.A., Takagishi, T. et al. (2017). *Physiological roles of zinc transporters: molecular and genetic importance in zinc homeostasis. The Journal of Physiological Sciences* 67 (2): 283–301. https://doi.org/10.1007/s12576-017-0521-4.

Hoffmann, L., Eggers, S.L., Allhusen, E. et al. (2020). *Interactions between the ice algae* Fragillariopsiscylindrus *and microplastics in sea ice. Environment International* 139: 105697. https://doi.org/10.1016/j.envint.2020.105697. Epub 2020 Apr 22. PMID: 32334123.

Howard, G.T., Crother, B., and Vicknair, J. (2001a). *Cloning, nucleotide sequencing and characterization of a polyurethanase gene (pueB) from* Pseudomonas chlororaphis. *International Biodeterioration & Biodegradation* 47 (3): 141–149. https://doi.org/10.1016/s0964-8305(01)00042-7.

Howard, G.T., Vicknair, J., and Mackie, R.I. (2001b). *Sensitive plate assay for screening and detection of bacterial polyurethanase activity. Letters in Applied Microbiology* 32 (3): 211–214. https://doi.org/10.1046/j.1472-765x.2001.00887.x.

Hutchins, D.A. and Fu, F. (2017). *Microorganisms and ocean global change. Nature Microbiology* 2 (6): 17058. https://doi.org/10.1038/nmicrobiol.2017.58.

Iiyoshi, Y., Tsutsumi, Y., and Nishida, T. (1998). *Polyethylene degradation by lignin-degrading fungi and manganese peroxidase. Journal of Wood Science* 44 (3): 222–229. https://doi.org/ 10.1007/bf00521967.

Jacquin, J., Cheng, J., Odobel, C. et al. (2019). *Microbial ecotoxicology of marine plastic debris: a review on colonization and biodegradation by the "Plastisphere" a new niche for marine microorganisms. Frontiers in Microbiology* 10: 865. https://doi.org/10.3389/fmicb.2019. 00865.

Jeon, H.J. and Kim, M.N. (2013). *Isolation of a thermophilic bacterium capable of low-molecular-weight polyethylene degradation. Biodegradation* 24 (1): 89–98. https://doi. org/10.1007/s10532-012-9560-y. Epub 2012 Jun 4. PMID: 22661062.

Krasowska, A. and Sigler, K. (2014). *How microorganisms use hydrophobicity and what does this mean for human needs? Frontiers in Cellular and Infection Microbiology* 4 https://doi. org/10.3389/fcimb.2014.00112.

Krueger, M.C., Harms, H., and Schlosser, D. (2015). *Prospects for microbiological solutions to environmental pollution with plastics. Applied Microbiology and Biotechnology* 99 (21): 8857–8874. https://doi.org/10.1007/s00253-015-6879-4. Epub 2015 Aug 30. PMID: 26318446.

Lee, B., Pometto, A.L., Fratzke, A., and Bailey, T.B. (1991). *Biodegradation of degradable plastic polyethylene by phanerochaete and streptomyces species. Applied and Environmental Microbiology* 57 (3): 678–685. https://doi.org/10.1128/AEM.57.3.678-685.1991.

Li, W.C., Tse, H.F., and Fok, L. (2016). *Plastic waste in the marine environment: a review of sources, occurrence and effects. Science of the Total Environment* 566–567: 333–349. https://doi.org/10.1016/j.scitotenv.2016.05.

Liu, W., Zhang, J., Liu, H. et al. (2021). *A review of the removal of microplastics in global wastewater treatment plants: characteristics and mechanisms. Environment International* 146: 106277. https://doi.org/10.1016/j.envint.2020.106277.

Ly, Y.T.-P., Johnson, L.A., and Jane, J. (1998). Soy protein as biopolymer. In: *Biopolymers from Renewable Resources: Macromolecular Systems — Materials Approach* (ed. D.L. Kaplan). Berlin, Heidelberg: Springer https://doi.org/10.1007/978-3-662-03680-8_6.

Mammo, F.K., Amoah, I.D., Gani, K.M. et al. (2020). *Microplastics in the environment: interactions with microbes and chemical contaminants. Science of the Total Environment*: 140518. https://doi.org/10.1016/j.scitotenv.2020.140518.

Mohanty, A.K., Misra, M., and Hinrichsen, G. (2000). *Biofibres, biodegradable polymers and biocomposites: an overview. Macromolecular Materials and Engineering* 276–277 (1): 1–24. https://doi.org/10.1002/(sici)1439-2054(20000301)276.

Naik, R.K. et al. (2019). *Microplastics in ballast water as an emerging source and vector for harmful chemicals, antibiotics, metals, bacterial pathogens and HAB species: a potential risk to the marine environment and human health. Marine Pollution Bulletin* 149: 110525. https://doi.org/10.1016/j.marpolbul.2019.110525.

Ng, E.L., Huerta Lwanga, E., Eldridge, S.M. et al. (2018). *An overview of microplastic and nanoplastic pollution in agroecosystems. Science of the Total Environment* 627: 1377–1388. https://doi.org/10.1016/j.scitotenv.2018.01.341. Epub 2018 Feb 20. PMID: 30857101.

Picó, Y. and Barceló, D. (2019). *Analysis and prevention of microplastics pollution in water: current perspectives and future directions. ACS Omega* 4 (4): 6709–6719. https://doi.org/10.1021/acsomega.9b00222.

Poerio, T., Piacentini, E., and Mazzei, R. (2019). *Membrane processes for microplastic removal. Molecules* 24 (22): 4148. https://doi.org/10.3390/molecules24224148.

Poliakoff, M., Fitzpatrick, J., Farren, T., and Anastas, P. (2002). *Green chemistry: science and politics of change. Science* 297 (5582): 807–810. Retrieved June 14, 2021, from http://www.jstor.org/stable/3831987.

Procter, J., Hopkins, F.E., Fileman, E.S., and Lindeque, P.K. (2019). *Smells good enough to eat: dimethyl sulfide (DMS) enhances copepod ingestion of microplastics. Marine Pollution Bulletin* 138: 1–6. https://doi.org/10.1016/j.marpolbul.2018.11.014.

Pruzzo, C., Vezzulli, L., and Colwell, R.R. (2008). *Global impact of vibrio cholerae interactions with chitin. Environmental Microbiology* 10 (6): 1400–1410. https://doi.org/10.1111/j.1462-2920.2007.01559.x. Epub 2008 Feb 27. PMID: 18312392.

Romera-Castillo, C., Pinto, M., Langer, T.M. et al. (2018). *Dissolved organic carbon leaching from plastics stimulates microbial activity in the ocean. Nature Communications* 9 (1) https://doi.org/10.1038/s41467-018-03798-5.

Rummel, C.D., Jahnke, A., Gorokhova, E. et al. (2017). *Impacts of biofilm formation on the fate and potential effects of microplastic in the aquatic environment. Environmental Science & Technology Letters* 4 (7): 258–267. https://doi.org/10.1021/acs.estlett.7b00164.

Russell, J.R., Huang, J., Anand, P. et al. (2011). *Biodegradation of polyester polyurethane by endophytic fungi. Applied and Environmental Microbiology* 77 (17): 6076–6084. https://doi.org/10.1128/AEM.00521-11. Epub 2011 Jul 15.

Sanchez, J.G., Tsuchii, A., and Tokiwa, Y. (2000). *Degradation of polycaprolactone at 50 °C by a thermotolerant* Aspergillus sp. *Biotechnology Letters* 22 (10): 849–853. https://doi.org/10.1023/a:1005603112688.

Sangeetha Devi, R., Rajesh Kannan, V., Nivas, D. et al. (2015). *Biodegradation of HDPE by* Aspergillus *spp. from marine ecosystem of Gulf of Mannar, India. Marine Pollution Bulletin* 96 (1–2): 32–40. https://doi.org/10.1016/j.marpolbul.2015.05.

Sarmah, P. and Rout, J. (2018). *Efficient biodegradation of low-density polyethylene by cyanobacteria isolated from submerged polyethylene surface in domestic sewage water. Environmental Science and Pollution Research International* 25 (33): 33508–33520. https://doi.org/10.1007/s11356-018-3079-7. Epub 2018 Sep 28. PMID: 30267347.

Shah, A.A., Kato, S., Shintani, N. et al. (2014). *Microbial degradation of aliphatic and aliphatic-aromatic co-polyesters. Applied Microbiology and Biotechnology* 98 (8): 3437–3447. https://doi.org/10.1007/s00253-014-5558-1. Epub 2014 Feb 13. PMID: 24522729.

Shen, M., Zhu, Y., Zhang, Y. et al. (2019). *Micro(nano)plastics: Unignorable vectors for organisms. Marine Pollution Bulletin* 139: 328–331. https://doi.org/10.1016/j.marpolbul.2019.01.004.

Singh, B. and Sharma, N. (2008). Mechanistic implications of plastic degradation. *Polymer Degradation and Stability* 93 (3): 561–584. https://doi.org/10.1016/j.polymdegradstab.2007.11.008.

Tiwari, M., Rathod, T.D., Ajmal, P.Y. et al. (2019). *Distribution and characterization of microplastics in beach sand from three different Indian coastal environments. Marine Pollution Bulletin* 140: 262–273. https://doi.org/10.1016/j.marpolbul.2019.01.055.

Tokiwa, Y. and Calabia, B.P. (2004). *Degradation of microbial polyesters. Biotechnology Letters* 26 (15): 1181–1189. https://doi.org/10.1023/B:BILE.0000036599.15302.e5. PMID: 15289671.

Tokiwa, Y. and Calabia, B.P. (2007). *Biodegradability and biodegradation of polyesters. Journal of Polymers and the Environment* 15 (4): 259–267. https://doi.org/10.1007/s10924-007-0066-3.

Tokiwa, Y. and Suzuki, T. (1977). *Hydrolysis of polyesters by lipases. Nature* 270 (5632): 76–78. https://doi.org/10.1038/270076a0.

Tokiwa, Y., Calabia, B., Ugwu, C., and Aiba, S. (2009). *Biodegradability of plastics. International Journal of Molecular Sciences* 10 (9): 3722–3742. https://doi.org/10.3390/ijms10093722.

Tribedi, P. and Sil, A.K. (2013). *Cell surface hydrophobicity: a key component in the degradation of polyethylene succinate by* Pseudomonas *sp. AKS2. Journal of Applied Microbiology* 116 (2): 295–303. https://doi.org/10.1111/jam.12375.

Tuson, H.H. and Weibel, D.B. (2013). *Bacteria–surface interactions. Soft Matter* 9 (17): 4368. https://doi.org/10.1039/c3sm27705d.

Vijaya, C. and Mallikarjuna, R.R. (2008). *Impact of soil composting using municipal solid waste on biodegradation of plastics. Indian Journal of Biotechnology* 7 (2): 235–239.

Vimal Kumar, R., Kanna, G.R., and Elumalai, S. (2017). *Biodegradation of polyethylene by green photosynthetic microalgae. Journal of Bioremediation & Biodegradation* 08 (01) https://doi.org/10.4172/2155-6199.1000381.

Von Moos, N., Burkhardt-Holm, P., and Köhler, A. (2012). *Uptake and effects of microplastics on cells and tissue of the blue mussel* Mytilus edulis L. *after an experimental exposure. Environmental Science & Technology* 46 (20): 11327–11335. https://doi.org/10.1021/es302332w.

Vroman, I. and Tighzert, L. (2009). *Biodegradable polymers. Materials* 2 (2): 307–344. https://doi.org/10.3390/ma2020307.

Yamada-Onodera, K., Mukumoto, H., Katsuyaya, Y. et al. (2001). *Degradation of polyethylene by a fungus,* Penicillium simplicissimum YK. *Polymer Degradation and Stability* 72 (2): 323–327. https://doi.org/10.1016/s0141-3910(01)00027-1.

Yang, J., Yang, Y., Wu, W.-M. et al. (2014). *Evidence of polyethylene biodegradation by bacterial strains from the guts of plastic-eating waxworms. Environmental Science & Technology* 48 (23): 13776–13784. https://doi.org/10.1021/es504038a.

Yang, Y., Yang, J., Wu, W.-M. et al. (2015). *Biodegradation and mineralization of polystyrene by plastic-eating mealworms: part 2. Role of gut microorganisms. Environmental Science & Technology* 49 (20): 12087–12093. https://doi.org/10.1021/acs.est.5b02663.

Yoshida, S., Hiraga, K., Takehana, T. et al. (2016). *A bacterium that degrades and assimilates poly(ethylene terephthalate). Science* 351 (6278): 1196–1199. https://doi.org/10.1126/science.aad6359.

Yuan, J., Ma, J., Sun, Y. et al. (2020). *Microbial degradation and other environmental aspects of microplastics/plastics. Science of the Total Environment* 715: 136968. https://doi.org/10.1016/j.scitotenv.2020.136968.

Zhang, K., Xiong, X., Hu, H. et al. (2017a). *Occurrence and characteristics of microplastic pollution in Xiangxi Bay of Three Gorges Reservoir, China. Environmental Science & Technology* 51 (7): 3794–3801. https://doi.org/10.1021/acs.est.7b00369.

Zhang, C., Chen, X., Wang, J., and Tan, L. (2017b). *Toxic effects of microplastic on marine microalgae* Skeletonemacostatum: *interactions between microplastic and algae. Environmental Pollution* 220: 1282–1288. https://doi.org/10.1016/j.envpol.2016.11.005.

Zhu, J. and Wang, C. (2020). *Biodegradable plastics: green hope or greenwashing? Marine Pollution Bulletin* 161: 111774. https://doi.org/10.1016/j.marpolbul.2020.111774.

13

Life Cycle Assessment (LCA) of Plastics

Kailas L. Wasewar[1], Sushil Kumar[2], Dharm Pal[3], and Hasan Uslu[4]

[1] *Advance Separation and Analytical Laboratory (ASAL), Department of Chemical Engineering Visvesvaraya National Institute of Technology (VNIT), Nagpur, Maharashtra, India*
[2] *Department of Chemical Engineering, Motilal Nehru National Institute of Technology (MNNIT) Allahabad, Prayagraj, UP, India*
[3] *Department of Chemical Engineering, National Institute of Technology, Raipur, India*
[4] *Food Engineering Department, Engineering Faculty, Niğde Ömer Halisdemir University, Niğde, Turkey*

13.1 Introduction

Plastics are known as synthetic polymers which consist mainly of carbon, hydrogen, oxygen, nitrogen, and chlorine. Plastics such as polyethylene (PE), polypropylene (PP), polyethyleneterephthalate (PET), polystyrene (PS), nylons, polyvinylchloride (PVC), etc., are derived from fossil sources such as coal, oil, and natural gas. These plastics often wind up as environment pollutants, which pose increasing ecological threats to the world due to their ineffective and inefficient disposals. In 2015, around 381 million tons of plastic was produced and it was cumulative as 7.81 billion tons by 2015. The plastics used are mainly discarded, incinerated, and recycled as methods of disposal. In the last 50 years, plastic usage has increased more than twenty fold, and it is predicted to be doubled in the next 20 years as (World Economic Forum 2016; Ahamed et al. 2021). The various applications of plastics have been multiplied many times since its introduction. In 2018, more than 350 million tons of polymers, excluding fibers, were globally manufactured (PlasticsEurope 2019).

In the current economic context, an environmental performance of various materials and services has been gaining attention and importance in recent years. Several tools such as environmental impact assessment (EIA), life cycle assessment (LCA), environmental risk assessment (ERA), material flow analysis (MFA), cost–benefit analysis (CBA), and the ecological footprint (EF) can be used to measure their performance. Usually, LCA is applicable for all products and services in different contexts; therefore, efforts have been made to maximize the LCA value in the current context of its applications. In day-to-day life, there have been continuous increases in consumption of plastics, which has had very significant environmental impacts (Rajendran et al. 2013). Plastic has significant impacts on environments, and the environmental impact category in the lifecycle of the plastics can be

Plastic and Microplastic in the Environment: Management and Health Risks, First Edition.
Edited by Arif Ahamad, Pardeep Singh, and Dhanesh Tiwary.
© 2022 John Wiley & Sons Ltd. Published 2022 by John Wiley & Sons Ltd.

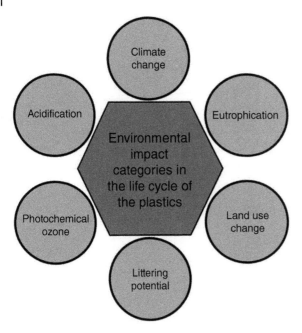

Figure 13.1 Environmental impact categories in the life cycle of plastics (UNEP 2020).

quantified in terms of climate change, eutrophication, acidification, photo-chemical ozone, littering potential, and land use change (Figure 13.1) (UNEP 2020). This chapter focuses on LCA of plastics for the issues of sustainability.

13.2 Plastics

Plastic is one of the most important resources that has many applications in our lives. Plastics are comparatively easy to manufacture and inexpensive, which has led to the large growth in the production of plastics in the last 50 years (Gu et al. 2017). There have been continuous increases in production of plastics due to significant increases in demand for wide applications. In 2011, the annual production of plastic was about 279 million tons, 280 million tons in 2014, 322 million tons in 2015, 335 million tons in 2016, and 350 million tons in 2017 (Shen and Worrell 2014; PlasticsEurope 2016; EP 2018; Bataineh 2020).

13.2.1 Types of Plastics

There are many classifications of plastics available, but most are classified as thermoplastics and thermosets based on their flow characteristics. Various types of polyethylenes (PEs), polypropylene (PP), polyvinyl chloride (PVC), polystyrene (PS), polyethylene terephthalate (PET), and others are examples of thermoplastics; whereas melamine, urea, and phenolic formaldehydes, and epoxy, vinyl ester, and alkyd resins are examples of thermoset plastics. (Aryan et al. 2018). The typical classification of plastics given by the Central Pollution Control Board (CPCB) India has been depicted in Figure 13.2 (CPCB 2018). The typical properties, advantages, and disadvantage of plastics are mentioned in Figures 13.3 and 13.4, respectively (CPCB 2018).

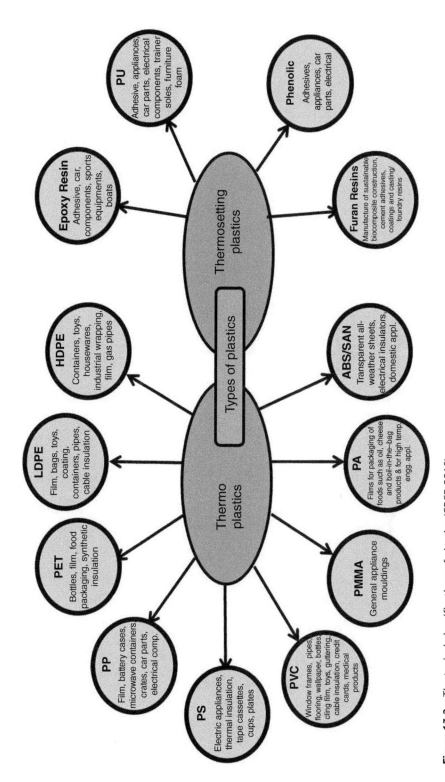

Figure 13.2 The typical classifications of plastics (CPCB 2018).

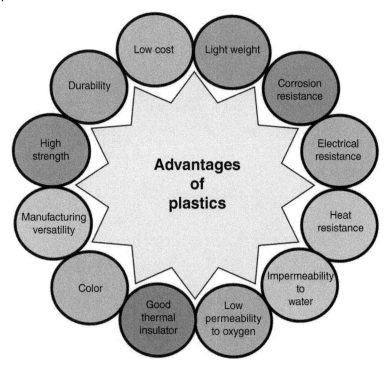

Figure 13.3 The typical advantages of plastics (CPCB 2018).

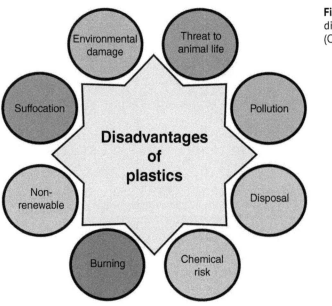

Figure 13.4 The typical disadvantages of plastics (CPCB 2018).

13.2.2 Plastic Bags

In 1957, sandwich bags were introduced, and now plastic bags have become an important part of our daily lives. Lightweight, portable, durable, easily available, insulating characteristics, and low costs are the main features of plastic bags that attract consumers and promote them for use. Therefore, plastic bags have many applications such as packing, carrying, and wrapping materials. Nearly, everyone in the world uses plastic bags; and almost all the stores (supermarkets, clothing stores, bookstores, restaurants, etc.) around the world offer plastic bags in their businesses (Hanun et al. 2019). Worldwide, single-use plastic bags have been one of the most consumed plastics. These single-use plastic bags (SUPBs) have been causing issues of environmental impact; while there are alternatives for SUPBs, such as cotton and paper bags, they don't perform as well as SUPBs (UNEP 2020).

Globally, around 348 million tons of plastics were produced in 2017, and of these plastics, around one-third were manufactured as single-use plastic products (PlasticsEurope 2018; UNEP 2018; UNEP 2020). In spite of the various features mentioned, these facilitate many problems associated with their use (EPHC 2010).

13.2.3 Plastic Waste

Plastic bags are the major contributors in the application of plastics in day-to-day life. Even though plastic bags have been banned for many applications with specific criteria, the generation of plastic bag waste is significant. As per the United Nations report, by 2030, the generation rate of plastic bag waste is predicted to increase, having 752 million people projected to live in densely populated cities (>10 million inhabitants) (United Nations 2018). Due to the non-biodegradable nature of plastics, the waste generated after use of plastic has been one of the most challenging issues to society (Aryan et al. 2018). Plastic waste remains for many years (500–1000 years) even after disposal in dumps and landfills, because they are chemically stable (Barnes et al. 2009). Annually, more than 8.8 million tons of plastic waste has been discarded in the oceans, and in India around 0.6 million tons of plastic waste have been dumped in rivers and oceans (Jambeck et al. 2015).

There have been increasing demands for more plastic for various industrial, health, transport, food, pharmaceuticals, and other applications which can lead to concerns of waste management issues, health related problems, and environmental impacts (Aryan et al. 2018; Bataineh 2020).

13.2.4 Recycling and Disposal

The major operations involved in plastic recycling are presented in Figure 13.5. These are pre-operations (collection, disassembly, break, sorting, washing, separation); mechanical recycling (dehalogenation, physical modification, re-manufacture); chemical recycling (modification, recycle products); incineration (energy recovery); and final disposal (landfill) (Zhang and Kang 2013). Recycling and disposal of used plastic materials is the end of life of plastic. Disposal is mainly landfilling, but this is not the most suitable approach. The recycling of plastic waste is a complex method in which various operations and steps are involved (Zhang and Kang 2013). The various steps in recycling of waste plastics

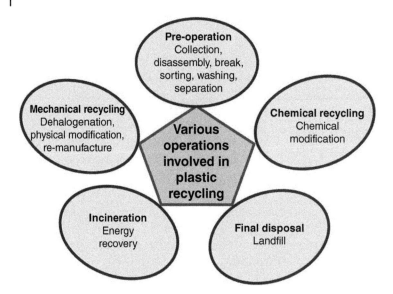

Figure 13.5 Various operations involved in plastic recycling (Zhang and Kang 2013).

mentioned, along with the physical and chemical changes, consume energy and resources, which are further responsible to generate new products with certain environmental impact (Zhang and Kang 2013).

Recycling plastic waste has many advantages, the major advantages being: (i) reduction in fossil fuel consumption, as the production of plastic utilized 4–8% of oil production globally, whereas 4% came from feedstock, and an additional 4% from conversion; (ii) reduction of energy and municipal solid waste (MSW); and (iii) reduction in the emissions of green house carbon dioxide (CO_2), nitrogen oxide (NOx), and sulfur dioxide (SO_2) (Perdon 2004; Bataineh 2020).

13.2.5 Indian Scenario

In India, plastic consumption has increased from five million tons per year (2005) to eight million tons per year (2008), and was expected to increase to 24 million tons in 2020 (Singh and Ruj 2015). This growth rate of plastic consumption is found to rise 2.5 times of the GDP of India (CPCB 2015). The huge amount of different types of plastics produced every year is due to an extraordinary growth of the plastic/polymer industries during the past several years (Zhang and Kang 2013).

Based on an investigation performed in India (60 cities), it was found that almost 15 342 tons of plastic waste is generated per day. Around 60% (9205 tons per day) of total plastic waste was recycled and around 6137 tons remained uncollected (CPCB 2015; Aryan et al. 2018). In India, about 0.6 million tons of plastic waste has been dumped in the rivers and oceans (Jambeck et al. 2015). Based on investigations of the CPCB India, around 94% of plastic waste has thermoplastics, which are recyclable (LDPE, PET, PVC, HDPE, PS, etc.), and the remaining 6% is thermosets and other plastics (FRP, SMC, thermocol, multi-layered, etc.) which is non-recyclable (CPCB 2018).

13.3   Life Cycle Assessment (LCA)

For the estimation of proper environmental impacts of various products and services, it is essential to analyze plastics in the life cycle perspective; in this way, problems shifting from one part of the life cycle to another may be avoided. Therefore, it is important to perform a comprehensive assessment in a life cycle perspective in order to avoid environmental problems shifting from one area to another (Finnveden et al. 2005; 2009).

The LCA originated back in 1969 as a resource and environmental profile analysis (REPA), and used for the assessment of beverage industry environmental impacts involving five steps (Figure 13.6) (Hunt et al. 1992; Gnansounou 2017). REPA were used for various industrial sectors including food products, beverages, plastic, chemicals, and others. There was gap of proper standardization, and hence in 1993 LCA guidelines were presented based on recommendations of European and North American Chapters of the Society of Environmental Toxicology and Chemistry (SETAC) and the International Organization for Standardization (ISO). The various procedures, guidelines report, technical reports, technical specifications, and case studies are presented by ISO as 14040 series and these are summarized in Table 13.1 (Gnansounou 2017).

One can use LCA as a tool for the assessment of environmental impacts in a product's life cycle, such as raw material, production, and application phases, and for waste management. LCA is intended for the comprehensive assessment of environmental impacts of various products and services. The LCA methodology to assess the environmental impacts has been somewhat established during the last decades. Currently, some other activities with respect to databases, consistency, quality assurance, and method harmonization also contribute to LCA, with new application areas being developed.

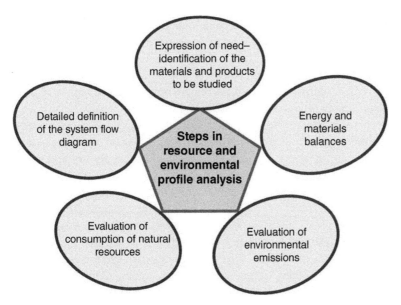

Figure 13.6   Steps involved in resource and environmental profile analysis (Hunt et al. 1992; Gnansounou 2017).

Table 13.1 A few ISO 14040 series documents for Life Cycle Assessment (Gnansounou 2017).

ISO Series	References	Description
ISO 14040	ISO (2006a)	International standard presenting the principles and framework of LCA
ISO 14044	ISO (2006b)	International standard providing requirements and guidelines for LCA
ISO/TR 14047	ISO (2003)	Technical Report related to life cycle impact assessment (LCIA)
ISO/TP 14048	ISO (2002)	Technical specifications related to data documentation format
ISO/TR 14048	ISO (2000)	Informative document with an example related to goal and scope definition and inventory analysis
ISO 14046	Klöpffer (2014)	Related to the water footprint
ISO 14067	ISO (2013)	Related to the carbon footprint of products
ISO TS 14071	ISO (2006b)	Critical review processes and reviewer competencies
ISO TS 14072	ISO (2014a,b)	Additional requirements and guidelines for organizations
ISO/AWI TR 14073	ISO (2016)	Water footprint: Illustrative examples of how to apply ISO 14046 (water footprint)

Globally, LCA is documented as a consistent and planned approach to evaluate the environmental impacts developed during the complete life cycle of a product, process, or activity. The typical structure of the LCA build model, life cycle inventory, and life cycle impact assessment is presented by Culaba (2018). Understanding and addressing the impact of these categories at an early stage of the processes by LCA is a major challenge. Although LCA may account for a complete scope of the development of many products, it may be considered as one solution to investigate the environmental impact in various directions. In addition, energy products and other byproducts may be made decisive with the help of LCA. As per the International Organization for Standardization (ISO) 14000 series, the LCA methodology technical framework comprises four segments; goal and scope definition, inventory analysis, impact assessment, and finally interpretation (Gnansounou 2017).

The LCA is categorized as both an attributional life cycle assessment (ALCA) and a consequential life cycle assessment (CLCA) (Finnveden et al. 2005; 2009; Zamagni et al. 2012). ALCA presents physical flow in the life cycle, which utilizes average data of each unit process and environmental consequences in CLCA, which needs a market-oriented approach (Earles and Halog 2011). The typical steps involved are from laboratory data to LCA for assessment of any system, which are laboratory protocol, plant flow chart with scale and reactor sizes, separate scale-up of each process steps, linkage of process steps, and performing LCA (Piccinno et al. 2016; Elginoz et al. 2020). The environmental impact assessed by LCA can be quantified based on various impact

Table 13.2 The impact categories most used for life cycle assessment (Gnansounou 2017).

	Impact Categories
1.	Depletion of abiotic resources
2.	Depletion of biotic resources
3.	Climate change
4.	Stratospheric ozone depletion
5.	Human toxicity
6.	Ecotoxicity
7.	Freshwater aquatic ecotoxicity
8.	Marine ecotoxicity
9.	Terrestrial ecotoxicity
10.	Photo-oxidant formation
11.	Acidification
12.	Eutrophication
13.	Impact of ionizing radiation
14.	Loss of biodiversity

categories, from resources to biodiversity, which have been listed in Table 13.2 (Gnansounou 2017).

Recycling and waste management options of plastics may be assessed, and the best possible option should be identified. The alternatives for single-use or other plastics may be decided on based on certain potential environmental impacts such as fit-for-purpose and more sustainable for replacement. LCA is the best approach to understand the various impacts and to decide suitable alternatives (UNEP 2020). The LCA approach may be used to precisely compare and identify the best waste management option(s) for plastics in the view of the environment (Aryan et al. 2018).

LCAs are able to enumerate the probable environmental impacts throughout the product life cycle (raw material to final disposal after use, including waste treatment). LCA is presented as a standard method for predicting the environment impact and given by the International Standard Organization (ISO 2006a,b). LCA is "a technique to evaluate the environmental loads, which are associated with a process, product, and any other activity by identifying and computing energy and materials used and/or released to the environment; in addition, to identify and estimate the opportunities to improve the environmental effects" (ISO 2006a,b).

The potential impacts on global or regional scales, such as climate, eutrophication, acidification, and resource use may be quantified by adapted LCA, which represents a potent tool to compare the environmental aspects in different products and services including technological systems. In addition, LCA also created decision-making perspectives of procurement and policy making in industry, and gained acceptance (USEPA 2006; UNEP 2020).

13.3.1 Phases of LCA

The LCA is carried out in four phases. According to the new terminology of International organization for standardization, these phases are: (i) goal and scope definition; (ii) inventory analysis; (iii) impact assessment; and (iv) interpretation (UNEP 2020).

13.3.1.1 Goal and Scope Definition

In this phase, the LCA's purpose is intended toward the user and application. The goal of LCA is to determine the type of assessment needs to be conducted, either (i) attributional, which includes only the processes, which are part of the life cycle investigation; or (ii) consequential with the wider perspectives, which includes the processes affected due to the supply and the product uses. According to the LCA goal, the functional unit, detail levels, impact categories, assumptions and limitations, and allocation process with system boundaries are also defined (UNEP 2020).

13.3.1.2 Inventory Analysis

The life cycle flow charts' construction and all the relevant input (material and energy) and output (wastes and emissions) data collections associated with the life cycle are carried out in the second phase of LCA, inventory analysis. After which these data are set with respect to the functional units which are defined in the first phase (UNEP 2020).

13.3.1.3 Impact Assessment

The third phase of LCA, impact assessment, is divided into two parts; (i) classification and (ii) characterization. In the first part of classification, the inventory results are allocated to the respective impact categories such as global warming. The characterization of impact assessment is performed in two-steps. In the first step, the assigned inventory results are multiplied with equivalence factors of different impacts, and summed up in the second step. The impact assessment provides the knowledge enhancement to analyze the inputs and outputs for the environmental importance. This betterment is useful in the collection of both improved data and inventory analysis (UNEP 2020).

13.3.1.4 Interpretation

In this last phase, interpretation, the final results are analyzed in accordance to the goal and scope defined. The conclusions with respect to the aim of impact assessment are drawn up and recommendations are made, and the limitations of the provided results are presented. The LCA conclusions should have compatibility with the goals and scope of the study (UNEP 2020).

As per the guidelines given by CPCB, India, various stages involved in LCA have been mentioned: goal scoping; inventory; impact assessment; classification; characterization; normalization; valuation; improvement assessment; interpretation (CPCB 2018).

13.3.2 Importance of LCA

The Central Pollution Control Board India, has set the importance of LCA as follows: LCA is useful to restrict the shifting environmental impact problems; if it is used with proper design, it helps in minimization of the secondary effects; it can reduce the environmental pollution and properly use the resources; and it enables us to get the true and total costs (CPCB 2018).

13.3.3 LCA for Plastics

In LCA, the assessment usually includes the entire life cycle of the considered system (product/process/activity) considering an acquisition of the energy and material, manufacture use, and waste/disposal management. In LCA, the environmental problem is associated with a product from "cradle to grave," examines all sequential steps in the life cycle of a system or product; extraction of raw materials, material production, fabrication of product, its use and reuse/recycling, whichever is applicable, and final disposition. (APR 2018).

A life cycle of petrochemical-based plastics is not found to be complete and they remain on the landscape for many years. Plastic products may be recycled three or four times, and after each recycling process the product quality becomes deteriorated and is finally dumped on a landfill site, which leads to environmental issues and burdens due to their non-biodegradable nature. (CPCB 2018). The problems associated with plastics may be characterized by two factors; environmental and societal. The environmentally sustainable impact of plastic may be understood by life cycle perspective and be used for comparison with other alternatives. Based on various LCA investigations, the virgin plastic/polymer in a ratio of 1:1 may be used to substitute a high-quality recyclable plastic/polymer waste.

13.4 Plastics Sustainability by LCA

The environmental impact of plastic bags and other plastic materials may be found in various aspects such as climate change, acidification, eutrophication, photochemical ozone, land use change, and littering potential (Figure 13.1) (UNEP 2020). The United Nationals Environmental Programme has made a few observations on environmental impact of plastic bags in their report (UNEP 2020).

The consumption of resources, pollution in air, water, and soil, and load on limited space available for landfill create various adverse impacts on the environment due to the use of plastics (SEC 2018). The various studies on the environmental impact of plastics have gained the attention of different political and social bodies for proper management of plastic waste. Apart from various impacts on the environment, the demand of plastic has been increasing and it has created concerning issues, as it involves the consumption of available energy sources of the earth in terms of fossil fuel, which has been depleting very quickly (CPCB 2018). There are only a few LCA studies which report plastics as environment friendly (Edwards and Fry 2011; Chaffee and Yaros 2007; Muthu and Li 2014b; Saibuatrong et al. 2017), while other studies promote paper and biodegradable bags (Khoo et al. 2010; Muthu and Li 2014c). The typical flow chart of LCA for plastic wastes is presented in literature (Treenate et al. 2018).

Many LCA studies were performed on plastics waste management as mechanical recycling, incineration, and landfilling in the early 1990s. Craighill and Powell (1996) carried out LCA for the variety of materials and evaluated individual environmental impacts in different categories such as global warming, eutrophication, and acidification. Over the past 15–20 years, various studies have been performed on plastic waste management with the perspective of life cycle. Various plastic materials and their waste methods as described in the particular study are presented by Bernardo et al. (2016).

The sorted and unsorted plastic waste was used for mechanical recycling, pyrolysis, and incineration with energy recovery, and LCA studies were performed to find the impact level and to compare various recycling options (Molgaard 1995). For the case of mixed plastic waste, the energy recovery option was the best when considering all impact categories, and in the case of sorted plastic waste, mechanical recycling was the best.

Ten different approaches for the management of waste PET bottles in Korea have been investigated using LCA. These approaches are listed as; landfill (with and without pyrolysis), incineration (with and without pyrolysis), open-loop recycle (with landfill and incineration), solvolysis (with landfill and incineration), and closed-loop polymer feedback (with landfill and incineration) (Song and Hyun 1999). Various emissions have been evaluated, and closed-loop recycling in combination with land filling is the best approach for disposal based on CO_2 emissions. Closed-loop recycling with incineration was found to be most beneficial in terms of SO_x and NO_x emissions, and energy requirement (Song and Hyun 1999).

Plastics have been widely used as packaging materials, and disposal or recycling is required after use to know their impact toward the environment. LCA of plastic packaging material in Italy has been performed for mechanical recycling, incineration with energy recovery, and landfilling of PE waste and PET waste bottles to understand the environmental impacts (Arena et al. 2003). To quantify the material and energy consumption and environmental emissions, various phases have to be investigated individually; these phases are collection, sorting, compaction, reprocessing, and disposal. For the case of all collected plastic waste, mechanical recycling was found to be the most favorable; and in the case of processed wastes, incineration with energy recovery was found to be most environment-friendly.

Various approaches for plastic waste treatments such as feedstock recycling, mechanical recycling, energy recovery, and landfilling were analyzed and compared by LCA in Italy (Perugini et al. 2003). It is compulsory to add the type of method for recovery of energy rather than projecting value from the gross calorific value. The emission and recovery of energy depends on the type of method used for recovery of energy (Perugini et al. 2004).

The various approaches of mechanical recycling (with and without hydrocracking and low temperature pyrolysis), incineration with recovery of energy, and landfilling have been investigated for quantifying the environmental impact of plastic packaging wastes by LCA (Perugini et al. 2005). Mechanical recycling with hydrocracking was found to be the superior approach as compared to other studied approaches in terms of oil and energy consumption, and mechanical recycling alone was found to be better in all other impact parameters. The study from Eriksson et al. (2005) was based on the assumption that recycled materials replace virgin materials, which may not be valid in all applications, since the level of degradation in recycled polymers depends on the number of thermal cycles, their previous applications, and processing conditions.

Plastic wastes obtained from discarded mobile phones have been investigated for various recycling approaches by LCA (Takahashi and Dodbiba 2007). Mechanical recycling was found to be the best potential approach for waste treatment as compared to incineration with energy recovery in terms of environmental impact. The treatment of brominated plastic waste was analyzed by LCA and the thermal treatment of staged-gasification was demonstrated as a more energy-efficient approach as compared to co-combustion (Bientinesi and Petarca 2008). LCA was also performed for assessment of environmental impact of plastic for hydrophilic catheters in medical care (Stripple and Westman 2008).

Plastic waste recycling LCA was studied to replace virgin plastics, wood, and energy based on greenhouse gas emissions (Astrup et al. 2009). It was observed that almost 700–1500 kg CO_2 equivalent per ton of plastic waste was the emission. Plastic waste can be used as a fuel in industry (e.g. cement kilns) if it is not sufficient to replace virgin plastic, which can be the second best option.

The comparison of mechanical recycling on waste PET bottles for remaking PET bottles with burning of uncollected/non-recycled PET wastes have been performed by LCA (Chilton et al. 2010). Significant reduction in all types of pollutant emissions was observed in the case of mechanical recycling, which may be due to avoiding the impacts of replacement of virgin PET bottles. Post-consumer PET bottle waste disposal in Mauritius have been studied by LCA to quantify the environmental impact (Foolmaun and Ramjeeawon, 2012). Different approaches such as incineration (with landfilling and with energy recovery), landfilling, and flake production (with and without landfilling) were considered, and incineration with energy recovery was found to be best, and landfilling was worst for environmental impact.

The environmental impacts of grocery bags has been recognized by LCA using an eco-functional approach in which a scoring system was demonstrated with the environmental impacts to evaluate and integrate the functionality of the various options (Muthu and Li 2014a).

Mechanical recycling of waste plastics is an environmental solution to the problem of waste plastic disposal, and has already become a common practice in industry (Gu et al. 2017). LCA has been effectively employed to help understand the environmental impact of mechanical recycling of waste plastics, and has been compared to different alternative disposal methods (Lazarevic et al. 2010; Wäger et al. 2011; Al-Maadeed et al. 2012; Wäger and Hischier 2015). Mechanical recycling is comparably better than incineration and landfills and it is environmentally friendly. Perugini et al. (2005) performed LCA for mechanical recycling of post-consumer PET and PE liquid containers in Italy, which requires almost 7.97 MJ energy per kg production of recycled PET flakes. This energy is very low as compared to 70 830 MJ/ton for virgin PET (Chilton et al. 2010).

Three types of garbage bags are made of PE, Bio-PE (produced from bio-ethanol), and poly (butylene adipate-co-terephthalate) (PBAT)/Starch bags have been compared for environmental impact by LCA (Saibuatrong et al. 2017). PE bags were found to be better than Bio-PE and PBAT/Starch in terms of all indicators for environmental impacts except the climate change potential.

Gu et al. (2017) investigated the LCA for environmental burdens arisen due to mechanical recycling of plastic materials in China. The waste plastics were collected from various sources (plastic product manufacturers, agricultural wastes, solid plastic wastes collected, and dismantled parts from waste electrical and electronic components). Three different routes (extrusion, fillers, and additives) have been used for processing the final product from waste plastics. The better environmental aspects were obtained in mechanical recycling as compared to virgin plastics (Gu et al. 2017).

Treenate et al. (2017) investigated the environmental impact of lubricant oil bottles (HDPE) in Thailand by performing LCA through the steps of acquiring, processing, using, and disposing of the product. Recycle and incineration were considered for environmental impact. Treenate et al. (2018) performed the LCA of polyethylene such as LDPE, LLDPE, and HDPE resins and other plastic bags by the cradle-to-grave approach. It was found that the major environmental impacts were brought by PE resins including production and raw material acquisitions, in addition to the energy consumption.

A plastic grocery bag production plant in Columbia was investigated by LCA for ozone layer depletion potential (Morales-Mendez and Silva-Rodríguez 2018). The study of LCA only for the production process may underestimate the environmental impacts of plastic bags and hence it is essential to use an integrated approach (The Danish Environmental Protection Agency 2018).

Polyethylene Terephthalate (PET) and Polyethylene (PE) are the two major plastic wastes, and their LCA have been performed with respect to recycling, landfilling (without biogas recovery), and incineration (with and without energy recovery) (Aryan et al. 2018). LCA was performed based on ISO 14040/44 using SimaPro 8.0.5 with impact categories of abiotic depletion of fossil fuels, global warming potential, human toxicity potentials, depletion of ozone layer, fresh water aquatic ecotoxicity, acidification, and eutrophication potential.

Hanun et al. (2019) performed the LCA investigation of $660.4 \times 838.2 \, mm \times 0.04 \, mm$ thick polyethylene plastic bags (#150) obtained from three hypermarkets in Kota Bharu, Kelantan, Malaysia. These bags were claimed to be biodegradable bags. ISO standards were employed for LCA study (ISO 2006a,b). The typical methodology as goal and scope, impact assessment, and inventory were employed. The parameters considered for impact assessment are climate change (global warming), freshwater aquatic ecotoxicity, human toxicity, marine aquatic ecotoxicity, and ozone layer depletion. OpenLCA software has been used for study, and the study showed high environmental impact due to high emissions of carbon dioxide and carbon monoxide.

In the last few years, the demand has increased for returnable packaging by various industrial sectors (food and beverages, pharmaceuticals, consumer goods, and automobiles) (Markets and Markets 2018). The re-use principle is important and was found to be a key role in the circular economy (Tua et al. 2019). Particularly in the packaging sector, the re-use concept is the initiation of useful change that is expected to bring the benefits to economics and the environment (EC 2015). The returnable packages are categorized as pallets, bottles, crates, drums and barrels, dunnage, intermediate bulk containers, and other items (Tua et al. 2019). Plastic crates are mostly used for distribution of vegetables and fruits; and in Italy, almost 36% of the reusable plastic crates have been used in thi way. Inspection, washing, and sanitization with chemicals and hot water are the steps involved in the reconditioning processes for several uses of plastic crates (Tua et al. 2019). The LCA of reusable plastic crates have been performed using 12 impact categories; ozone depletion, climate change, human toxicity of both cancer and non-cancer effects, particulate matter, acidification, photochemical ozone formation, terrestrial eutrophication, marine eutrophication, freshwater eutrophication, resources depletion, and freshwater ecotoxicity (Tua et al. 2019).

HDPE, LDPE, PP woven, single-use recycled paper, and biodegradable bags were compared in Spain by LCA and correlated for marine litter impacts (Civancik-Uslu et al. 2019). For the case of plastic crates, reusable plastic crates are better than the single-use cardboard boxes in food packaging applications (Abejón et al. 2020).

The recycling of PET bottles has the benefit of reducing greenhouse gas emissions by 1.5 tons CO_2 equivalent per ton of recycled PET, which may be around 27% reduction in emission along with savings in energy requirements (Patel et al. 2000; Shen et al. 2010). Recycled PET has many environmental benefits as compared to virgin PET (Hopewell et al. 2009; Chilton et al. 2010). Based on allocation methods for PET plastic recycling, around 40–85% savings in fuel energy and 25–75% savings in global warming is achieved as additional benefits (Bataineh 2020).

Bataineh (2020) carried out LCA to obtain the environmental impact of mechanical recycling of the postconsumer high-density polyethylene (HDPE) and polyethylene terephthalate (PET) in Jordan. The environmental impact has been assessed based on total energy requirements, energy sources, atmospheric pollutants, waterborne pollutants, and solid waste for mechanical recycling of PET and HDPE resin from the postconsumer plastic. Based on this study, 40–85% savings in non-renewable energy and 25–75% savings in greenhouse gas emission were observed. Total energy requirements for recycled PET flakes are found in the range of 14–17% of that of virgin PET flakes, and 57% of virgin resins are obtained based on the "system expansion" method. In the case of HDPE, recycled HDPE pellets need the energy of 12–13% of virgin HDPE resins using "cut-off" recycling method, while these require 62% based on the "system expansion" method (Bataineh 2020).

Ahamed et al. (2021) performed the LCA of single-use and reusable bags for environmental impact in Singapore city. High-density polyethylene (HDPE), kraft paper, and biodegradable plastic-based single-use bags, and reusable bags made from cotton and polypropylene non-woven were considered for study. The negative environmental impacts of plastic bags have been significantly contributed by characteristics of usage, production process, and emissions. The lowest impact on the environment was observed for the reusable polypropylene non-woven bag (PNB) and the single use HDPE plastic bag (HPB). The trend for global warming potential was obtained as: kraft paper > cotton woven ~ biodegradable polymer > HDPE plastic > polypropylene non-woven.

13.5 Discussion and Conclusion

Plastics are omnipresent materials in all major societies of the world. Due to the useful characteristics, these are found to contribute decisively in the improvement of the standard of living of societies, even though many drawbacks are associated with their uses. In the case of plastics, their end of life is a key concern (Bernardo et al. 2016). LCA is used to monitor and measure the impact of industrial processes on the environment and human health. Since LCA models show the complex relationship between the environment and production, they are used quite frequently (Ruban 2012). The typical life cycle impact and damage categories of plastics are mentioned in Table 13.3 (CPCB 2018).

Over the last few decades, application of LCA has increased to industrial process and products in many sectors, particularly to chemical activities. The new considerations in LCA have permitted a complete understanding of all the unit operations which are included in a production unit, i.e. the influence of operating parameters and the synergistic effects between different unit operations (Julio et al. 2017). However, this approach is only

Table 13.3 Typical life cycle impact categories of plastics (CPCB 2018).

	Life cycle impacts		
Climate change	**Resource depletion**	**Ecological quality**	**Human health**
Global warming	Abiotic resource depletion	Land transformation and use	Photo chemical smog
	Mineral and fossil fuels	Water depletion	Ozone depletion
	Land transformation and use	Eutrophication	Ionizing radiation
	Water depletion	Acidification	Human toxicity
	Eutrophication	Ecotoxicity	Respiratory effects
		Photo chemical smog	Nuisance
		Ozone depletion	Indoor environmental quality
		Ionizing radiation	

achievable on previously designed processes, by which a sufficient amount of data essential to realizing the LCA, are obtained. The typical phases of LCA are represented and discussed by Uihlein and Schebek (2009).

Based on the available studies on the LCA of plastic and its waste to quantify the environmental impacts and compare various options of recycling and disposal, it can be concluded that to reduce the environmental impact of plastic it is not possible by just to choose, ban, recommend, or prescribe the alternative material. It will only change by changing consumer behavior to increase the reuse rate and to avoid littering (UNEP 2020).

Recycling of plastic waste was found to be one of the best alternatives for utilization of plastic waste to reduce the environmental impact as per the LCA study. If alternative energy were used instead of thermal energy, recycling industries may be more attractive, sustainable, and environmentally friendly (Aryan 2018). Over the last 25 years, significant studies have been performed on LCA of plastic waste, which has analyzed waste management in various perspectives, mostly of the environment. Table 13.3 represents various investigations of plastic materials and waste methods used in different studies (Bernardo et al. 2016).

In the case of LCA investigations on recycling of various types of plastic waste material as PE and PET, it is observed that the PE waste has lower impact on the environment as compared to PET waste. Hence, recycling of PE may be encouraged at all levels, including domestic. Based on LCA investigations, the disposal of waste plastics by incineration has more potential in terms of saving in energy and reduction in acidification. Hence, these technologies may be considered in India, as plastic waste of PE occurs far more than PET wastes (Aryan 2018). As per the report of The Association of Plastic Recyclers prepared by Franklin Associates, a Division of Eastern Research Group, the significant reduction in total energy, water consumption, solid waste, global warming acidification, eutrophication, and smog were observed based on LCA study for recycle of PET, HDPE, and PP as compared to virgin resins (APR 2018). The LCA studies for plastic show that the results of investigations are based on many system boundary conditions such as transportation distances, different geographical coverage, the weight of products, the recycling infrastructure, and the rate of mismanaged waste (Abejón et al. 2020; Ahamed et al. 2021).The integrated life cycle based critical evaluation would aid in focused waste prevention initiatives, efficient resources management, environmental footprint reduction, and policy decision making (Ahamed et al. 2021).

References

Abejón, R., Bala, A., Vázquez-Rowe, I. et al. (2020). *When plastic packaging should be preferred: life cycle analysis of packages for fruit and vegetable distribution in the Spanish peninsular market. Resources, Conservation and Recycling* 155: 104666.

Ahamed, A., Vallam, P., Iyer, N.S. et al. (2021). *Life cycle assessment of plastic grocery bags and their alternatives in cities with confined waste management structure: a Singapore case study. Journal of Cleaner Production* 278: 123956, 1–11.

Al-Maadeed, M., Madi, N.K., Kahraman, R. et al. (2012). *An overview of solid waste management and plastic recycling in Qatar. Journal of Polymers and the Environment* 20: 186–194.

APR (The Association of Plastic Recyclers) (2018). Life cycle impacts for postconsumer recycled resins: PET, HDPE, and PP. Franklin Associates, A Division of Eastern Research Group (ERG).

Arena, U., Mastellone, M.L., and Perugini, F. (2003). *Life cycle assessment of a plastic packaging recycling system. The International Journal of Life Cycle Assessment* 8 (2): 92.

Aryan, Y., Yadav, P., and Samadder, S.R. (2018). *Life cycle assessment of the existing and proposed plastic waste management options in India: a case study. Journal of Cleaner Production* https://doi.org/10.1016/j.jclepro.2018.11.236.

Astrup, T., Fruergaard, T., and Christensen, T.H. (2009). *Recycling of plastics: accounting of green house gases and global warming contribution. Waste Management and Research* 27: 763–772.

Barnes, D.K., Galgani, F., Thompson, R.C., and Barlaz, M. (2009). *Accumulation and fragmentation of plastic debris in global environments. Philosophical Transactions of the Royal Society B: Biological Sciences* 364 (1526): 1985–1998.

Bataineh, K.M. (2020). *Life-cycle assessment of recycling postconsumer high-density polyethylene and polyethylene terephthalate. Advances in Civil Engineering* 2020, Article ID 8905431, 15 pages. doi:https://doi.org/10.1155/2020/8905431: 1–15.

Bernardo, C.A., Simões, C.L., and Pinto, L.M.C. (2016). *Environmental and economic life cycle analysis of plastic waste management options; a review. AIP Conference Proceedings* 1779: 140001. https://doi.org/10.1063/1.4965581.

Bientinesi, M. and Petarca, L. (2008). *Comparative environmental analysis of waste brominated plastic thermal treatments. Journal of Waste Management* 29, 3: 1095–1102. https://doi.org/10.1016/j.wasman.2008.08.004.

Chaffee, C. and Yaros, B.R. (2007). *Life Cycle Assessment for Three Types of Grocery Bags - Recyclable Plastic; Compostable, Biodegradable Plastic; and Recycled*. Recyclable Paper. Boustead Consulting & Associates Ltd.

Chilton, T., Burnley, S., and Nesaratnam, S. (2010). *A life cycle assessment of the closed-loop recycling and thermal recovery of post-consumer PET. Resources, Conservation and Recycling* 54 (12): 1241–1249.

Civancik-Uslu, D., Puig, R., Hauschild, M., and Fullana-i-Palmer, P. (2019). *Life cycle assessment of carrier bags and development of a littering indicator. Science of the Total Environment* 685: 621e630.

CPCB (Central Pollution Control Board). (2015). Report of central pollution control board on "status of compliance by CPCB with municipal solid wastes (management and handling) rules, 2000." http://www.cpcb.nic.in/divisionsofheadoffice/pcp/MSW_Report.pdf (accessed 18 July 2017).

CPCB (Central Pollution Control Board) (2018). *Life Cycle Assessment Study of Plastic Packaging Products.* Central Pollution Control Board India.

Craighill, A. and Powell, J.C. (1996). *Life cycle assessment and economic evaluation of recycling: a case study. Resources, Conservation and Recycling* 17: 75–96.

Culaba A. (2018). *Life Cycle Analysis of Plastic: Case Study of Plastic Carrying Bags.* De La Salle University Power Point Presentation. https://www.nast.ph/index.php/downloads/category/142-visayas-regional-scientific-meeting?download=578:4-dr-culaba-life-cycle-analysis-case-study-of-plastic-carrying-bags (accessed 26 October 2021).

Earles, J. and Halog, A. (2011). *Consequential life cycle assessment: a review. The International Journal of Life Cycle Assessment* 16 (5): 445–453.

Edwards, C. and Fry, J. M.(2011). *Life cycle assessment of supermarket carrier bags: a review of the bags available in 2006.* Evidence report, Bristol, the United Kingdom. https://assets.publishing.service.gov.uk/government/uploads/system/uploads/attachment_data/file/291023/scho0711buan-e-e.pdf (accessed 17 November 2019).

Elginoz, N., Khatami, K., Owusu-Agyeman, I., and Cetecioglu, Z. (2020). *Life cycle assessment of an innovative food waste management system. Frontiers in Sustainable Food Systems* 4: 23.

EPHC (Environmental Protection and Heritage Council) (2010). *Plastic Bags.* www.ephc.gov.au/taxonomy/term/54.

Eriksson, M., Carlsson, R., Frostell, B. et al. (2005). *Municipal solid waste management from a systems perspective. Journal of Cleaner Production* 13: 241–252.

Finnveden, G., Johansson, J., Lind, P., and Moberg, Å. (2005). *Life cycle assessment of energy from solid waste – part 1: general methodology and results. Journal of Cleaner Production* 13: 213–229.

Finnveden, G., Hauschild, M.Z., Ekvall, T. et al. (2009). *Recent developments in life cycle assessment. Journal of Environmental Management* 91 (1): 1–21.

Foolmaun, R.K. and Ramjeeawon, T. (2012). *Disposal of post-consumer polyethylene terephthalate (PET) bottles: comparison of five disposal alternatives in the small island state of Mauritius using a life cycle assessment tool. Environmental Technology* 33 (5): 563–572.

Gnansounou, E. (2017). Fundamentals of life cycle assessment and specificity of biorefineries. In: *Life-Cycle Assessment of Biorefineries* (eds. E. Gnansounou and A. Pandey), 41–75. Elsevier.

Gu, F., Guo, J., Zhang, W. et al. (2017). *From waste plastics to industrial raw materials: a life cycle assessment of mechanical plastic recycling practice based on a real-world case study. Science of the Total Environment* 601–602: 1192–1207.

Hanun, R.S.F., Sharizal, A.S., Mazlan, M. et al. (2019). *Life cycle assessment (LCA) of plastic bag: current status of product impact. International Journal of Advanced Science and Technology* 28 (18): 94–101.

Hopewell, J., Dvorak, R., and Kosior, E. (2009). *Plastics recycling: challenges and opportunities. Philosophical Transactions of the Royal Society B: Biological Sciences* 364 (1526): 2115–2126.

Hunt, G.R., Sellers, D.J., and Franklin, E.W. (1992). Resource and environmental profile analysis: a life cycle environmental assessment for products and procedures. *Environmental Impact Assesment Review* 12 (3): 245–269.

ISO (International Standard) (2000). Environmental Management—Life Cycle Assessment—Examples of Application of ISO 14041 to Goal and Scope Definition and Inventory Analysis. Technical report ISO/TR 14049, (E), Switzerland.

ISO (International Standard) (2002). Environmental Management—Life Cycle Assessment—Data Documentation Format. Technical specification ISO/TS 14048, (E), Switzerland.

ISO (International Standard) (2003). Environmental Management—Life Cycle Impact Assessment—Examples of Application of ISO 14042 ISO/TR 14047, (E), Switzerland.

ISO (International Standard) (2006a). ISO 14040:2006. Environmental Management – Life Cycle Assessment – Principles and Framework. International Organization for Standardization, Geneva, Switzerland.

ISO (International Standard) (2006b). ISO 14044:2006 Environmental Management – Life Cycle Assessment – Requirements and Guidelines. International Organization for Standardization, Geneva, Switzerland.

ISO (International Standard) (2013). Greenhouse Gases—Carbon Footprint of Products—Requirements and Guidelines for Quantification and Communication. Technical specification ISO/TS 14067, Switzerland.

ISO (International Standard) (2014a). Environmental Management—Life Cycle Assessment—Requirements and Guidelines for Organizational Life Cycle Assessment. Technical specification ISO/TS 14072, Switzerland.

ISO (International Standard) (2014b).Environmental Management—Life Cycle Assessment—Critical Review Processes and Reviewer Competencies: Additional Requirements and Guidelines to ISO 14044:2006b. Technical specification ISO/TS 14071, Switzerland.

ISO (International Standard) (2016). Environmental Management—Water Footprint—Illustrative Examples on How to Apply ISO 14046. Technical report ISO/AWI TR 14073, May, Switzerland.

Jambeck, J.R., Geyer, R., Wilcox, C. et al. (2015). *Plastic waste inputs from land into the ocean.* Science 347 (6223): 768–771.

Julio, R., Albet, J., Vialle, C. et al. (2017). *Sustainable design of biorefinery processes: existing practices and new methodology. Biofuels, Bioproducts and Biorefining* 11 (2): 373–395.

Khoo, H.H., Tan, R.B.H., and Chng, K.W.L. (2010). *Environmental impacts of conventional plastic and bio-based carrier bags. International Journal of Life Cycle Assessment* 15: 284e293.

Klöpffer, W. (ed.) (2014). *Background and Ffuture Pprospects in Llife Ccycle Aassessment,* Part of the LCA Compendium – The Complete World of Life Cycle Assessment Book Series (LCAC). Springer.

Lazarevic, D., Aoustin, E., Buclet, N., and Brandt, N. (2010). *Plastic waste management in the context of a European recycling society: comparing results and uncertainties in a life cycle perspective. Resources, Conservation and Recycling* 55 (2): 246–259.

Markets and Markets™ Inc. (2018). Returnable Packaging Market by Product Type (Pallets, Crates, Intermediate Bulk Containers, Drums & Barrels, Bottles, Dunnage), Material (Plastic, Metal, Wood, Glass, Foam), End-Use Industry, Region-Global Forecast to 2023. Market Research Report. Available online: https://http://www.marketsandmarkets.com/Market-Reports/returnable-packaging-market-231944920.html (accessed on 20 May 2019).

Mølgaard, C. (1995). *Environmental impacts by disposal of plastic from municipal solid waste. Resources, Conservation and Recycling* 15: 51–63.

Morales-Mendez, J.-D. and Silva-Rodríguez, R. (2018). *Environmental assessment of ozone layer depletion due to the manufacture of plastic bags. Heliyon* 4: e01020.

Muthu, S.S. and Li, Y. (eds.) (2014a). Chapter 1: Basic Introduction to Shopping Bags and Eco-function. In: *Assessment of Environmental Impact by Grocery Shopping Bags: An Eco-functional Approach*, EcoProduction Series, 1–6. Springer https://doi.org/10.1007/978-981-4560-20-1.

Muthu, S.S. and Li, Y. (eds.) (2014b). Chapter 3: Life Cycle Assessment of Grocery Shopping Bags. In: *Assessment of Environmental Impact by Grocery Shopping Bags: An Eco-functional Approach*, EcoProduction Series, 15–54. Springer https://doi.org/10.1007/978-981-4560-20-7_3.

Muthu, S.S. and Li, Y. (eds.) (2014c). Chapter 7: Eco-functional Assessment of Grocery Shopping Bags. In: *Assessment of Environmental Impact by Grocery Shopping Bags: An Eco-functional Approach*, EcoProduction, Series, 99–113. Springer https://doi.org/10.1007/978-981-4560-20-7_7.

Patel, M., von Thienen, N., Jochem, E., and Worrell, E. (2000). *Recycling of plastics in Germany. Resources, Conservation and Recycling* 29 (1–2): 65–90.

Perdon, S. (2004). Chapter 1: sustainable development in practice: case studies for engineers and scientists. In: *Introduction to Sustainable Development* (eds. A. Azapagic, S. Perdon and R. Clift), 3–25. Hoboken, NJ: Wiley.

Perugini, F., Mastellone, M.L., and Arena, U. (2003). *Life cycle assessment of a plastic packaging recycling system. International Journal of Life Cycle Assessment* 8 (2): 92–98.

Perugini, F., Mastellone, M.L., and Arena, U. (2004). *The environmental aspects of mechanical recycling of PE and PET: a life cycle assessment study. Progress in Rubber, Plastics and Recycling Technology* 20: 69–84.

Perugini, F., Mastellone, M.L., and Arena, U. (2005). *A life cycle assessment of mechanical and feedstock recycling options for management of plastic packaging wastes. Environmental Progress* 24 (2): 137–154.

Piccinno, F., Hischier, R., Seeger, S., and Som, C. (2016). *From laboratory to industrial scale: a scale-up framework for chemical processes in life cycle assessment studies. Journal of Cleaner Production* 135: 1085–1097. https://doi.org/10.1016/j.jclepro.2016.06.164.

PlasticsEurope (2016). Plastics – The Facts: An analysis of European plastics production, demand and waste data. http://www.plasticeurope.org/documents/document/plasticsthefacts2016finalversion.pdf (accessed 16 October 2017).

PlasticsEurope (2018). Plastics – The Facts: An analysis of European plastics production, demand and waste data. Brussels, Belgium.

PlasticsEurope (2019). Plastics – the facts: an analysis of European plastics production, demand and waste data. https://www.plasticseurope.org/application/files/9715/7129/9584/FINAL_web_version_Plastics_the_facts2019_14102019.pdf. (accessed 7 January 2019).

Rajendran, S., Hodzic, A., Scelsi, L. et al. (2013). *Plastics recycling: insights into life cycle impact assessment methods. Plastics, Rubber and Composites* 42 (1): 1–11.

Ruban, A. (2012). Life cycle assessment of plastic bag production, Master Thesis. Department of Earth Sciences, Geotryckeriet, Uppsala University, Villavägen 16, SE-752 36 Uppsala, Sweden.

Saibuatrong, W., Cheroennet, N., and Suwanmanee, U. (2017). *Life cycle assessment focusing on the waste management of conventional and bio-based garbage bags. Journal of Cleaner Production* 158: 319e334.

SEC (Singapore Environmental Council). (2018). *Consumer plastic and plastic resource system ecosystem in Singapore.* A position paper by the Singapore Environmental Council. http://sec.org.sg/seaa/wp-content/uploads/2018/08/DT_PlasticResourceResearch_29Aug_FinalPrint.pdf (accessed 14 Augest 2019).

Shen, L. and Worrell, E. (2014). Chapter 13: plastic recycling. In: *Handbook of Recycling* (eds. E. Worrell and M. Reuter), 179–190. Elsevier.

Shen, L., Worrell, E., and Patel, M.K. (2010). *Open-loop recycling: a LCA case study of PET bottle-to-fibre recycling fibre recycling resources. Resources, Conservation and Recycling* 55 (1): 34–52.

Singh, R.K. and Ruj, B. (2015). *Plastic waste management and disposal techniques: Indian scenario. International Journal of Plastics Technology* 19 (2): 211–226.

Song, H.S. and Hyun, J.C. (1999). *A study on the comparison of the various waste management scenarios for PET bottles using the life-cycle assessment (LCA) methodology. Resources, Conservation and Recycling* 27 (3): 267–284.

Stripple, H. and Westman, R. (2008). *Development and environmental improvements of plastics for hydrophilic catheters in medical care: an environmental evaluation. Journal of Cleaner Production* 16 (16): 1764–1776. https://doi.org/10.1016/j.jclepro.2007.12.006.

Takahashi, K. and Dodbiba, G. (2007). *Assessing different recycling options for plastic wastes from discarded mobile phones in the context of LCA. Journal of Resources Processing* 54: 29–34.

The Danish Environmental Protection Agency. (2018). Life cycle assessment of grocery carrier bags. The danish environmental protection agency, Copenhagen, Denmark. https://www2.mst.dk/Udgiv/publications/2018/02/978-87-93614-73-4.pdf (accessed 20 March 2019).

Treenate, P., Ruangrit, C., and Chavalparit, O. (2017). *A complete life cycle assessment of high density polyethylene plastic bottle. IOP Conference Series: Materials Science and Engineering* 222: 012010. https://doi.org/10.1088/1757-899X/222/1/012010.

Treenate, P., Ruangrit, C. and Chavalparit, O. (2018). Life cycle management for plastic waste management: a life cycle assessment of polyethylene bag in Thailand, 7th International Workshop, Advances in Cleaner Production – Academic Work, "cleaner production for achieving sustainable development goals" Barranquilla – Colombia – June 21st and 22nd.

Tua, C., Biganzoli, L., Grosso, M., and Rigamonti, L. (2019). *Life cycle assessment of reusable plastic crates (RPCs). Resources* 8 (2): 110–125. https://doi.org/10.3390/resources8020110.

Uihlein, A. and Schebek, L. (2009). *Environmental impacts of a lignocellulose feedstock biorefinery system: an assessment. Biomass and Bioenergy* 33 (5): 793–802.

UNEP (United Nations Environment Programme). (2018). Single-use plastics: a roadmap for sustainability.

UNEP (United Nations Environment Programme). (2020). Single-use plastic bags and their alternatives – recommendations from life cycle assessments.

United Nations. (2018). United Nations, department of economic and social affairs, population division (2018). The world's cities in 2018 - data booklet (ST/ESA/SER.A/417). https://www.un.org/en/events/citiesday/assets/pdf/the_worlds_cities_in_2018_data_booklet.pdf (accessed 26 October 2021).

USEPA (United States Environmental Protection Agency) (2006). Solid waste management and grenhouse gases: A life-cycle assessment of emissions and sinks (3rd edition), USA.

Wäger, P.A. and Hischier, R. (2015). *Life cycle assessment of postconsumer plastics production from waste electrical and electronic equipment (WEEE) treatment residues in a central European plastics recycling plant. Science of the Total Environment* 529: 158–167.

Wäger, P.A., Hischier, R., Eugster, M., and Eugster, M. (2011). *Environmental impacts of the Swiss collection and recovery systems for waste electrical and electronic equipment (WEEE): a follow-up. Science of the Total Environment* 409 (10): 1746–1756.

World Economic Forum. (2016). The New Plastics Economy: Rethinking the Future of Plastics. World Economic Forum, Geneva, Switzerland. http://www3.weforum.org/docs/WEF_The_New_Plastics_Economy.pdf. Date accessed: 7 January 2019.

Zamagni, A., Guinée, J., Heijungs, R. et al. (2012). *Lights and shadows in consequential LCA. The International Journal of Life Cycle Assessment* 17 (7): 904–918.

Zhang, B. and Kang, M. (2013). *Life-Cycle Assessment for Plastic Waste Recycling Process: Based of the Network Evaluation Framework.* In: *20th CIRP International Conference on Life Cycle Engineering*, 377–382. Singapore.

14

Role of Education and Society in Dealing Plastic Pollution in the Future

Nalini Singh Chauhan[1] and Abhay Punia[2]

[1] *Department of Zoology, Kanya Maha Vidyalya, Jalandhar, Punjab, India*
[2] *Department of Zoology, Guru Nanak Dev University, Amritsar, Punjab, India*

14.1   Introduction

Humans have benefited through the use of polymers since the prehistoric Mesoamericans first shaped unprocessed plastic into hoops, toys, and bands around 1600 BCE (Hosler et al. 1999). With the expansion if the development of modern plastics, man's dependence on plastic has greatly expanded, and 15 new polymer groups were produced in the first half of the decade. Plastic supply and demand have increased dramatically across the world because of massive urbanization and socioeconomic growth in various countries. The percentage of plastic used by various industries is as follows; 18% in packaging, 8% in building materials, 7% in PPA fibers, and the rest in other industries (Figure 14.1).

Due to the lower recycling potential of plastic as well as a level of technical assistance, the extent of plastic recovery from waste continues to be a challenge. The majority of it is dumped in landfills, charred in incinerators, or flushed into the oceans. Environmental degradation, food chain disruption, habitat breakdowns, fuel cost, and economic decline are all effects of these massive quantities of plastic waste. Plastic waste won't just pollute the air and soil (Li et al. 1995; Barnes et al. 2009; Steelys Drinkware 2013), or threaten an individual's well-being (Elliott et al. 1996; Yamamoto and Yasuhara 1999; Maffini et al. 2006; Crinnion 2010), but it also pollutes the water (Laist 1987; Howarth 2013; Perkins 2014; Schwartz 2014), disturbs the food chain (Thompson et al. 2009; Rochman et al. 2015), threatens biological richness (Derraik 2002; Grant and Ryder 2009; Gregory 2009), and results in waste of a lot of energy (Cho 2012; Hong Kong Cleanup 2012; European Commission 2013; Themelis and Mussche 2014; StudyMode 2015). Recycling, banning the sale of certain goods, and imposing taxes or fines are all being utilized to eliminate the harmful effects of plastic pollution (Lopez and Martin 2015). In cities across China, India, and other countries, migrant workers gather and recycle the bulk of these wastes. Recycling is the process of converting waste into useful new products and is encouraged all over the world as part of the circular economy model. However, effective recycling continues to be

Plastic and Microplastic in the Environment: Management and Health Risks, First Edition.
Edited by Arif Ahamad, Pardeep Singh, and Dhanesh Tiwary.
© 2022 John Wiley & Sons Ltd. Published 2022 by John Wiley & Sons Ltd.

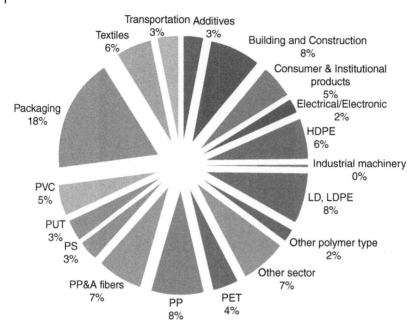

Figure 14.1 Primary plastic production. *Source:* David et al. (2019).

a challenge for a variety of reasons, including the difficulty of managing waste streams (USEPA 2017). The recycling of plastic waste has developed into a well-established underground economy that employs unskilled workers and the urban poor. The inappropriate processing and burning of waste for energy restoration makes the workers vulnerable to toxic heavy metals and pollutes the air, water, and soil.

14.2 Consumption

Plastics are unquestionably an integral part of the commodities used in modern culture. Plastics are used in almost every aspect of daily life, including clothing, accessories, food, and health care devices. Well over one-third of the plastic usage accounts for the packaging of common goods, with more than one-third in construction materials such as plastic pipes and vinyl cladding. More than forty million tons of plastics are processed into textile fiber such as nylon, polyester, and acrylic, which is then used in the clothing industry around the globe. A large amount of PET plastic is being used in poly-cotton blend apparel along with many other high-performance fabrics which are entirely made of plastics including polyesters, fluoropolymers, and nylons. These polymers are referred to as single use plastics (SUPs) since they are designed to be used just once.

In the developed world, the trends of usage of the five most commonly used plastics in various sectors seem to be stable. However, in developing nations, the trends of plastic usage may vary a bit, for example, in India, the packaging sector is responsible for 42% of resin utilization (Mutha et al. 2006). Plastics are used in vast numbers in industrial applications, furniture, and toy manufacturing. In developed countries, plastics are increasingly being used as a means to replace products including paper, metals, wood, and glass due to reduced unit costs and improved performance specifications. Plastics have seen many popularities as a commodity and have proven to be flexible in a variety of styles and shapes. Plastics possess a variety of special characteristics, including the ability to be used at a broad spectrum of temperatures, chemical and light resistance, and the ability to be worked as a hot melt. Although at a moderate average expansion rate of 5%, a continuity of this pattern means that by 2050, the planet would use at least 1800 million tons of plastics annually (Table 14.1). The increase in public demand for plastics is largely to blame for this projected development.

Table 14.1 Top 20 countries ranked by mass of mismanaged plastic waste in 2010 and 2025, with percent increase in coastal population from 2010 to 2025. MMT; million metric tons.

| | | Year 2010 | | Year 2025 | |
Rank	Country	Mismanaged plastic waste [MMT/year]	Country	Mismanaged plastic waste [MMT/year]	% pop. change since 2010
1	China	8.82	China	17.81	3.7%
2	Indonesia	3.22	Indonesia	7.42	11.9%
3	Philippines	1.88	Philippines	5.09	26.0%
4	Vietnam	1.83	Vietnam	4.17	13.3%
5	Sri Lanka	1.59	India	2.88	18.7%
6	Thailand	1.03	Nigeria	2.48	45.1%
7	Egypt	0.97	Bangladesh	2.21	18.5%
8	Malaysia	0.94	Thailand	2.18	5.4%
9	Nigeria	0.85	Egypt	1.94	25.0%
10	Bangladesh	0.79	Sri Lanka	1.92	9.0%
11	South Africa	0.63	Malaysia	1.77	23.6%
12	India	0.60	Pakistan	1.22	26.6%
13	Algeria	0.52	Burma	1.15	11.1%
14	Turkey	0.49	Algeria	1.02	18.4%
15	Pakistan	0.48	Brazil	0.95	10.6%
16	Brazil	0.47	South Africa	0.84	7.2%
17	Burma	0.46	Turkey	0.79	16.2%
18	Morocco	0.31	Senegal	0.74	44.3%
19	Korea, North	0.30	Morocco	0.71	14.1%
20	United States	0.28	North Korea	0.61	5.0%

Source: "Plastic waste inputs from land into ocean," by J. Jambeck et al. 2015.

14.3   Global Dimension of Plastic Pollution

Global production of plastic has increased at an average of 9% per year since the 1950s. In 2018, global production reached 360 million metric tons, up from 1.7 million metric tons in 1950 (Geyer et al. 2017; PlasticsEurope 2019). According to the MacArthur Foundation, the production of plastic will double during the next two decades (Elias 2018) (Figure 14.2). These SUPs have also been shown to disintegrate slowly when exposed to UV light, heat, or mechanical stress into minute particles called secondary microplastics (MPs), having diameters only around 5 mm (Rillig 2012; Schopel and Stamminger 2019). MP pollution has been steadily increasing because they are more readily disseminated and are more difficult to contain than the larger plastics (Thompson et al. 2004, 2005). Moreover, recent researchers have revealed the presence of these MPs in freshwater lakes and rivers, as well as in drinking water (McCormick et al. 2016; Ivleva et al. 2017; Karthik et al. 2018; Schymanski et al. 2018; Wu et al. 2019).

The amount of plastic waste that reaches the oceans, mainly derived from terrestrial sites, is an extremely disturbing phenomenon (Thompson et al. 2004, 2005; Barnes et al. 2009; Gregory 2009; Teuten et al. 2009). Tourist activities, sewerage overflow, dumping sites along coastlines, unregulated disposal, and accidental industrial spillage are the main causes that led to the entry of plastic pollution into the oceans over the past 10 years. This waste is estimated to be between 0.2 and 0.3%, as taken from a Greenpeace survey (Allsopp et al. 2006). Due to the limitation in the availability of landfill sites in most countries,

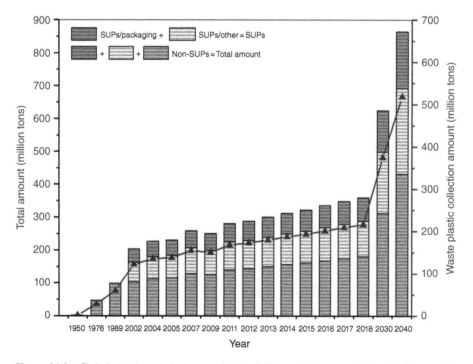

Figure 14.2   Global plastic and single use plastic (SUP) production growth trends. *Sources:* Geyer et al. (2017) and Advisors et al. (2019).

different solid waste management methods should be used for material recycling and energy recovery. In Korea, for example, domestic trash from everyday life and commercial operations declined dramatically, from 1.3 kg per individual per day in 1994 to 1.04 kg in 2002, and total material recycling has surpassed landfilling for the first time in 2002. Hong Kong generates a large amount of solid waste on a regular basis (Environmental Protection Department [EPD] 2015). To treat such large volumes of garbage, Hong Kong has three landfill sites; West New Territories, South East New Territories, and North East New Territories. They too have a finite amount of space which will be limited for the next two decades. Landfill expansions are minimal and the installation of an incinerator would entail additional consultations.

The plastic pollution and deleterious consequences are particularly severe and ubiquitous in megacities like Japan (PWMI 2014), Taiwan (Walther 2015), the UK (GHK 2006; Howarth 2013), and Hong Kong (Environmental Protection Department 2013), where economic activities are prosperous and utilization of plastic is higher (Table 14.2). The unsustainable use and unplanned disposal of plastic products has globally threatened the economies, human health, and ecosystems. The increased health hazards are more visible in developing Asian and African nations due to low recycling rates. Since solid waste includes large levels of harmful substances such as heavy metals, chemicals, and chronic organic toxins, their excessive processing has resulted in substantial aquatic and air contamination, land destruction, and medical repercussions in developing nations. The industrial distribution and trade of e-waste has also created transboundary environmental governance issues globally. The United States, the United Kingdom, and the European Union are the top exporters of waste, including e-waste and plastic waste to developing nations like China, India, and Nigeria, which benefit from low-cost workers and disposal cost (Chi et al. 2011; Luo et al. 2013). Only about one-quarter of this is tracked and safely recycled, allowing the remainder to pollute the environment.

14.4 Plastic Pollution in Natural Environments

Disposal of plastic waste, just as all consumer materials, adds to the rise of municipal waste and contributes to urban litter. The amount of waste materials in urban litter is increasing, including thermoplastic items and a large number of other discarded materials. Plastic litter will persist for a long time because thermoplastics are not easily biodegraded in the ecosystem (Andrady 2003). The exact lifespan of discarded plastics is highly subjective depending on the chemical composition of the substance, the properties of the system in which it is placed, and even how "degradation" is described or calculated. The environmental or biological breakdown of compostable plastics can occur within months, but for conventional plastics, it can take years (Andrady 2003). Plastic's durability allows it to have a broad variety of uses, but it often raises environmental issues as it is ultimately discarded as trash. It is not surprising that fisheries waste ends up in the marine environment, given that most fishing equipment is made of plastic, as are fish packaging boxes and other supplies. By 2025, MSW's total volume will be more than two times what it is now (Hoornweg and Bhada-Tata 2012). Oceans are now receptacles for at least 8 billion kilograms of plastics per year; equivalent to a waste truck that releases plastic every minute into the ocean. The

Table 14.2 Current plastic waste problem in mega-cities.

	Japan	Taiwan/Taipei	UK	Hong Kong
Status quo of plastic wastes	Around 9.3 million tons discharged per year	Nearly 7200 kg plastic waste collected on 18 beaches	5 million tons of plastic consumed per year	Around 730 000 tons of plastic waste discharged per year
Major source of plastic wastes	Domestic/packaging	Domestic/restaurants	Packaging	Municipal/shopping bags
Plastic waste management strategies	Incineration	—	—	Landfilling
Affects of plastic wastes				
(a) Health	Respiratory infection	Food chain contamination	—	—
(b) Environment	Air pollution	Water pollution	—	Land pollution
(c) Biodiversity	—	—	Endanger biodiversity	—
(d) Energy	—	—	—	Energy waste and economic loss

Source: Chow et al. (2017).

growth rates mentioned mean that the oceans will receive the equivalent of one truckload of plastic every 15 seconds, day and night, by 2050. Until 2050, the volume of plastic in our oceans will exceed the mass of fish unless we severely cut the plastic production and dumping. Once, the ocean plastics remained in the form of small and smaller particles, for centuries. Other human impacts are added to this massive contamination such as overfishing, acidification, and an increase in ocean temperature. Given the clear link between land-based emissions sources and marine contamination, the plastic industry must work closely with water and sewerage authorities to discuss issues and adjust plastic treatment and waste management strategies accordingly. The plastic industry can be encouraged to design biodegradable plastic, to produce materials having the longer lifespan, to recycle plastic waste, to assist plastic removal from the environment, etc.

14.5 Role of Education for Plastic Waste Management

The responsible behavior of the future adult toward plastic waste management may be achieved through education. Given the long-term nature of reducing plastic waste, it is critical that children and young people gain adequate education about the effect of plastic pollution, and measures to reduce its influence by everyday actions since they serve future generations. Furthermore, as the country's potential backbone, they will motivate more people to become conscious of this urgent threat (Kong et al. 2014). Individual behavior and a lack of adequate facilities (such as litter bins) are two major contributors to the excessive level of plastic trash in the ecosystem. Individual's undesirable behaviors, such as dumping cigarettes in the sidewalk, are frequently adopted and accepted societal norms. As a result, substantial modifications in social practices, such as the decrease or removal of widespread usage of single-use water bottles, will need a huge number of individuals to adjust their behavior (Shove et al. 2012). People that cause the most to plastic pollution, such as by littering irresponsibly, may be less linked to both the concerns (i.e., they overlook plastic waste due to a lack of awareness or option) and the aggregation site (e.g. on aquatic bodies, roadsides). However, a large number of individuals rate plastic pollution as a significant ecological concern (Ipsos 2018; IFAT poll/The Recycler 2018). Thus, increasing knowledge and responsibility for resolving the problem of plastic waste should encourage better behavioral patterns. Policy developments to take the plastic related challenges have increased at both the global and national levels during the last five years (Bourguignon 2018; DEFRA 2019).

Littering is a behavioral challenge that must be tackled solely by schooling. Increased consumer understanding of the environmental effects of litter needs more effort and energy, as this is the most effective remedy. Classrooms and outdoor experiences may be used to teach students about how much plastic they use on a daily basis, the impact of plastic on the environment (especially the aquatic environment), how human existence is intertwined with aquatic ecosystems, how aquatic pollution may harm them, the necessity of minimizing, reusing, and recycling plastic, and lastly how implacable plastic is. Pupils' involvement in plastic waste management activities may be further carried out by field training so that they can monitor and share their findings of plastic pollution through school competitions and engagement in lake, river, and beach cleanup efforts. In general, these programs should focus on avoiding plastic littering.

Plastic waste reduction must take into account choices such as minimizing, reuse, and recycling which necessitates greater effort in the existing waste management system, as well as improved equipment and education. Sometimes people can be seen reusing plastic bags or tossing plastic waste into recycle bins. While awareness of plastic recycling is given, there is a shortage of suitable recycling bins for various forms of plastics, which make segregating plastics at the source easier. As a result, a modern plastic recycling bin may be a viable option as well as an educational platform for putting awareness into practice and improving the efficiency of plastic recycling. A poster or video may be displayed near the PRB to show bin users useful recycling details, such as the plastic recycling process afterward. For local schools, a teacher's guidebook called "Reduce Your Waste and Recycle Your Plastics" was released by the Hong Kong government to properly educate students about plastic recycling (EPD 2011). According to a study conducted in the United States, good education combined with recycling awareness will encourage pro-recycling action rather than relying on financial incentives (De Young 1990). As a result, it is essential to support bin users to appreciate their role in recycling so that they will continue to recycle. Students may attend seminars or workshops on plastic education to improve their awareness of plastic recycling. Inter-hall contests may also be formed to inspire and promote recycling participation. Plastic recycling ambassador programs should be held to provide students with the awareness they need to further promote the recycling idea to their peers. Long et al. (2013) found that peer sharing among students resulted in positive reinforcement of recycling behaviors. The PRB, in conjunction with education and promotion, can improve the rate of plastic recycling and public knowledge of environmental issues in the future.

The participant's knowledge has a huge effect on their behavior and desire to make a difference, so increasing their perception of such a pervasive problem is a crucial step toward improving their behavior. Studies in Texas (Bradley et al. 1999), Malaysia (Aminrad et al. 2013), and Istanbul (Aminrad et al. 2013) all found similar results (Ergen et al. 2015). Participants who became more aware expressed a greater willingness to contribute and create a change, either by assisting others in becoming aware or by self-involvement, despite the fact that the municipality has the main responsibility of limiting the spread of pollution. However, it is also the responsibility of the municipality to have safer recycling environments, more campaigns, and stringent rules to prevent more factories and workshops from violating the law. In Singapore, a government program involving significant penalties and remedial work orders has proven to be a highly successful anti-littering tactic. Quality recycling should be studied and implemented to deal with the massive amount of plastic waste. However, only a small percentage of the population is aware of plastic recycling and reprocessing programs.

To minimize plastic waste, the public's awareness of plastic waste management must be improved by improving people's perceptions, beliefs, and actions toward plastic waste management. The management of plastic waste is divided into four categories: (i) the 4Rs- reduce, reuse, recycle, and regenerate; (ii) two strategies – landfilling and incineration; (iii) the four stages to recycling procedures – cleaning, isolation, processing, and compression; and (iv) understanding of the plastics life cycle. However, according to the World Economic Forum (2016), only 14% of packaging plastic is recycled, while more than 80% is thrown away as litter, with 40% being landfilled, 14% being set ablaze, and 32% being released into the natural surroundings.

14.6   Significance of Plastic Waste Education

Education is defined as a powerful tool for developing new awareness, skills, and values in order to accomplish a healthy world and a better standard of living. (UNCED 1992; Nagra 2010). Education is a learning process which helps learners to procure new scientific facts, values, and information, and thus leads to change in knowledge, behavior, and attitudes. The common examples of PET are beverage drink bottles and food containers whose usage rate is higher in the academic life. During a research undertaken by a professor at Hong Kong's Education Institute, it was observed that there was an urgent need for educating learners about environmental issues and sustainable initiatives for conserving it, as these students would take the job of educators for the coming generations. Pro-environmental activities on campus and in student's dorms offered realistic possibilities for environmental education and provided long-term studies on the efficacy of teaching strategies. Several studies have supported that the amount of proper education people have appears to be strongly associated with their environmental awareness and the promotion of positive behaviors toward the environment (Mobley et al. 2009). In addition, Scott and Willits (1994) also reported that the more one is educated, the more likely he or she is to be involved in responsible behavior toward the environment. Japan, Taiwan, the United Kingdom, and Hong Kong have implemented environmental education for plastic waste management, but it isn't enough yet.

Thus, the importance of education and information in our lives cannot be overstated. More individuals carefully choosing the plastic items they consume or how they dispose of trash would gradually affect public behavior, resulting in a major drop in plastic pollution. There is still much room for progress in terms of altering people's perceptions, habits, and awareness about plastic waste management, much of which is dependent on the efficacy of educational strategies.

14.7   Role of NGOs and General Public

Despite the fact that dismantling freezers can release mercury, oil, and other chemicals, and expel CFC emissions, the workers doing this job seems to be unaware of these hazards. Most of the people are aware that flaming plastic is bad for the atmosphere, but they are largely ignorant of its environmental and health consequences. The public is generally unaware of the health risks and strategies to be followed and are also unaware of the dangers of informal recycling. This absence of knowledge about the ecological and health-related concerns associated with plastic pollution has increasingly become an important barrier in communicating with people about the problem, and it continues to be a barrier in successful implementation of strategies and legislations. Education seems to be the only tool available to combat public ignorance about such a serious environmental problem at this critical juncture, especially among youth, who constitute the most important community in the battle against plastic waste in terms of potential educated workers, industry innovators, and legislators. Thus, increasing people's understanding of such a prevalent issue is a critical step in changing their behavior, as the awareness will have a significant impact on the attitudes and willingness of the public to make a change. Local youth efforts like #FridaysForFuture, which began in August 2018 when Greta Thunberg, a Swedish-born 15-year-old girl, campaigned for three weeks against a lack of support for the climate problem, might be a fantastic source of inspiration. She documented her strike with images on Instagram and Twitter, and it rapidly became a worldwide phenomenon. The hashtags #FridaysForFuture and #Climatestrike went viral, and students and adults from all around the globe began to demonstrate outside their parliaments and local municipal halls, encouraging kids from over 140 nations and addressing international leaders on September 23, 2019 at the United Nation conference (Fridays for Future 2019). Similar results showing that community behaviors can significantly minimize the plastic hazards when people are made aware were discovered in research done in Texas (Bradley et al. 1999), Malaysia (Aminrad et al. 2013), and Istanbul (Ergen et al. 2015). The 1977 Tbilisi Intergovernmental Conference on Environmental Education also described that education will transform conduct by "providing social groups and individuals with an opportunity to be actively engaged at all levels in working toward resolution of environmental concerns" (Hungerford and Volk 1980). The community can thus be involved by encouraging and providing guidance for recycling through daily publishing of numerous guidebooks and magazines, as well as their door-to-door distribution for separating recyclables and trash. Moreover, daily columns dedicated to resource recycling must be included in the newspapers offering advice to the public. United Nations Environment developed a significant initiative, the "massive open online course" (MOOC) on Marine Litter, in collaboration with the Open University of the Netherlands, with the goal of teaching students, regardless of their field of work or place, how to implement successful and inspirational activities to their own local environment through action-oriented teaching (Open Universiteit 2019).

In developing countries such as China, public sector efforts in e-waste control are very important and the lack of interest of the public in policy making for waste management is very challenging for the Chinese government (Lu et al. 2015). There is a greater need for generating awareness among general masses along with their involvement in waste management strategies, and for these NGOs can play a very efficient role. The goal of environmental education is to transform the minds of people for assisting in social movements and individuals in acquiring a collection of beliefs and emotions of environmental interest, as well as inspiration for active involvement in environmental development and protection. In India various NGOs have collaborated with the government to recycle informal waste, but still more work is required for educating people on the proper management of solid waste. NGOs and Government Organized Non-Governmental Organizations (GONGOs) may also assist small businesses in developing monitoring strategies so that their operations meet workplace health and safety requirements, even though they are not approved or licensed by government. Local recycling businesses should be encouraged and strengthened so that they may supply more informal waste collection. This will assure the steady supply of recycling waste to these small enterprises and thus increase their recycling capacity. NGOs may collect ground research data for developing manual recycling practices that are safer and thus reduce the chances of contamination. The role of NGOs is to address the "governance" vacuum in traditional recycling practices. Several NGOs, such as Greenpeace and the Basel Action Network, have highlighted the tremendous medical threats posed by informal plastics recyclers. In China, NGOs have emerged as significant non-state players in the political arena (Mol and Carter 2006). There is need for increased grassroots efforts along with more participation of general masses in helping government in delivering education related to different solid waste management policies (Qu et al. 2013; Lu et al. 2015). NGOs may act as good resources that will help the implementation of on-the-ground work with the help of government, legislations, and authentic recycling methods. More local analysis, including interviews and surveys from a social viewpoint, is needed to understand the dynamics of the plastic waste management, as well as how the plastic waste merchants operate on a daily basis, what obstacles they encounter, and what solutions will be most helpful in facilitating their secure and healthy work. Local government can play an important role for delivering education related to the health risks of plastic pollution in a manner that encourages the exchange of information and collaboration among the community. Methods should not be replaced, but rather strengthened by constructive improvements that can be tracked and tested by NGOs or scientific agencies. Thus, local plastic waste can be handled through regulations that work along with social realities. Owing to the well-developed informal economy, which has proven to be more productive than formal methodologies, it is currently impractical to consider the informal sector to be formalized or organized. Supporting the informal sector on a social and political level involves the government legally acknowledging the service and jobs that informal garbage collection and recycling provides. There is no question that empowering the informal sector would necessitate a reorganization of agencies among private businesses, local governments, and national policymakers.

14.8 Conclusion

To summarize, plastic pollution challenges are gradually becoming more severe all over the globe, threatening the wildlife, environment, and human health. Different nations have different educational systems, but there is still room for progress in terms of changing public perceptions, habits, and awareness about plastic waste management. Different teaching methods may increase student's knowledge dramatically, and the most significant reform can occur at the societal level, where several options exist for reducing plastic waste by easy improvements in everyday habits.

Interdisciplinary research involving different fields such as biology, ecotoxicology, medical sciences etc. should be aimed at identifying plastics having higher risks and replacing them with more ecologically friendly alternatives.

References

Advisors, D., Wit, D. W., Hamilton, A., et al. (2019). *Solving plastic pollution through accountability*. World Wide Fund for Nature report. https://www.worldwildlife.org/publications/solving-plastic-pollution-through-accountability (accessed 26 October 2021).

Allsopp, M., Walters, A., Santillo, D., and Johnston, P. (2006). *Plastics debris in the worlds oceans*, Greenpeace survey. http://http://oceans.greenpeace.org/raw/content/en/documentsreports/plastic_ocean_report (accessed 26 October 2021).

Aminrad, Z., Zakariya, S., Hadi, A., and Sakari, M. (2013). *Relationship between awareness, knowledge and attitudes towards environmental education among secondary school students* in Malaysia. *World Applied Sciences Journal* 22 (9): 1326–1333. https://doi.org/10.5829/idosi. wasj.2013.22.09.275.

Andrady, A.L. (ed.) (2003). *Plastics and the Environment*. West Sussex, England: Wiley. ISBN: 0-471-09520-6.

Barnes, D.K.A., Galgani, F., Thompson, R.C., and Barlaz, M. (2009). *Accumulation and Fragmentation of Plastic Debris in Global Environments*, vol. 364, 1985–1998. Philosophical Transactions of the Royal Society B; Biological Sciences.

Bourguignon, D. (2018). Single-use plastics and fishing gear. EU legislation in Progress. European Union: Brussels. http://www.europarl.europa.eu/RegData/etudes/ BRIE/2018/625115/EPRS_BRI(2018)625115_EN.pdf (accessed 26 October 2021).

Bradley, J., Waliczek, T., and Zajicek, J. (1999). *Relationship between environmental knowledge and environmental attitude of high school students. The Journal of Environmental Education* 30 (3): 17–21. https://doi.org/10.1080/00958969909601873.

Chi, X.W., Streicher-Porte, M., Wang, M.Y.L., and Reuter, M.A. (2011). *Informal electronic waste recycling: a sector review with special focus on China. Waste Management* 31 (4): 731e742.

Cho, R. (2012). *What happens to all that plastic?* State of the planet. The Earth Island Institute, Columbia University: NYC.

Chow, C.F., So, W.M.W., Cheung, Y.T.Y. et al. (2017). Plastic waste problem and education for plastic waste management. In: *Emerging Practices in Scholarship of Learning and Teaching in a Digital Era*, 125–140. Singapore: Springer https://doi.org/10.1007/978-981-10-3344-5_8.

Crinnion, W.J. (2010). *Toxic effects of the easily avoidable phthalates and parabens. Alternative Medicine Review: A Journal of Clinical Therapeutic* 15 (3): 190–196.

David, A.,Thangavel, Y. D., Sankriti, R. (2019). Recover, recycle and reuse: An efficientway to reduce the waste. *International Journal of Mechanical and Production Engineering Research and Development (IJMPERD)*, ISSN (P), 2249–6890.

De Young, R. (1990). *Recycling as appropriate behavior: a review of survey data from selected recycling education programs in Michigan. Resources, Conservation and Recycling* 3 (4): 253–266.

DEFRA (Department for Environment, Food & Rural Affairs) (2019). Gove takes action to ban plastic straws, stirrers, and cotton buds to slah plastic waste. Gov.uk. https://www.gov.uk/ government/news/gove-takes-action-to-ban-plastic-straws-stirrers-and-cotton-buds (accessed 26 October 2021).

Derraik, J.G.B. (2002). *The pollution of the marine environment by plastic debris: a review. Marine Pollution Bulletin* 44 (9): 842–852.

Elias, S.A. (2018). *Plastics in the ocean. Encyclopedia of the Anthropocene* 1: 133–149.

Elliott, P., Shaddick, G., Kleinschmidt, I. et al. (1996). *Cancer incidence near municipal solid waste incinerators in Great Britain. British Journal of Cancer* 73 (5): 702–710.

Environmental Protection Department (EPD) (2011). *Reduce your Waste and Recycle your Plastics Campaign teachers' Guidebook*. Hong Kong, China: Government Printer.

Environmental Protection Department (EPD). (2013). *Monitoring of solid waste in Hong Kong. Waste Statistics for 2013.*

Environmental Protection Department (EPD) (2015). *Monitoring of Solid Waste in Hong Kong-Waste Statistics for 2013*. Hong Kong, China: Government Printer.

Ergen, A., Baykan, B., and Turan, S. (2015). *Effect of materialism and environmental knowledge on environmental consciousness among high school students: a study conducted in Istanbul*

province. International Journal of Human Sciences 12 (1): 511. https://doi.org/10.14687/ijhs.
v12i1.3130.

European Commission (2013). Green paper on a European strategy on plastic waste in the
environment.

Fridays for Future. (2019). *Statistics.* List-countries. https://www.fridaysforfuture.org.

Geyer, R., Jambeck, R.J., and Law, L.K. (2017). *Production, use and fate of all plastics ever made.*
Science Advances 3: 1–5.

GHK in association with Recoup (2006). UK plastics waste; A review of supplies for recycling,
global market demand, future trends and associated risks. Waste & Resources Action
Programme.

Grant, R. and Ryder, B. (2009). Drowning in plastic: The great Pacific garbage patch is twice the
size of France. www.telegraph.co.uk/news/earth/environment/5208645/Drowning-in-plastic-
The-Great-Pacific-Garbage-Patch-is-twice-the-size-of-France.html (accessed 26 October 2021).

Gregory, M.R. (2009). *Environmental implications of plastic debris in marine settings-*
entanglement, ingestion, smothering, hangers-on, hitch-hiking and alien invasions.
Philosophical Transactions of the Royal Society B 364: 2013–2025.

Hong Kong Cleanup (2012). Plastic bags fact sheet. http://www.hkcleanup.org/en/content/
plastic-bags-fact-sheet (accessed 26 October 2021).

Hoornweg, D. and Bhada-Tata, P. (2012). *What a Waste: A Global Review of Solid Waste*
Management. Washington, DC: World Bank.

Hosler, D., Burkett, S.L., and Tarkanian, M.J. (1999). *Prehistoric polymers: rubber processing in*
ancient mesoamerica. Science 284: 1998–1991. https://doi.org/10.1126/science.284.5422.1988.

Howarth, S. (2013). *Why plastic waste is such a problem and the future of bioplastics.* Huffpost
Tech. www.huffingtonpost.co.uk/simon-howarth/why-plastic-waste-issuch-a-
problemb3346167.html (accessed 26 October 2021).

Hungerford, H.R. and Volk, T.L. (1980). *Changing learner behavior through environmental*
education. Journal of Environmental Education 21 (3): 8–21.

IFAT poll/The Recycler (2018). IFAT poll/The Recycler Survey reveals British qualms over
plastic waste. https://www.therecycler.com/posts/survey-reveals-british-qualms-over-
plastic-waste (accessed 26 October 2021).

Ipsos, MORI (2018). *Public concern about plastic and packaging waste is not backed up by*
willingness to act. Ipsos MORI. https://www.ipsos.com/ipsos-mori/en-uk/public-concern-
about-plastic-and-packaging-waste-not-backed-willingness-act (accessed 26 October 2021).

Ivleva, N.P., Wiesheu, A.C., and Niessner, R. (2017). *Microplastic in aquatic ecosystems.*
Angewandte Chemie International Edition 56 (7): 1720–1739.

Jambeck, J.R., Geyer, R., Wilcox, C. et al. (2015). *Plastic waste inputs from land into the ocean.*
Science 347 (6223): 768–771. https://doi.org/10.1126/science.1260352.

Karthik, R., Robin, R.S., Purvaja, R. et al. (2018). *Microplastics along the beaches of southeast*
coast of India. Science of the Total Environment 645: 1388–1399.

Kong, D., Ytrehus, E., Hvatum, A.J., and Lin, H. (2014). *Survey on environmental awareness of*
Shanghai college students. Environmental Science and Pollution Research 21 (23): 13672–
13683. https://doi.org/10.1007/s11356-014-3221-0.

Laist, D.W. (1987). *Overview of the biological effects of lost and discarded plastic debris in the*
marine environment. Marine Pollution Bulletin 18 (6): 319–326.

Li, C.T., Lee, W.J., Mi, H.H., and Su, C.C. (1995). *PAH emission from the incineration of waste oily sludge and PE plastic mixtures. Science of the Total Environment* 170 (3): 171–183.

Long, J., Harre, N., and Atkinson, Q. (2013). *Understanding change in recycling and littering behavior across a school social network. American Journal of Community Psychology* 53 (3–4): 16–17.

Lopez, J. and Martin, M. (2015). Social perceptions of single-use plastic consumption of the Balinese population. Doctoral dissertation, Novia University of Applied Sciences, Vasaa, Finland. https://www.theseus.fi/bitstream/handle/10024/93403/Lopez_Javier. pdf?sequence=1 (accessed 26 October 2021).

Lu, C.Y., Zhang, L., Zhong, Y.G. et al. (2015). *An overview of e-waste management in China. The Journal of Material Cycles and Waste Management* 17 (1): 1e12.

Luo, C.L., Liu, C.P., Wang, Y. et al. (2013). *A review of development of an e waste collection system in Dalian, China. Journal of Cleaner Production* 52: 176e184.

Maffini, M.V., Rubin, B.S., Sonnenschein, C., and Soto, A.M. (2006). *Endocrine disruptors and reproductive health: The case of bisphenol-A. Molecular and Cellular Endocrinology* 25 (254): 179–255, 86. https://doi.org/10.1016/j.mce.2006.04.033.

McCormick, A.R., Hoellein, T.J., London, M.G. et al. (2016). *Microplastic in surface waters of urban rivers: concentration, sources, and associated bacterial assemblages. Ecosphere* 7 (11): e01556.

Mobley, C., Vagias, W.M., and DeWard, S.L. (2009). *Exploring additional determinants of environmentally responsible behavior: The influence of environmental literature and environmental attitudes. Environment and Behavior* 42 (4): 420–447.

Mol, A.P.J. and Carter, N.T. (2006). *China's environmental governance in transition. Environmental Politics* 15 (02): 149e170.

Mutha, N., Patel, M., and Premnath, V. (2006). *Plastics materials flow analysis for India. API Resources, Conservation and Recycling* 47: 222–244.

Nagra, V. (2010). *Environmental education awareness among school teachers. The Environmentalist* 30 (2): 153–162.

Open Universiteit (2019). Massive open online course on marine litter. https://www.ou.nl/-/ unenvironment-mooc-marine-litter (accessed 26 October 2021).

Perkins, S. (2014). *Plastic waste taints the ocean floors. Nature, (News Features)* 16581.

Plastic Waste Management Institute (PWMI) JAPAN (2014). *Plastic products, plastic waste and resource recovery* [2012] PWMI. *Newsletter* 43: 4.

PlasticsEurope (2019). Plastics – The Facts 2019: An Analysis of European Plastics Production, Demand and Waste Data (PlasticsEurope 2018).

Qu, Y., Zhu, Q.H., Sarkis, J. et al. (2013). *A review of development on e-waste collection system in Dalian, China. Journal of Cleaner Production* 52: 176e184.

Rillig, M.C. (2012). Microplastic in terrestrial ecosystems and the soil? *Environ. Sci. Technol.* 46 (12): 6453–6454.

Rochman, C.M., Tahir, A., Williams, S.L. et al. (2015). *Anthropogenic debris in seafood: plastic debris and fibers from textiles in fish and bivalves sold for human consumption. Scientific Reports* 5: 14340.

Schopel, B. and Stamminger, R. (2019). *A comprehensive literature study on microfibres from washing machines. Tenside Surfactants Detergents* 56 (2): 94–104.

Schwartz, J. (2014). New research quantifies the oceans' plastic problem. New York Times.

Schymanski, D., Goldbeck, C., Humpf, H.U., and Furst, P. (2018). *Analysis of microplastics in water by micro-Raman spectroscopy: release of plastic particles from different packaging into mineral water. Water Research* 129: 154–162.

Scott, D. and Willits, F.K. (1994). *Environmental attitudes and behavior: a Pennsylvania survey. Environment and Behavior* 26: 239.

Shove, E., Pantzar, M., and Watson, M. (2012). *The Dynamics of Social Practice: Everyday Life and how it Changes*, –London. Sage.

Steelys Drinkware (2013). *The growing global landfill crisis.* http://steelysdrinkware.com/growing-global-landfill-crisis (accessed 26 October 2021).

StudyMode (2015). *Hong Kong landfill problem.* http://www.studymode.com/essays/Hong-Kong-Landfill-Problem-962598.html (accessed 26 October 2021).

Teuten, E.L., Saquing, J.M., Knappe, D.R.U. et al. (2009). *Transport and release of chemicals from plastics to the environment and to wildlife. Philosophical Transactions of the Royal Society B* 364: 2027–2045. https://doi.org/10.1098/RSTB.2008.0284.

Themelis, N. J. and Mussche, C. (2014). 2014 energy and economic value of municipal solid waste (MSW), including Non-recycled Plastics (NRP), currently landfilled in the fifty states.

Thompson, R.C., Olsen, Y., Mitchell, R.P. et al. (2004). *Lost at sea: where is all the plastic? Science* 304 (5672): 838. https://doi.org/10.1126/science.1094559.

Thompson, R., Moore, C., Andrady, A. et al. (2005). *New directions in plastic debris. Science* 310: 1117.

Thompson, R.C., Moore, C.J., Saal, F.S.V., and Swan, S.H. (2009). *Plastics, the Environment and Human Health: Current Consensus and Future Trends*, vol. 364(1526), 2153–2166. Philosophical Transactions The Royal Society.

UNCED (United Nations Conference on Environment and Development) (1992). Promoting education and public awareness and training, Agenda 21. http://www.undocuments.net/a21–36.htm (accessed 26 October 2021).

USEPA (U.S. Environmental Protection Agency) (2017). Trash-free waters: frequently asked questions about plastic recycling and composting. https://www.epa.gov/trash-free-waters/frequently-questionsabout-plastic-recycling-and-composting (accessed 26 October 2021).

Walther, B. (2015). *Nation engulfed by plastic tsunami.* Taipei Times. http://www.taipeitimes.com/News/editorials/archives/2015/01/09/2003608789/2 (accessed 26 October 2021).

World Economic Forum (2016). *The New Plastics Economy: Rethinking the Future of Plastics.* World Economic Forum.

Wu, Y.M., Guo, P.Y., Zhang, X.Y. et al. (2019). *Effect of microplastics exposure on the photosynthesis system of freshwater algae. Journal of Hazardous Materials* 374: 219–227. https://doi.org/10.1016/j.jhazmat.2019.1004.1039.

Yamamoto, T. and Yasuhara, A. (1999). *Quantities of bisphenol a leached from plastic waste samples. Chemosphere* 38 (11): 2569–2576.

Index

Page numbers in *italic* refer to figures; page numbers in **bold** indicate tables. The abbreviation MP/MPs (microplastic/s) is used throughout the index. US spelling is used in the index.

Plastic and Microplastic in the Environment: Management and Health Risks, First Edition.
Edited by Arif Ahamad, Pardeep Singh, and Dhanesh Tiwary.
© 2022 John Wiley & Sons Ltd. Published 2022 by John Wiley & Sons Ltd.